AF769761

VAUBAN UNDER SIEGE

HISTORY OF WARFARE

General Editor

KELLY DEVRIES
Loyola College

Founding Editors

THERESA VANN
PAUL CHEVEDDEN

VOLUME 41

VAUBAN UNDER SIEGE

Engineering Efficiency and Martial Vigor in the War of the Spanish Succession

BY

JAMEL OSTWALD

TUTA SUB AEGIDE PALLAS · 1683 ·

BRILL

LEIDEN · BOSTON
2007

Cover illustration: Planches du second memoire de l'attaque et deffense des places de Mr. de Vauban [early 18th century] (Courtesy of the Society of the Cincinnati).

This book is printed on acid-free paper.

Library of Congress Cataloging-in-Publication Data

A C.I.P. record for this book is available from the Library of Congress.

ISSN 1385–7827
ISBN-13: 978 90 04 15489 6
ISBN-10: 90 04 15489 2

© Copyright 2007 by Koninklijke Brill NV, Leiden, The Netherlands.
Koninklijke Brill NV incorporates the imprints Brill,
Hotei Publishing, IDC Publishers, Martinus Nijhoff Publishers and VSP.

All rights reserved. No part of this publication may be reproduced, translated, stored in a retrieval system, or transmitted in any form or by any means, electronic, mechanical, photocopying, recording or otherwise, without prior written permission from the publisher.

Authorization to photocopy items for internal or personal use is granted by Koninklijke Brill provided that the appropriate fees are paid directly to The Copyright Clearance Center, 222 Rosewood Drive, Suite 910 Danvers MA 01923, USA.
Fees are subject to change.

PRINTED IN THE NETHERLANDS

For my parents Sharon and Mel Ostwald, and my wife Liz

CONTENTS

Acknowledgements ix
List of Maps, Figures and Tables xi
Notes xiii

Chapter One Introduction 1

Chapter Two The Perfect Siege of Ath 1697 21

Chapter Three Efficiency and the Perfect Siege 46

Chapter Four The Transitional War of the Spanish Succession 92

Chapter Five Implementing the Paradigm Siege 123

Chapter Six Contesting the Paradigm Siege 173

Chapter Seven Vigor: An Alternative Paradigm 215

Chapter Eight Succeeding with Vigor 253

Chapter Nine The Vigorous Future 309

Appendices:
- Appendix A: Allied Campaign Lengths in Flanders 325
- Appendix B: Siege Dataset Methodology 326
- Appendix C: Siege Lengths Dataset 340
- Appendix D: Non-Siege Positional Tactics 348
- Appendix E: Length Estimates, Selected Sieges 361
- Appendix F: Siege Artillery 367

Bibliography 371
Index 383

ACKNOWLEDGEMENTS

This work would not have been possible without the assistance and support of many individuals and institutions, and I owe them all my deepest thanks.

My mentors at Ohio State University provided the necessary background and support for me to embark on this project. These include the historian of Louis XIV's France John Rule, a Renaissance scholar of all periods and places but particularly 16th century Mediterranean naval warfare Joe Guilmartin, quantitative historian Randy Roth, and Geoffrey Parker, who provided (among other things) the needed impetus to just finish the damn thing when I was mired in my own never-ending quest for ever-greater efficiency. The research for this project was also supported by grants from the Ohio State University Department of History, the OSU Graduate School, and the OSU Center for Medieval and Renaissance Studies.

John Lynn of the University of Illinois, Urbana-Champaign also deserves many thanks for his unfailing support of a Buckeye who shared an interest in the Giant of the *Grand Siècle*. While a postdoctoral fellow at George Mason University I also received unflagging support from Jack Censer, Mills Kelly and Mack Holt, as well as the crucial time and resources needed to flesh out several aspects of the project.

History is built on its sources, and this work would not have been possible without support from the holdings and staffs at the following institutions: the research collections of the Service Historique de l'Armée de Terre, Archives de Guerre at Vincennes; the Bibliothèque nationale in Paris (and their amazing Gallica digitization project at http://gallica.bnf.fr); the British Library in London; the Public Record Office (National Archives) at Kew; the Koninklijke Bibliotheek at The Hague; the Algemeen Rijksarchief (now Nationaal Archief) at The Hague; the Library of Congress in Washington D.C.; and most warmly Ellen Clark at the Society of the Cincinnati Library in D.C., a truly invaluable and unique resource for all students of 18th century military history. I must also thank the staff and supporters of the various university libraries I have frequented over the past many years, including the Ohio State University Libraries, the George

Mason University Library, and especially the general holdings and Special Collections of the University of Michigan, Ann Arbor.

The following individuals have been friends to provide valuable suggestions and insight into their own periods, places and topics of study: Ben Trotter, Derek Croxton, Olaf van Nimwegen, Michiel de Jong, Guus Veenendaal, George Satterfield, Steve Walton, Ron Martin, Mark Danley, Erik Lund, Pat Speelman, Jean-Pierre Rorive and most of all John Stapleton. My new colleagues in the Department of History, Political Science, Philosophy and Geography at Eastern Connecticut State University have also warmly supported my efforts to delve into the minutiae of far-off wars from long-forgotten times.

I would also like to thank those associated with Brill for seeing this project through its various stages, including Kelly DeVries of Loyola College and Editor Marcella Mulder.

Finally, I must express my deepest gratitude to family and friends that have provided encouragement and support over the many years that this project has taken. These include Karen Straits, Shawn Steene, the Mindels and Mindegros, my parents and sister, and most important of all, my wife Liz, without whom none of this would have been possible.

LIST OF MAPS, FIGURES AND TABLES

Maps

Map 1	The Low Countries Theater	20
Map 2.1	The Low Countries Theater in the Nine Years War	39
Map 2.2	The Fortress of Ath 1697	40
Map 2.3	Fortification Terms	41
Map 2.4	The Siege of Ath 1697	42
Map 2.5	Ath Phase I: Attacking the Covered Way	43
Map 2.6	Ath Phase II: Attacking the Ravelin	44
Map 2.7	Ath Phase III: Attacking the Bastions	45
Map 3	The Efficiency of Vauban's Batteries at Ath	91
Map 4.1	Low Countries Sieges 1702–1712	117
Map 4.2	Fortifications and the Low Countries in 1700	118
Map 4.3	Allied Conquests in the Low Countries 1702–1711	119
Map 4.4	The Low Countries Campaigns 1702–1705	120
Map 4.5	The Low Countries Campaign of 1706	121
Map 4.6	The Low Countries Campaigns 1707–1712	122

Figures

Figure 1.1	Length of Allied Campaign Seasons in Flanders	3
Figure 6.1	Distribution of Siege Length Estimate Errors	205
Figure B.1	A Comparison of Low Countries Siege Lengths, by Source	330
Figure B.2	A Frequency Comparison of Siege Lengths, by Source	332
Figure B.3	Revised Spanish Succession Siege Lengths	334
Figure D.1	Theoretical Results of Positional Tactics	357

Tables

Table 1.1	Percent of Allied Campaigns Spent Besieging, Flanders theater	14
Table A.1	Allied Campaign Lengths in Flanders	325
Table C.1	Siege Lengths, War of the Spanish Succession	341–343
Table D.1	Positional Tactics Used to Capture a Town	351
Table D.2	Non-Siege Tactics in the Flanders Theater	358–359
Table E.1	Estimates of Siege Durations	362–364
Table F.1	Siege Artillery in Selected Flanders Sieges	368

NOTES

Dates are given as Month/Day/Year, with the years in New Style unless noted by "O.S." for Old Style. This Julian system was used in England until 1752. Beginning in 1700 the Julian calendar was eleven days behind the new Gregorian calendar.

The spelling of proper nouns is a surprisingly tricky affair, so I should briefly note my treatment of them here. As this book is targeted at an English-reading audience, I have retained the seemingly-random spelling of all contemporary quotes originally written in English—spelling was far from standardized in the period, so much so that using *sic* would make many quotes unreadable. Even personal names lacked standardized spellings; here I have attempted to spell all names with modern orthography. In the spirit of a multinational work, I have endeavored to employ the form currently used in their country of origin. For example, I refer to the Dutch commander Ouwerkerk rather than the Francophone Auverquerque (how he often signed his name) or the Anglophone version Overkirk (as he was known to the English). Similarly, I have usually referred to place names by their modern spelling, except in quotations. In Belgium, I usually use Flemish orthography for Flemish towns (Oostende rather than Ostend) and French spelling for Walloon places (Mons instead of Bergen). The main exceptions are place-names with well-known Anglicizations, such as Ypres rather than Ieper, and The Hague rather than Den Haag (or more formally 's Gravenhage). In any case, the differences are rarely large enough to muddle comprehension. The reader should not read nationalistic biases into the selection of one form over another.

The maps were created by the author in Adobe Illustrator (and in previous incarnations, AutoCAD) with the assistance of Mike Palumbo and the staff at Eastern Connecticut State's Center for Instructional Technology.

CHAPTER ONE

INTRODUCTION

Throughout the twentieth century, military historians have acknowledged the often-pivotal role of siegecraft in warfare. A necessity since prehistory, fortress walls delineated boundaries, gave succor to the weak and protected the holder's resources, making them obvious targets in any conflict. The romantic vision of medieval castles and the more prosaic reality of urban citadels ought to have ensured that they would continue to play a dominant role in the innumerable wars fought by early modern statesmen. Their relevance depended instead on the balance between these walls' resistance and the force of new offensive techniques and technologies, particularly the 13th century arrival of gunpowder. Responding to this chemical imbalance, defensive architects soon created a defensive system tailored to neutralize the besieger's new-found advantages. This disparity between the power of the defense and the frailty of the attack has been considered a prime cause of the widespread indecisiveness in Europe's 16th and 17th century wars, which in turn crippled efforts to centralize early modern governance in royal hands. One individual, Sébastien le Prestre de Vauban, is said to have played a central part in reducing the defense's dominance while setting the stage for a Military Enlightenment later in the century. Assessing the impact of his legacy on this grand narrative can contribute to a better understanding of this balance and its influence on the early modern world.

1. *Siege Warfare's Pivotal Place*

Before the age of mechanization and motorized transport, warfare was inherently time-conscious. Without the combustion engine, the age's only means of overland cargo transit was horsepower, whose requirements forced a highly circumscribed campaign season. In order to provide the tons of supplies (food, weapons, munitions, equipment) needed by armies with tens of thousands of men, large-scale military operations could normally be sustained only with the onset of spring weather and had to be concluded before winter.[1] Unpaved roads would then be transmuted into mud and waterways blocked with ice, while the horses that armies depended on (including cavalry mounts as well as teams for wagons and towing barges upriver) could not survive the strenuous campaigning on their vitamin-deficient diet of dry fodder (hay, oats, and straw) and thus had to wait for the spring's green grass.[2] Wintertime also gave warring powers a respite in which to rebuild their forces and hammer out diplomatic fine points in peace negotiations. The seasonality of campaigning did vary somewhat by theater: the Mediterranean climates of Iberia and Italy allowed campaigning far earlier and later in the year, but in Spain the severe heat also demanded a respite during hot summer months (usually July and August). In the Low Countries and northern France, the quintessential theater for siege warfare, campaign seasons in the War of the Spanish Succession (the focus of this study) averaged only six months out of the year, usually starting in May and ending in early November, as Figure 1.1 illustrates.[3]

In this time-sensitive context, the later an offensive commander entered the field and the longer a siege lasted, the less time available for future actions. The garrison's duty was to bring any approaching army to a standstill and force it to conduct a difficult siege. Its job was made easier since advancing armies were unlikely to leave an unmolested garrison along their path of advance, for fear of it sallying out to threaten their communication with the rear.[4] Capturing

[1] Military logistics has seen an explosion of recent studies in the past few decades. A sampling includes: Géza Perjés, "Army Provisioning, Logistics and Strategy in the Second Half of the 17th Century," *Acta Historica Academiae Scientiarum Hungaricae* 16 (1970), pp. 7–51; Jean Milot, "Un problème opérationnel du XVIIe siècle illustré par un cas régional," *Revue du Nord* 53, no. 209 (1971), pp. 269–285; Geoffrey Parker, *The Army of Flanders and the Spanish Road, 1567–1659*, 2nd edition (New York, 2004); Martin van Creveld, *Supplying War: Logistics from Wallenstein to Patton*, (New York, 1977); John Lynn (ed.), *Feeding Mars: Logistics in Western Warfare from the Middle Ages to the Present*, (Boulder, CO, 1993); James Scott Wheeler, "Logistics of the Cromwellian Conquest of Scotland, 1650–1651," *War and Society*, 10(1) 1992, pp. 1–18; and Olaf van Nimwegen, *De subsistentie van het leger: Logistiek en strategie van het Geallieerde en met name het Staatse leger tijdens de Spaanse Successieoorlog in de Nederlanden en het Heilige Roomse Rijk (1701–1712)*, (Amsterdam, 1995).

[2] David Chandler, *The Art of Warfare in the Age of Marlborough*, (London, 1994 reprint of 1976 original), p. 14.

[3] For the data used to create this graph, see Appendix A. This is based off of Allied campaign seasons, which were usually longer than the time French armies could remain in the field.

[4] Clausewitz, the famous Prussian theoretician of war, summarizes this 'passive'

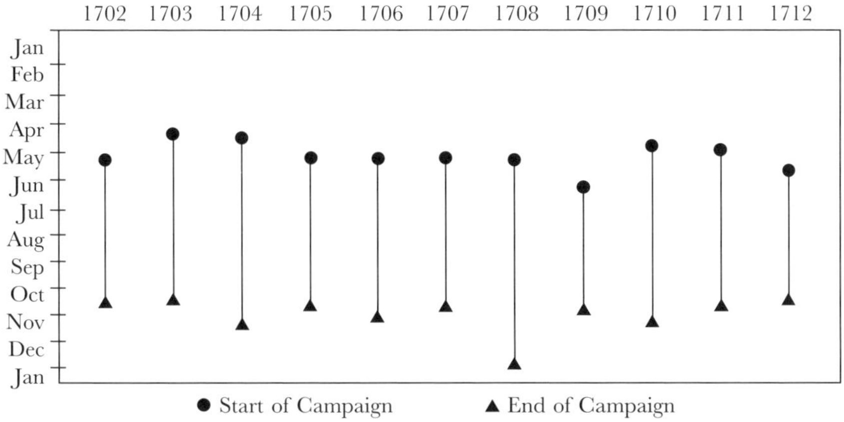

Figure 1.1: Length of Allied Campaign Seasons in Flanders

a besieged town did not always mean success in this context, for even if a well-defended fortress were to eventually fall, it could have halted enemy operations for months. In a war of attrition, the costs incurred by the besiegers would be calculated in money, blood, and powder. Well before the Spanish Succession war, the English general George Monck, fighting in the Netherlands, English and Irish theaters in the middle of the century, best elucidated the impact of lengthy attacks on the momentum of a campaign:

> Long sieges ruine armies; empty the purse, and most commonly it falleth out so, that it hindreth armies from better imployments; and after a long siege, though things fall out according to a commanders desire, he will have little reason to brag of his victory, when he vieweth his expences, his time, and his army. The malice of a great army is broken, and the force of it spent in a great siege.[5]

As Monck intimates, the significance of a siege for those seeking to conduct a rapid war of annihilation was measured most immediately in days. A slower tempo of campaigning gave the advantage to the defensive side in such a conflict, allowing him to run out the clock

function in Carl von Clausewitz, *On War*, edited and translated by P. Paret and M. Howard (Princeton, 1984), pp. 394ff., book 6 chapter 10.

[5] George Monck, duke of Albemarle, *Observations upon Military and Political Considerations*, (London, 1671), p. 119.

on the campaign season. This generated what Thomas Arnold refers to as the "friction fortifications exerted on the pace and effectiveness of offensive operations."[6] With the campaign clock always ticking, too many sieges would doom a campaign to indecision and interminably protract the war.

2. *A Fluctuating Offensive-Defensive Balance*

Defensive fortifications then, served not only as force-multipliers, but also served to slow down the advance of a relentless enemy. The ease with which they were captured—how effectively they consumed a limited campaign season—has fluctuated wildly over the centuries as new defensive and offensive technologies struggled for dominance.[7] The early modern period (c. 1400–c. 1775) witnessed several swings in this balance between the attack and defense. At the dawn of the period, a 14th and 15th century "gunpowder revolution" brought picturesque medieval castle walls crashing down—the thunderous roar of bombards heralded a dramatic shift in the offensive-defensive balance of power.[8] In response to this new threat of gunpowder

[6] Thomas Arnold, "Fortifications and the Military Revolution: The Gonzaga Experience, 1530–1630," in C. Rogers, *The Military Revolution Debate*, pp. 219–221, quote on 220. This pattern is well established in the literature: John Childs, *The Nine Years' War and the British Army, 1688–1697: The Operations in the Low Countries*, (Manchester, 1991), p. 91; John Lynn, *Giant of the Grand Siècle: The French Army, 1610–1715*, (Cambridge, 1996), pp. 549ff.; Parker, "In Defense," pp. 348–350.

[7] For a few of the many general surveys of siegecraft across the ages, consult: Martin Brice, *Forts and Fortresses: From the Hill Forts of Prehistory to Modern Times*, (London, 1990/1999); William Seymour, *Great Sieges of History*, (London, 1991); Bruce Allen Watson, *Sieges: A Comparative Study*, (Westport, CT, 1993); and the more academic James Tracy (ed.), *City Walls—The Urban Enceinte in Global Perspective*, (Cambridge, 2000).

[8] The pace and exact nature of this change remains contentious. Recent studies of the emerging gunpowder era can be sampled in Alain Salamagne, "L'attaque des places-fortes au XVe siècle à travers l'exemple des guerres anglo- et franco-bourguignonnes," *Revue historique*, 585 (1993), pp. 65–113; Clifford Rogers, "The Military Revolution of the Hundred Years War," in C. Rogers (ed.), *The Military Revolution Debate: Readings on the Military Transformation of Early Modern Europe*, (Boulder, CO: Westview Press, 1995), pp. 55–94; and Kelly DeVries, "The Impact of Gunpowder Weaponry on Siege Warfare in the Hundred Years War," in I. Corfis and M. Wolfe (eds.), *The Medieval City under Siege*, (Woodbridge, 1995). For recent summaries of medieval siegecraft, see more broadly Jim Bradbury, *The Medieval Siege*, (Woodridge, 1992); Bernard Bachrach, "Medieval Siege Warfare: A Reconnaissance," *The Journal of Military History*, 58(1) 1994, pp. 119–133; and Richard Jones, "Forti-

artillery, Italian architects developed in the beginning of the 16th century a fortification style referred to as the *trace italienne* to counteract the powerful cannons' ability to smash the thin, tall medieval walls that had been designed to prevent escalade.[9] These designers decreased the height of the walls in order to provide a less visible target, while increasing their thickness to provide greater solidity—both as greater proof against gunpowder artillery and to support the defender's own new weaponry. Angled bastions were also added to provide converging fields of fire at every potential point of attack. Defended by a garrison armed with gunpowder weapons of its own, fortifications once again dominated military campaigns.

The *trace italienne*'s success led to a period of defensive predominance that lasted through the 16th century and beyond. The design's widespread adoption in the Netherlands during their 80-year revolt against Spain led Geoffrey Parker to declare that "even a small, unimportant town might resist capture for several months provided it had the *trace italienne*."[10] Christopher Duffy, the first modern historian to give his full attention to early modern siegecraft, dedicated all three of his survey works to the development of artillery fortresses.[11] Parker later combined his own research with Michael Roberts' postulated battlefield 'military revolution,' and the resulting 1988 book proposed a revised Military Revolution founded on these new 16th century fortification designs. "[W]hen constructed as part of an integrated system, fortifications *alla moderna* dominated the conduct of warfare."[12] James Wood's analysis of the early French Wars of Religion

fications and Sieges in Western Europe c. 800–1450" in M. Keen (ed.), *Medieval Warfare: A History*, (Oxford, 1999), pp. 163–185.

[9] Reginald Blomfield, *Sebastien Le Prestre de Vauban, 1633–1707*, (New York, 1971 reprint of 1938 original), pp. 14–26; John R. Hale, "The Early Development of the Bastion: An Italian Chronology c. 1450–1534" in J.R. Hale (ed.), *Europe in the Late Middle Ages*, (Evanston, IL, 1965); Duffy, *Siege Warfare*, pp. 25ff.; Anne Blanchard, *Les ingénieurs du "Roy" de Louis XIV à Louis XVI: Etude du corps des fortifications*. (Montpellier, 1979), pp. 40–42.

[10] Parker, *The Army of Flanders*, p. 10.

[11] Christopher Duffy, *Fire and Stone: The Science of Fortress Warfare, 1660–1860* (London, 1975); Duffy, *Siege Warfare: The Fortress in the Early Modern World 1494–1660* (London, 1979); and Duffy, *The Fortress in the Age of Vauban and Frederick the Great 1660–1789*, (London, 1985).

[12] Parker, "In Defense of the Military Revolution," in C. Rogers (ed.), *The Military Revolution Debate*, pp. 348–349. His Military Revolution thesis was first presented in "The 'Military Revolution,' 1560–1660—a Myth?" *Journal of Modern History* 48 (June 1976), pp. 195–214, expanded in his *The Military Revolution: Military Innovation and the*

(1562–1576) also concluded that "even great victories on the battlefield would prove empty if there was no means to silence [a fortress's] defending artillery and to batter the breaches that enabled the infantry to mount their assaults."[13] His examination of several extremely bloody sieges confirmed that this task was difficult in an era of defensive superiority, for "the moderately up-to-date defensive works of a single medium-sized city could resist even the most determined bombardment . . ."[14] Another recent student of the sixteenth century highlighted more generally the "growing inefficacy of the siege assault" when firearms and cannon were fully incorporated into the defense—garrisons had adopted the new weapons as quickly as their attackers.[15] Those scholars who downplay the revolutionary nature of the artillery fortress also accept the defensive advantages of fortresses. Simon Pepper and Nicholas Adams, for example, emphasize the resistance that obsolete and transitional *trace italienne* fortresses could put up when facing diminutive 16th century siege trains.[16] The protracted, bloody struggles of this century were prompted by a swing towards the defense.

This balance continued to benefit the defense in the 17th century. David Parrott's recent tome on the French military during Louis XIII's reign reinforces the point's validity for the first half of the century, informing us that the French "with few exceptions [were] attuned to thinking in terms of sieges as the normal means of waging warfare."[17] Derek Croxton's investigation of the Thirty Years' War campaigns along the Rhine in the 1640s also finds a large number of sieges alongside many field battles.[18] George Rothrock

Rise of the West, 1500–1800, (New York, 1988, second edition 1996) and reprinted in Rogers (ed.), *The Military Revolution Debate*, pp. 37–54. He responded to a number of critiques in "In Defense of the Military Revolution," Rogers (ed.), *The Military Revolution Debate*, pp. 337–365 and added as a separate chapter in the second edition of his *The Military Revolution.*

[13] James Wood, *The King's Army: Warfare, Soldiers and Society during the Wars of Religion in France, 1562–1576*, (Cambridge, 1996), p. 155.

[14] Wood, *The King's Army*, p. 272.

[15] David Eltis, *The Military Revolution in Sixteenth-Century Europe*, (London, 1995), p. 136.

[16] Simon Pepper and Nicholas Adams, *Firearms and Fortifications: Military Architecture and Siege Warfare in Sixteenth-Century Siena* (Chicago, 1986), e.g. pp. 170–171.

[17] David Parrott, *Richelieu's Army: War, Government and Society in France, 1624–1642*, (Cambridge, 2001), pp. 57–59.

[18] Derek Croxton, *Peacemaking in Early Modern Europe: Cardinal Mazarin and the Congress of Westphalia, 1643–1648* (Selinsgrove, 1999).

was one of the first 20th century historians to embrace rather than excuse the fact that "field battles were rarely fought, and the preponderance of military effort was absorbed in the attack and defense of fortified places or in maneuvers to pose the threat of such an attack."[19] David Chandler, the most prolific recent biographer of the famous John Churchill, First Duke of Marlborough, was forced to admit that despite his preference for decisive battles, "few periods of military history have been more dominated by siege warfare than the 60-odd years between 1680 and 1748."[20] John Lynn's research also recognizes positional warfare as the "most common form of military operation," reaching its apogee during the Sun King's reign (1661–1715).[21] John Childs emphasized the importance of pitched battles in the Low Countries during the 1688–1697 Nine Years War (also known as the War of the League of Augsburg or the War of the Grand Alliance), but still declared the siege as "the principal military and political operation."[22] The Earl of Orrery's well-known 1677 quip that Europeans fought more like foxes than lions encapsulates in a single phrase the widely-recognized prevalence and predominance of siegecraft in early modern warfare. Even in military operations far from the artillery fortress' heartland, sieges still played a critical role in individual campaigns. East of Parker's *trace italienne* 'heartland,' the influence of fortifications was also keenly felt well into the 18th century.[23] In the relative backwater of Civil War England, Charles Carlton's accounting finds that sieges still comprised a third of all combats (which included many small skirmishes) and that they resulted in more casualties than the much better known field battles.[24] A more recent survey states matter-of-factly that "The

[19] George Rothrock, "Preface" in G. Rothrock, trans. *A Manual of Siegecraft and Fortification*, (Ann Arbor, 1968), p. v.

[20] Chandler, *The Art of Warfare in the Age of Marlborough*, p. 234.

[21] John Lynn, *The Wars of Louis XIV, 1667–1714*, (New York, 1999), quote on p. 71; on p. 63 it is labeled "the more common form of conflict during this period;" also p. 369. See also Lynn, *Giant*, pp. 530–532.

[22] Childs, *The Nine Years' War*, p. 87.

[23] For their role in poorly-fortified eastern Europe, see M.S. Anderson, *The War of the Austrian Succession, 1740–1748*, (London, 1995), p. 37; and Dennis Showalter, *The Wars of Frederick the Great*, (London, 1996), p. 4.

[24] Charles Carlton, *Going to the Wars: The Experience of the British Civil Wars, 1638–1651*, (New York, 1992), pp. 154ff. For studies that emphasize the backwardness of British fortifications and siegecraft vis-à-vis continental Europe (particularly during the British Civil Wars), see Christopher Duffy, *Siege Warfare: The Fortress*

characteristic military action of the British and Irish Civil Wars was an attack upon a fortified strongpoint."[25] A recent survey of early modern war and society reflects the common view of the entire period: "warfare was reduced to a seemingly interminable succession of sieges."[26]

3. *Vauban's Legacy and Siege warfare's Decline*

This early period of modulation between attack and defense is identified by the technologies associated with them (gunpowder, the *trace italienne*), but the next shift—a return once again to offensive ascendancy—is summed up by an individual rather than an artifact. According to the laudatory literature, this French military engineer took advantage of decades of military experience to almost single-handedly reverse this trend of long, expensive sieges inaugurated by the spread of the artillery fortress. Vauban was born (1633) in the rugged Morvan region of Burgundy to a family of poor provincial nobles, and soon volunteered to fight first against and then for the young Louis XIV. Quickly proving his abilities as a military engineer, he surpassed his master Louis-Nicolas chevalier de Clerville, gained the King's personal trust to conduct sieges as he saw fit, and was rewarded with the position of *commissaire-général des fortifications* (1678) to oversee the realm's defenses. The culmination of his career was his promotion to *maréchal de France* (Marshal of France) in 1703,

in the Early Modern World 1494–1660, chapter 6; Peter Harrington, "English Civil War Fortifications," *Fort: The International Journal of Fortification and Military Architecture* 15 (1987), pp. 39–60; Geoffrey Parker, *The Military Revolution*, pp. 26–32; James Burke, "The New Model Army and the problems of siege warfare, 1648–51," *Irish Historical Studies* 27, no. 105 (May 1990), pp. 7–29; Rolf Loeber and Geoffrey Parker, "The military revolution in seventeenth century Ireland," in J. Ohlmeyer (ed.), *Ireland from Independence to Occupation 1641–1660*, (Cambridge, 1994), pp. 66–88; Ronald Hutton and Wylie Reeves, "Sieges and Fortifications," in J. Kenyon and J. Ohlmeyer (eds.), *The Civil Wars: A Military History of England, Scotland and Ireland, 1638–1660*, (Oxford, 1998), pp. 195–233; and James Burke, "Siege Warfare in Seventeenth Century Ireland," in Pádraig Lenihan (ed.), *Conquest and Resistance: War in seventeenth-century Ireland*, (Leiden, 2001), pp. 257–291. For a less-convincing counterargument, see Mark C. Fissel, *English Warfare, 1511–1641*, (London, 2001), pp. 183ff.

[25] Hutton and Reeves, "Sieges and Fortifications," in J. Kenyon and J. Ohlmeyer (eds.), *The Civil Wars*, p. 195.

[26] Frank Tallett, *War and Society in Early Modern Europe, 1495–1715*, (London, 1992), p. 52.

a late recognition of how much the Sun King owed to his faithful servant.

Though Vauban died in 1707, his legacy lasted far beyond his death. His stone and brick fortresses shaped campaigns long after he had been buried and outline France's hexagon today; some of them still stand, having survived even World War II bombardments (e.g. Cherbourg and Brest).[27] He also played a pivotal role in shaping the administrative organization of the French engineering corps.[28] Much has also been made of Vauban's strategic vision, particularly his creation of a rationalized, fortified *frontière de fer* shielding France, the famous *pré carré* (translated alternately as a dueling ground or a squared circle). Within this French school of historiography, the attention given to fortification design has revolved around Vauban's 'three systems,' all modern scholars insisting that the great French engineer would never have endorsed such a rigid partition of fortification styles.[29] The territorial boundaries Louis XIV conquered and defended with the aid of Vauban's abilities last to this day.

But his biographers, particularly Anglo-American commentators, identify his most significant bequest to siegecraft in his systematization of the siege attack, and declare him responsible for another radical shift from defensive dominance to offensive supremacy.[30] Unlike all other previous engineering authors, Vauban never wrote a treatise on fortification design. The biographer Blomfield perceived his

---

[27] My thanks to Dr. John Stapleton for these references. On Vauban's legacy as embodied in the fortified works themselves, see Duffy, *The Fortress in the Age of Vauban*, pp. 94–96. The effects of the 20th century assaults on Neuf-Brisach are illustrated in the town's Musée Vauban, while the various publications of the Association Vauban have traced his influence across the ages.

[28] On the legacy Vauban bequeathed to the administration of French fortifications and their engineers, see Ben Scott Trotter, *Marshal Vauban and the Administration of Fortifications under Louis XIV (to 1691)*, Ohio State University Ph.D. Dissertation, Columbus, OH (1993).

[29] Paul Lazard, *Vauban, 1633–1707*. (Paris, 1934), pp. 377–394; Reginald Blomfield, *Sebastien Le Prestre de Vauban*, p. 163; Henry Guerlac, "Vauban: The Impact of Science on War," in Peter Paret (ed.), *Makers of Modern Strategy from Machiavelli to the Nuclear Age* (Princeton, 1986), pp. 80–83; Anne Blanchard, *Vauban*, (Paris, 1996), pp. 388–404; Alain Salamagne, "Vauban et les fortifications du Quesnoy," *Revue historique des armées* 1986, pp. 45–51.

[30] For one example, see Duffy, *The Fortress in the Age of Vauban*, p. 96. More briefly, Ronald Martin, "The Army of Louis XIV," in P. Sonnino et al. (eds.), *The Reign of Louis XIV*, (New Jersey, 1991), p. 118; and F.J. Hebbert and George Rothrock, *Soldier of France: Sebastien Le Prestre de Vauban, 1633–1707*, (New York, 1989), pp. xvii, xix.

real merit: "The truth was that till Vauban came, there was nobody in the French army who understood 'grande finesse' as Vauban puts it. People wrote freely and voluminously on fortifications and defense, but nobody seems to have studied the not less important matter of attack."[31] Over the course of his fifty years of military service, he besieged fifty towns or more but found himself besieged only once, at the week-long attack on Oudenaarde (Audenarde) in 1677. Historians have tended to use a stereotyped 'rhetoric of siege history' to describe these Vauban-influenced sieges, emphasizing the clockwork regularity imposed by seemingly-omnipotent military engineers on the attacks.[32] According to one French writer, Vauban and his pupils created an attack regulated by "rules" which combined to form a "liturgy," attaining "quasi-perfection," "developing as in one of those ballets of Benserade orchestrated by Lully."[33] Chandler highlighted the systematization of the attack: "Vauban had virtually imposed a series of standards on both attack and defense," elsewhere crediting him with "perfect[ing] the techniques of the siege itself—with the laborious but almost mathematically certain 'sapping forward' by means of approach and parallel trenches, and the clever siting of batteries."[34] To stress the predictability of this process, the whole procedure is frequently given the inexorability of a drama—each act unfolding according to classical theory.[35] More recent applications of culture to early modern military history have stressed the intellectual context of the period, particularly the Scientific Revolutionary and Enlightenment predilections for geometry, Newtonian mechanics

---

[31] Blomfield, *Sebastien Le Prestre de Vauban*, p. 98. See also Rothrock (ed.), *A Manual*, p. viii; Chandler, *The Art of Warfare in the Age of Marlborough*, pp. 275–276.

[32] This term adapts John Keegan's discussion of the popular "rhetoric of battle history" described in the introduction to his seminal *The Face of Battle*, (New York, 1976), pp. 35ff.

[33] Y.J. Saint-Martin, "Le Pseudo-Siège de Toulon en 1707," *Provence historique* 176 (1994), pp. 199–200. Franco Cardini described 18th century siege warfare as a "stereotyped form, half liturgy and half accounting." *La culture de la guerre X^e^–XVIIIe siècles*, translation of 1982 Italian original (Paris, 1992), p. 253.

[34] David Chandler, *Marlborough as Military Commander*, (New York, 2000 reprint of 1973 original), p. 81 and David Chandler (ed.), *Military Memoirs of Marlborough's Campaigns, 1702–1712. Captain Robert Parker, Royal Regiment of Foot of Ireland, and the Count of Mérode-Westerloo, Field Marshal of the Holy Roman Empire*, (Hamden, CT, 1998), p. 234.

[35] Michel Parent and Jacques Verroust, *Vauban*, (Paris, 1971), pp. 110–113; Joan DeJean, *Literary Fortifications: Rousseau, LaClos, Sade*, (Princeton, 1984), pp. 26–29; Pernot, "Vauban, le siège devenu réglé," 256.

and rationalism.[36] Martha Pollak stressed the mechanical nature of Vaubanian siegecraft, a parallel to the clockwork universe discovered by Newton:

> Vauban was considered a theoretical, systematic and machinating genius . . . His tables of calculations gave the impression of strategic unassailability; since he calculated not only the dimensions of every element of the fortification, but also the length of time it would take the enemy to gain individual layers of the fortification, every stage of the siege could be predicted in advance. Vauban reduced the defense and attack of fortresses to double-entry bookkeeping, where the two columns balance each other precisely. The accountability of the smallest part of the defense, fortification and provisioning in Vauban reflect the earlier attempts by military theorists to set up a machine which can be expected to operate by itself, but which results—both in Vauban and his predecessors—in an obsession with the smallest detail.[37]

A man of his time, he purportedly made siege warfare "scientific."

Given such perfection, it is no surprise that Vauban is widely held to be responsible for determining the course of siege warfare for the next century or more. Not only did he provide a systematic discussion of the siege attack (the basis for tactical doctrine), but his legacy also resided in the large corps of French engineers he had trained in his methods; their experience would be critical in disseminating and explaining his ideas to posterity. Eighteenth century engineers and military authors had no choice but to come to grips with Vauban's legacy, either by accepting it or rejecting it.[38] Guerlac explained that Vauban's legacy "was followed with but little variation during the eighteenth century."[39] John Childs more recently concurred that

---

[36] For two examples which focus on the later 18th century and its implications for field tactics in particular, see Azar Gat, *The Origins of Military Thought from the Enlightenment to Clausewitz*, (Oxford, 1991); and Lee Kennett, "Tactics and Culture: The Eighteenth-Century Experience," *International Colloquy on Military History, No. 5* (1981), pp. 152–159. John Lynn's *Battle: A History of Combat and Culture*, (Boulder, CO, 2003) adopts a similar approach.

[37] Martha Pollak, *Military Architecture, Cartography and the Representation of the Early Modern European City: A Checklist of Treatises on Fortification in the Newberry Library*, (Chicago, 1991), p. xxxiv.

[38] For a survey of trends in 18th century siegecraft, see Duffy, *The Fortress in the Age of Vauban*, chapters 4 and 5.

[39] Guerlac, "Vauban," 79. The ossification of Vaubanian historiography is evidenced by the very fact that Guerlac's work remains widely-cited despite its age. Similarly, the most recent biographies of Vauban add nothing new to this aspect of their subject—the very aspect which he is most famous for. In addition to

"European engineers rapidly imitated the system of three parallels [one of Vauban's main innovations], and a 'siege in form' became synonymous with an attack according to the Vauban method."[40] The unequalled expert on early modern siegecraft, Christopher Duffy, concluded more forcefully:

> Vauban's impressive contribution to the defence is eclipsed by his still greater achievement as a taker of fortresses. The later engineers were allured, exercised and finally frustrated by the quest to undo his work by restoring the defence to an equilibrium with the Vauban-style attack.[41]

Janis Langins' recent study of eighteenth century French engineers declares much the same, recounting in great detail how his progeny labored under the shadow of the master.[42] In fact, his principles became veritable doctrine for future engineers. As Azar Gat explained: "Vauban's highly-renowned *De l'attaque et de la defense des places*, published in numerous editions, was the standard work for students of fortifications and siegecraft until the second half of the nineteenth century."[43] Paul Lazard, drawing on his military service as a colonel in France's *Génie* (engineering corps) at the beginning of the 20th century, contended that the great engineer's methods were not significantly modified until the outbreak of World War I.[44] In short, "Vauban . . . established a nearly-infallible routine which was accessible to ordinary mortals who were willing to take the trouble to become versed in it."[45]

Blanchard, consult also Bernard Pujo, *Vauban*, (Paris, 1991); and Hebbert and Rothrock, *Soldier of France: Sebastien Le Prestre de Vauban, 1633–1707*.

[40] Childs, *Warfare in the Seventeenth Century*, p. 148. Speaking of the siege of Menin in 1706, Hebbert and Rothrock (212) wrote that "the attack was carried on in the now generally accepted way, for Vauban's opponents had adopted his methods." See also Blomfield, *Sebastien Le Prestre de Vauban*, pp. 163–165.

[41] Duffy, *The Fortress in the Age of Vauban*, p. 96.

[42] Janis Langins, *Conserving the Enlightenment: French Military Engineering from Vauban to the Revolution*, (Cambridge, MA, 2003).

[43] Gat, *The Origins of Military Thought: From the Enlightenment to Clausewitz*, p. 35.

[44] P. Lazard, *Vauban, 1633–1707*, (Paris, 1934), p. 396.

[45] Duffy, *The Fortress in the Age of Vauban*, p. 96. For other examples, see M.S. Anderson, *War and Society in Europe of the Old Regime 1618–1789*, (New York, 1988), p. 88; Jean-Pierre Bois, "Armes, tactiques et batailles d'Azincourt à Fontenoy" in *Armes et Alliances en Europe*, (Nantes, 1992), p. 50; and Bois, *Maurice de Saxe*, (Paris, 1992), pp. 186, 222–226, where it is noted that Saxe saw little room for improvement in Vauban's attack.

Hence, by the dawn of the 18th century Vauban appears to have completely altered the face of early modern siegecraft, reversing the dominance of the century-old *trace italienne* design and overturning as well the pattern of long, bloody, uncertain sieges of the 16th and early 17th centuries. His swift conquests of Spanish and Dutch towns in the War of Devolution (1667–1668), the Dutch War (1672–1679), the War of Reunions (1683–1684), and the Nine Years' War (1688–1697) won him fame throughout Europe; a contemporary adage boasted that a town besieged by Vauban was one taken, while a town defended by him was one saved. In this new era patient besiegers could be certain to capture even a strong artillery fortress in only a month or two, assuming their manpower and supplies held out and a relief force did not force them to lift the siege. The Vauban legacy thus eliminated the eternal sieges of a previous age.

4. *Challenging Vauban*

How far Vauban made the pendulum swing towards the offensive pole is a matter of debate. The well-established biographical literature of another giant of the age, the Duke of Marlborough, presents their impatient subject as a precursor to that greatest of military commanders Napoleon Bonaparte, thanks to his perpetual quest for decisive battle. Sieges therefore appear through Napoleonic-tinted lenses a less-than-ideal replacement for field battle.[46] As a consequence, such historians reject Vauban's efforts as futile and consider absurd the very idea that a war of positions could somehow avoid indecision. Though writing about Vauban, Henry Guerlac's early formulation of the period's strategic culture set the tone for much of the rest of the century: "The strategic imagination of all but a few exceptional commanders was walled in by the accepted axioms of a war of sieges . . . . [they] accepted unconditionally this doctrine

[46] In addition to Chandler's many works (originally written in the 1970s and reprinted in the 1990s), 20th century biographies of the Duke include: Christopher Atkinson, *Marlborough and the Rise of the British Army*, (New York, 1921); H. Belloc, *The Tactics and Strategy of the Great Duke of Marlborough*, (London, 1933); Winston Churchill (descendant of the Duke), *Marlborough: His Life and Times*, 2 vols. (New York, 1947 reprinted 2003); Maurice Ashley, *Marlborough*, (London, 1956); Ivor Burton, *The Captain-General: The Career of John Churchill, Duke of Marlborough, from 1702 to 1711*, (London, 1968); and J.R. Jones, *Marlborough*, (Cambridge, 1993).

of the strategic primacy of the siege."[47] In this literature, the rhetoric of scientific sieges is also invoked, though with a disapproving tone. Not surprisingly, Marlburists pay particular attention to the "tedious" and "leisurely" nature of sieges, even those conducted *à la Vauban*.[48] Table 1.1 provides one justification for their claim, indicating that even in the War of the Spanish Succession a quarter of the Allies' Flanders campaigns were spent attacking fortresses. Adding to this the time needed to prepare for and clean up after the attack, offensive siegecraft took up more than one-third of all the campaign time available to the Allies, who held the initiative in the theater for most of the war.

Table 1.1: Percent of Allied Campaigns Spent Besieging, Flanders theater[49]

Year	# Days in campaign	# Days in sieges (OT-C)	# Days in sieges (I-BL)	% of Campaign in Sieges (OT-C)	% of Campaign in Sieges (I-BL)
1702	172	34	70	20%	41%
1703	194	13	44	7%	23%
1704	214	2	2	0.9%	0.9%
1705	168	10	17	6%	10%
1706	176	45	109	26%	62%
1707	161	0	0	0%	0%
1708	223	115	139	52%	62%
1709	127	83	104	65%	82%
1710	209	148	197	71%	94%
1711	179	21	70	12%	39%
1712	149	42	15	28%	10%
Total	1972	513	767	–	–
Mean	179	47	70	26%	39%

[47] Henry Guerlac, "Vauban: The Impact of Science on War," in P. Paret (ed.), *Makers of Modern Strategy from Machiavelli to the Nuclear Age*, (Princeton, 1986), p. 74. Originally published in the 1943 edition of *Makers of Modern Strategy*, it remained one of only three essays unchanged in the otherwise revised 1986 edition.

[48] Chandler discusses siege lengths in *The Art of Warfare in the Age of Marlborough*, pp. 245–246.

[49] See Appendix A for the sources for the campaign lengths; the siege data is derived from Appendix C; Appendix B describes the two measures for siege length used (from open trenches to the capitulation—OT-C—and from investment to when the besiegers finally left the site—I-BL). The average of percentages is calculated by dividing the total number of days spent in sieges into the total number of days campaigning, rather than an average of the yearly percentages. The real ratio may

Géza Perjés referred to this structural stalemate as an early modern "crisis of strategy," where politicians and military commanders were unable to achieve their strategic objectives because they could never completely defeat the enemy before the weather forced them to go into winter quarters and give the reeling foe a respite in which to recover and rebuild. Chandler and other more recent historians have adopted and expanded upon this concept.[50] How accurately both of these schools reflect the reality of siegecraft is the focus of this study.

5. *The Impact of Vauban's Legacy*

The significance of studying the tactical details of trenchworks, batteries and assaults goes far beyond an improved understanding of the nature of early modern warfare. The unique nature and great frequency of fortress warfare and urban combat had wide-ranging influences over early modern society.[51] Discussions of the period's 'limited' warfare have to ignore the innumerable occasions in which civilians found themselves caught in the crossfire. Battles were fought on plains (literally, 'field' battles), but sieges were waged for control of densely-populated urban areas—sieges brought the war into the inhabitant's hearth and home. Of most immediate interest to the townspeople, the ease with which a town could be captured influenced the experience of war for the participants (combatants and non-combatants alike). Formidable town walls might defend those inside from the casual depredations of marauding bands, but it also made them a target for military operations. In a time of defensive dominance,

---

actually be higher than shown, since the time an army spent idle waiting for the siege preparations to be completed prior to investment is not included. In the 1710 campaign the number of days the Allies spent besieging had to be modified because the Allies attacked both Aire and Saint-Venant at the same time. Therefore only the days spent at the siege of Aire were counted, since it was the longer of the two sieges and both were invested on the same day. If we were to measure the days of labor exerted in sieges however, we would include Saint-Venant's 24 days (I-C).

[50] Perjés, "Army Provisioning," especially pp. 45–46. Chandler, *The Art of Warfare in the Age of Marlborough*, pp. 13ff. Parker echoes this theme in *The Military Revolution*, p. 16.

[51] For an early exposition of some of these influences, see Duffy, *Siege Warfare*, chapter ten, The Fortress and Humankind. See also the wide-ranging collection of Ivy Corfis and Michael Wolfe (eds.), *The Medieval City under Siege*, (Woodbridge, 1995).

the strength of a besieged fortress might push both sides to extremes. The garrison would be encouraged to hold out to the bitter end, while the besiegers became frustrated with a slow siege and either settled down to a blockade in order to starve out the town, or else vented their anger by setting the town on fire for its recalcitrance. On the other hand, a town with weak fortifications might encourage the attacker to forego a formal siege and take the place by storm instead. If their assault succeeded, the fate of both garrison and townspeople would rely on how successfully the enemy commander could control his troops' bloodlust. Thus their lives and deaths were directly caught up in the last argument of kings.

The offensive-defensive balance dictated in large part how successful military campaigns would be and how much money those operations would cost. With the entire early modern state's fiscal apparatus dedicated to funding the period's lumbering martial juggernauts, these military factors determined the burden early modern society would bear. Sieges, as Monck's earlier quote recognized, were a particularly expensive way to wage war.[52] The eruption of logistical and administrative studies of early modern militaries in the past few decades has underlined the crushing expenses exacted by their military instruments and the difficulties with which these governments raised, equipped and maintained them.[53] For the French, building modern fortresses and renovating old ones cost Louis more than 105 million *livres*, or 2.5% of total royal expenditures.[54] The cost of fortifications to French society were in fact much higher than this, for this figure accounts for costs incurred during only half of his

[52] Lynn mentions the matter in passing in *Wars of Louis XIV*, p. 78 and in more detail in *Giant*, pp. 573–574. See also John Childs, *Warfare in the Seventeenth Century*, pp. 105–107.

[53] Among others (to limit ourselves to monographs): Parker, *The Army of Flanders* and *The Military Revolution*, chapter 2; John Lynn's relevant chapters (3–6) in *Giant*; the works in his edited *Feeding Mars: Logistics in Western Warfare from the Middle Ages to the Present*, (Boulder, CO, 1993); James Wood, *The King's Army*, chapter 11; David Parrott's *Richelieu's Army*, chapter 4; and Guy Rowlands, *The Dynastic State and the Army under Louis XIV: Royal Service and Private Interest 1661–1701*, (Cambridge, 2002). This is also one of the main themes of the recent volume edited by Philippe Contamine, *War and Competition between States*, (Oxford, 2000).

[54] Jean-Pierre Rorive, *La guerre de siège sous Louis XIV en Europe et à Huy*, (Brussels, 1997), pp. 39–40. Localities bore an even greater share of the burden. For an example from the French Wars of Religion, see Michael Wolfe, "Walled towns during the French wars of religion (1560–1630)," in J. Tracy, et al., *City Walls: The Urban Enceinte in Global Perspective*, (Cambridge, 2000), pp. 328–337.

reign, and does not include the many expenses paid for directly by the localities, in both cash and labor. Capturing and defending these defensive works also cost money, measured in lost lives, consumed powder and shot, and burst cannon. The costs of short sieges paled in comparison with the longest sieges, and both inflated wartime expenses further by bogging down armies for year after year—costs skyrocketed the longer armies remained immobilized in the environs of a particular town. With little to show for a single campaign other than a few conquered places, peace negotiations could easily break down (or never get started) and more military operations would be needed the next year, which required reestablishing unit strengths throughout the winter and spring, as well as disbursing funds for the costs of yet another campaign. Out of such costly stasis arose the hope that a decisive battle, though bloody, could save both lives and money in the long-term by cutting years off the length of an otherwise indecisive war.[55] The theory never worked for Louis XIV or his opponents, particularly during the two final slogging marathons of 1688–1697 and 1701–1714.

Enormous armies (increasing in size throughout Louis' reign)[56] and the Herculean tasks they performed forced war expenditures to spiral out of control, from half of the French King's expenses at the beginning of his reign to 90% of the royal "budget" in his last wars.[57]

---

[55] Russell Weigley presents a caricature of this battle-seeking philosophy in *The Age of Battles: The Quest for Decisive Victory from Breitenfeld to Waterloo*, (Bloomington, 1991).

[56] See John Lynn, "Recalculating French Army Growth during the Grand Siècle, 1610–1715," *French Historical Studies* 18(4) Autumn 1994, reprinted in Rogers (ed.), *The Military Revolution Debate*, pp. 117–148.

[57] Rorive, *La guerre de siège sous Louis XIV*, pp. 39–40. On early modern military financing generally, see P.G.M. Dickson and John Sperling, "War Finance, 1689–1714," in J.S. Bromley (ed.), *The New Cambridge Modern History, Vol. 6: The Rise of Great Britain and Russia, 1688–1715/25* (Cambridge, 1971), pp. 284–315. The most recent surveys of French finances during the period can be found in Richard Bonney, *Jean-Roland Malet: premier historien des finances de la monarchie française.* (Paris, 1993); as well as his "The Eighteenth Century II: The Struggle for Great Power Status and the End of the Old Fiscal Regime," in R. Bonney (ed.), *The rise of the fiscal state in Europe, c. 1200–1815*, (Oxford, 1995), pp. 315–392. These figures (e.g. 71% of the royal "budget" being expended on the war department in the early years of the war) are based on the financial histories of two 18th century historians, Jean-Roland Malet, an aide to the French controller-general Nicolas Desmaretz, and Véron de Fourbonnais. For the English, see D.W. Jones, *War and Economy in the Age of William III and Marlborough*, (Oxford, 1988); John Brewer, *The Sinews of Power: War, Money and the English State, 1688–1783*, (New York, 1989), and most recently James Scott Wheeler,

Finding the funds to pay for such massive undertakings in an era of almost perpetual armed conflict severely challenged every early modern state, no matter how well endowed it was with natural resources, how many peasants it could conscript, how many mercenaries it could hire, or how much money it could raise from its own tax base and financiers. Credit and cash, the sinews of early modern war, have been identified as the crucial connection between traditional military history and the broader histories of the composite monarchies that sought to consolidate their own authority while struggling with one another for dominance.[58] In a period of almost constant warfare, such insatiable demands could virtually enslave political leaders, forcing the often-undirected development of "fiscal-military" regimes.[59] The massive financial demands of waging war spurred monarchists and republicans alike to find innovative ways to fund their armies, forces that, more often than not, were involved in assaulting an enemy stronghold.[60] For early modern France in particular, scholars have already illustrated how war became the enemy of centralizing monarchs. The interminable need for immediate cash elevated the pursuit of creative, short-term financing above fiscal and political reforms that would improve the Crown's long-term budgetary stability. Reform-minded financial ministers such as Jean-Baptiste *le grand* Colbert saw their projects collapse with the declaration of war: the Crown was forced to sell even more venal offices and hire more autonomous tax-collectors instead of regaining royal control by buying back these administrative posts. Supporting this war effort for a decade or more further stretched the resources of Louis

The making of a world power: war and revolution in seventeenth-century England, (Stroud, 1999). Online datasets hosted by Richard Bonney for various European countries can be found at www.le.ac.uk/hi/bon/ESFDB/.

[58] Both Michael Roberts' initial Military Revolution and Geoffrey Parker's more recent permutation stressed the increasing army sizes, which in turn necessitated new methods of generating revenue to support the troops. David Parrott and Guy Rowlands argue, to different degrees, for a French royal army supported predominantly by the service ethos of its noble officer corps. Although historians might disagree over how many men were paid, how they were paid, when the most important military changes occurred, and whether military expenses drove administrative and fiscal innovations or the reverse, they all agree that states (and their sources of revenues) usually teetered on the brink of insolvency as a result.

[59] For a recent overview of this literature, see Gwynne Lewis's "'Fiscal States': Taxes, War, Privilege and the Emergence of the European 'Nation State', c. 1200–1800," *French Historical Studies*, 15(1) 2001, pp. 51–63, especially p. 54.

[60] Parrott argues this point in *Richelieu's Army*, p. 550.

XIV's state to the limit.[61] Elaborating the shifting balance between offensive and defensive tactics is of critical importance therefore for our understanding of the role war played in the development of the early modern state.

[61] For examples of such French financial gymnastics, see Gary McCollim, *The formation of fiscal policy in the reign of Louis XIV: the example of Nicolas Desmaretz, controller general of finances (1708–1715)*, Ph.D. dissertation (Columbus, OH, 1979); and Rowlands, *The Dynastic State.*

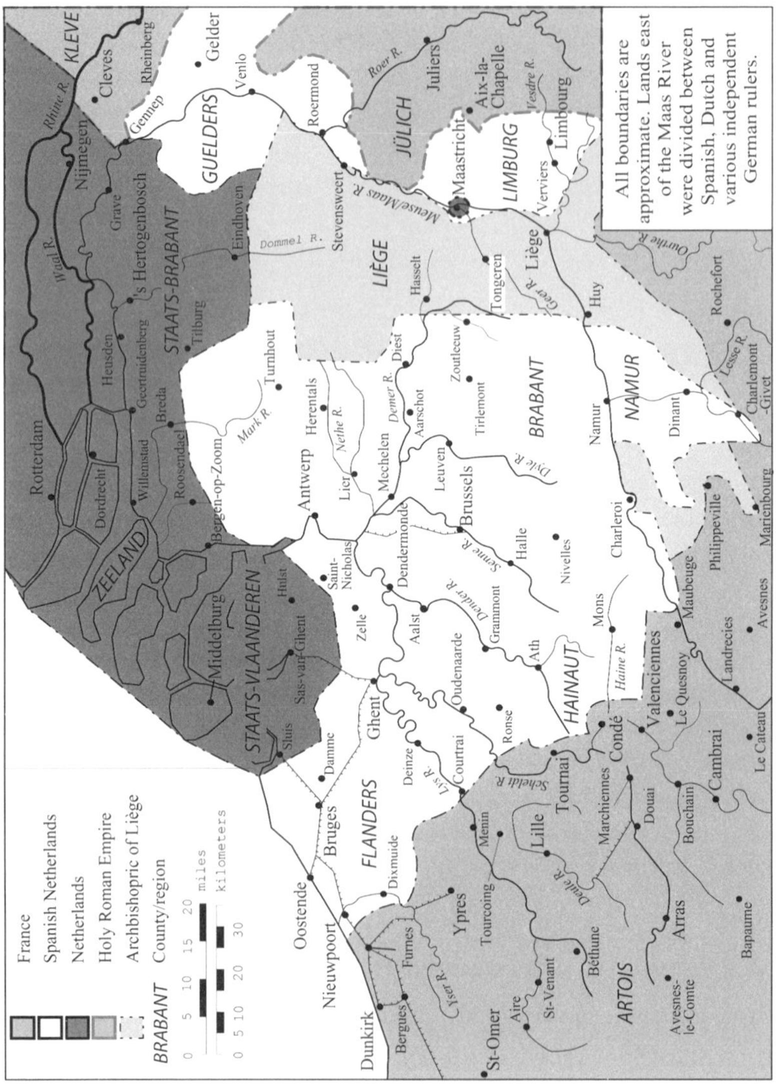

Map 1: The Low Countries Theater

CHAPTER TWO

THE PERFECT SIEGE OF ATH 1697

The epitome of Louisquatorzian siegecraft was the 1697 attack on Ath. As with many places in the cockpit of Europe, the town's sovereignty changed several times over the course of the late seventeenth century. Long under Spanish control, its garrison abruptly abandoned the town upon the approach of a French force in June 1667, one of Louis' many effortless victories during the unnaturally short War of Devolution. It would remain in French hands until it was returned, with substantial improvements, to the Spaniards at the Treaty of Nijmegen in 1678, along with several fortresses in the *pré carré*, a double barrier of fortresses defending France's northern border from the Channel to the Meuse (see Map 2.1 at the end of the chapter). When the Nine Years' War broke out in 1688, Ath witnessed the usual depredations of partisans and troops billeted in garrison, but it only became the focus of both belligerents in 1697.

Ath itself might never have been besieged in 1697 had the peace negotiations started in 1695 and convened more formally at Rijswijk (Ryswick) the next year taken hold. Waging war while discussing peace was common in an era of fluid coalition warfare, and the rarity of decisive military victories only reinforced both sides' willingness to keep fighting. When one of Louis XIV's enemies, Vittorio II Amadeo, Duke of Savoy abandoned his allies for neutrality in late 1696, Louis saw this as an opportunity to push the issue in the Spanish Netherlands the next year. He hoped, with the help of a reinforced army in the Low Countries, to give the Allies one more illustration of his military potency in order to force them to acknowledge his new-found pre-eminence. France's numerical superiority gave them the hope that the siege of Ath would be the straw to break the proverbial camel's back. Louis' forces entered the field in mid-April, and preparations for the siege were made amid recurrent rumors of an impending suspension of arms.[1]

---

[1] French archives relating to the siege are found in Service Historique de l'Armée

Rumors of peace dissipated as a force of 12,000 French cavalry arrived before Ath on the morning of 16 May from three directions—they quickly secured the main roads, river crossings, abbeys and buildings within a several-mile radius of the town. The main siege force left its camp at Helchin the same day, crossed the Scheldt River and camped about six miles from its intended target; two other armies moved to support the flanks of the besieging army. With all the troops assembled in their bivouacs, *maréchal de France* Nicolas Catinat commanded a besieging force of fifty battalions and as many squadrons, divided into three main camps around the town, with Vauban as the chief engineer. These two men, among the only low-ranking nobles to attain the elevated position of *maréchal de France* (Catinat in 1693 and Vauban in 1703), were particularly close and well disposed for the cooperation required of a successful siege. Vauban was seconded by his close companion Jean de Mesgrigny and further supported by more than sixty hand-picked engineers.[2]

de Terre (SHAT), Archives de Guerre (AG), series A[1], volumes 1394, 1400 and 1401. Among the most important published sources regarding the attack on Ath is an anonymous journal of the siege, published in English translation as "A Journal of the Siege of Ath conducted by Monsieur de Vauban" in Charles Goulon, *Memoirs of Monsieur Goulon, being a Treatise on the Attack and Defence of a Place*, (London, 1745), pp. 91ff. Goulon's abridged version is an edited and rearranged version from a fuller original French manuscript that has been published as *Relation du siège de la ville d'Ath en 1697*, (Mons, 1910). One historian attributed the work to one of Vauban's aides-de-camp Ferry, who is mentioned in the text in the third person; see also Gudin de Vallerin, "Les Ingénieurs de Vauban à Bazoches et le Journal du Siège d'Ath," *Bulletin de la Société nivernaises des lettres, sciences et arts*, vol. 28: 601–604. Other primary accounts can be found in various contemporary newsletters (*Le Mercure galant, Europische Mercurius, Gazette d'Amsterdam*); in Lecestre (ed.), *Mémoires du chevalier de Quincy*, vol. 1, pp. 41–59; in Edward d'Auvergne, *The History of the Campagne in Flanders for the Year 1697*, (London, 1698); and in Vauban's correspondence reproduced in Albert de Rochas d'Aiglun (ed.), *Vauban, sa famille et ses écrits, ses oisivetés et sa correspondance*, (Paris, 1910), vol. 2, pp. 449–465. Later histories and memoirs, which largely follow the Goulon journal, include: Charles Sevin marquis de Quincy, *Histoire militaire du règne de Louis le Grand, Roy de France*, (Paris, 1726), vol. 3, pp. 296–308; Pierre Alexandre Joseph Allent, *Histoire du corps impérial du génie . . . depuis l'origine de la fortification moderne jusqu'à la fin du règne de Louis XIV*, (Paris, 1805), pp. 345–351; Antoine-Marie Augoyat, *Aperçu historique sur les fortifications, les ingénieurs et sur le corps du Génie en France*, (Paris, 1860), vol. 1, pp. 216–220; F. Delvaux, "Sièges subis par la ville d'Ath," *Annales du Cercle royal d'histoire et d'archeologie d'Ath*, 24 (1938), 285–304; and Childs, *The Nine Years' War*, pp. 46–50.

[2] On Mesgrigny, see M. de Pinard, *Chronologie historique militaire*, (Paris, 1761–1764), vol. 4, pp. 396–398; and Michèle Virol, *Vauban: De la gloire du roi au service de l'État*, (Seyssel, 2003), pp. 287ff. For lists of the engineers serving at the siege, see AG A[1] 1400 #155 and #156.

Two other French *maréchaux*, Louis François, duc de Boufflers and François de Neufville, duc de Villeroi commanded separate forces (said to total 140,000 men) with orders to observe the enemy's movements and cover the siege army. Against Catinat's 40,000 Frenchmen, the motley garrison numbered a mere 3,600 soldiers of Spanish, Walloon, Italian, German, and Dutch origin. Assisting them in a technical capacity were thirty engineers and a like number of cannoniers to oversee the garrison's thirty-two pieces of artillery. All these assorted troops were governed by the sluggish sixty-five year old Ferdinand-Gaston-Lamorald de Croy, comte de Roeux. The more active Marquis de Conflans had been given orders to command the assorted regiments inside the town but was too late. He and several other Confederate officers who were absent from their regiments already inside the town were thus captured before they could throw themselves into the covered way. This surprise investment derived from a combination of Allied uncertainty and numerical inferiority—from their central position French armies had threatened Brussels, Oudenaarde and Ath all at the same time, forcing a dispersal of a smaller number of Allied troops. With their opponent's attention focused on the more important cities of Oudenaarde and Brussels, it was relatively easy to invest Ath uncontested. Reacting to the news of its investment, public Allied accounts expected the undersized garrison to withstand only fifteen to eighteen days of open trenches.[3]

With the town cut off from outside assistance, preparations for the siege were begun in earnest. On the defenders' side, Roeux's inactivity left direction of the defense to a young Prince Anton Günther of Anhalt-Zerbst.[4] Following the niceties of early modern siegecraft, passports were negotiated for the town's women of status to evacuate the town. With fewer mouths to feed and fewer companions to divert their attentions, the garrison could continue its other preparations. Warned by its outlying sentinels of investment, it set fire to the buildings surrounding the town the next day, a

[3] *Gazette d'Amsterdam*, 23 May 1697, Nouvelles extraordinaires. In Paris, people predicted that it would only last eight days. *Gazette d'Amsterdam*, 27 May, though it was later reported that Vauban claimed it would take twenty-five days (30 May edition, from Paris, 24 May).

[4] AG A[1] 1401 #2, Montueil to the Secretary of War Louis-François-Marie, marquis de Barbesieux, 1 June.

standard preparation intended to deny the enemy cover on their approach towards the fortress. As the defenders worked to shore up its fortifications and organize its defenses, the besiegers were equally busy. The trace of the siege lines surrounding the town had to be marked off, the various regimental quarters set up, and bridges built to allow communication between the three siege camps, separated as they were by the Dender and Leuze rivers. The garrison fired its cannon at these distant targets, but to no effect. Vauban reflected on what the garrison's preparations suggested about his opponent: their cannonfire was ineffectual while revealing the exact range of their guns, and Roeux further illustrated his incapacity by needlessly burning all the buildings within half a cannon-shot of the fortress, even on those fronts where no one would even consider attacking. Yet at the same time he failed to destroy the many hedges and gardens that the besiegers could use for cover.[5] But other matters needed to be attended to before Vauban could confirm his poor opinion of the governor. Civilian surgeons from Valenciennes, Cambrai and other neighboring cities were ordered to report to the siege camp to assist the army surgeons and apothecaries with the wounded.[6] The transportation of the supplies and munitions for the siege required some 4,000 wagons and their civilian driver teams recruited from the surrounding regions. Twenty thousand peasants from Tournai to Valenciennes were also conscripted into helping construct the lines of circumvallation crawling across the plains and through the forests in order to isolate the town more securely and ward off potential relief forces. This lengthy field fortification, consisting of eight-foot thick breastworks fronted by a six-foot deep ditch, bristled with wooden palisade stakes and was studded with v-shaped redans every 800 feet. The last line of defense in case of a relief effort, it would be manned by most of the fifty battalions and fifty squadrons encamped just inside its perimeter and just beyond the range of the garrison's cannon. Constructing the lines and trenches also required amassing an enormous amount of wood in advance, in addition to the equipment needed to shift tons of earth. Cavaliers, whose service during sieges was usually limited to defending against sorties and serving in

[5] Rochas d'Aiglun (ed.), *Vauban*, vol. 2, pp. 450–451; *Relation du siège d'Ath*, p. 21; additional critiques on p. 31.

[6] On the surgeons, *Gazette d'Amsterdam*, 30 May, from Brussels, 27 May.

the observation army, were often charged with finding nearby thickets of trees, stripping their branches and lashing them together into six-foot long fascine bundles as well as weaving wicker gabions, two-and-a-half foot tall baskets with a similar diameter. These siege materials would join the more mundane tools of siegecraft, the thousands of picks, shovels, axes, mallets, stakes, sandbags, wooden planks, and the miles of rope that had to be transported to the site. All these and more would be needed to build the trenches and platforms necessary for a siege.

At the same time as the siege lines were being constructed and supplies conveyed to the site, small detachments of French officers carefully snuck towards the counterscarp, beating back lurking enemy parties in order to deprive the garrison of information as well as to facilitate a closer look at the defenses. Vauban was among the scouts, comparing these reconnaissances with his knowledge of the fortifications as he had designed them thirty years earlier. The town itself had been founded by a twelfth century count of Hainaut (Henegouwen), who constructed it in a plain at the confluence of the Dender and Leuze rivers. By 1700 the burg had grown to a moderate size, housing a population of some 6,000 inhabitants, while main roads connected it with the similarly-sized Oudenaarde to the northwest, and to the much larger regional centers of Brussels, Mons (Bergen), and Tournai (Doornik).[7] The town's medieval walls and château were modernized by the Spanish in 1540 and Vauban in turn had these walls razed in 1668. He replaced them, over the course of the next six years, with an expanded octagonal trace sporting eight angled bastions—this modern trace is illustrated in Map 2.2, while a glossary of fortification terms can be found in Map 2.3.[8] With this design,

[7] Emmanuel Fourdin, *Inventaire analytique des Archives de la Ville d'Ath de 1260 à 1860*, (Brussels, 1873), 2 vols. For the population, see Denis Morsa, "L'urbanisation de la Belgique (1500–1800). Taille, hiérarchie et dynamique des villes," *Revue du Nord*, 79 (1997). As is often the case in a region of multilingual inhabitants and in a period of non-standard orthography, the names differ: the Dender (Dendre) is often referred to in French as the Denre or Tenre, while the Leuze rivulet is sometimes referred to in French as the Villette (Le Villet), or the Irchonwelz.

[8] There are numerous plans of the fortress of Ath, ranging from maps published in both contemporary (Goulon, De Fer, Bodenehr) and modern works (Van Belle, *Relation du siège de la ville d'Ath en 1697*), to archival sketches, to the famous plan-relief (scale model) constructed after its capture in 1697, currently on display in the basement of Lille's Musée des Beaux-Arts. Although usually reliable, the map insert in J.W. Wijn, *Het Staatsche Leger*, vol. 2 is misleading in several respects. De Fer's maps tend to have serious flaws.

Ath became one of the few fortresses to come close to the idealized 'regular' star shape most associated with the modern *trace italienne* style, its only departure a flattening on its western side. Surrounding this curtain wall (*enceinte* in French) was a ditch that added additional height relative to the sunken floor, making it thirty feet tall to someone standing at the bottom—consult the profile of Map 2.3.[9] Normally only a few feet of water flowed through this channel as the Dender and Leuze meandered through the town and its fortifications, but when there was a threat of siege a sluice gate held back the water as an added defense, retaining up to eight feet in the ditch during the siege. Dominating the fortress were the angled *bastions* made famous by the Italian-style trace. The distance between any two of them was approximately 600 feet, the effective range of musketfire, so that each bastion could be supported by the musketeers from its two neighboring bastions. The bastions on the western front were seriously compromised by the Mont Feron (or Monferron) heights immediately to the west, which provided any enemy a perch from which to see directly into the town itself. Accordingly, Vauban had constructed on each of these four bastions a *cavalier*—an elevated breastwork in the middle of the bastion that could shield its defenders from enfilading fire. Down in the ditch between each pair of bastions, a narrow, low-lying chevron-shaped outwork sheltered the wet ditch and stretch of curtain wall directly behind it. Some 150 feet in front of these *tenailles* were triangular *ravelins*, each massive island faced with masonry ('well-revetted,' as contemporaries liked to say) and with room to hold several hundred defenders and a few small-caliber cannon.[10] One hundred and twenty

[9] Despite the fact that most of the architectural features of *trace italienne* fortifications were developed in Italy, most names used in the late seventeenth century were either French or direct translations from French. At the end of the sixteenth century Sir Roger Williams defended this already-established practice to his English readers "Some will condemne mee for my strange names of fortifications, they ought to pardon me: for my part, I knowe no other names than are giuen by the strangers, because there are fewe or none at all in our language." *A Briefe Discourse of Warre*, in John Evans (ed.), *The Works of Sir Roger Williams*, (Oxford, 1972), p. 41. The German engineer in Dutch service Johan Landsberg similarly justified his usage of French to his fellow Germanophones: *Nouvelle manière de fortifier les places, qui demontre en même tems les défauts qui se rencontrent dans la construction des ouvrages faits par les plus fameux ingénieurs modernes, et les moyens faciles & assurez pour les corriger dans toutes les parties des fortifications, tant des villes, que des citadelles, par une nouvelle méthode qui n'a point été mise au jour*, (The Hague, 1712), p. 11.

[10] Both English and French used the terms *démi-lune* ('half-moon') and *ravelin* (rav-

feet beyond these outworks was the ditch's outer retaining wall known as the *counterscarp*, which served as the foundation of the *covered way*.[11] This open-air walkway snaked around the outer perimeter of the town, covered the defenders' movements from the besieger's view and offered *banquettes* (steps that allowed one to see over the parapet) for the soldiers to fire from. Though the covered way was a continuous pathway, it was also designed to isolate any section captured by the enemy. Besiegers would usually first capture the most exposed 'points' of the covered way (the *salient angles*), but small breastworks alternating like a zipper along the counterscarp's length obstructed their view along its full extent. Defenders could preserve the rest of the outer perimeter by retreating behind these traverses, and would be supported by the dozens of troops who sheltered in the assembly areas (the *place d'armes* or place of arms) located in the *'re-entrant' angles* nestled between and behind the salients. Though most of Ath's bastions had only the covered way to shield them, Vauban had added two reinforced bulwarks (*counterguards*) to the covered way in front of the Luxembourg and Flanders bastions for additional protection. Before such works could be tackled, however, besiegers had to first cross the dead ground between the surrounding countryside and the counterscarp. Just beyond the crest of the covered way was the *glacis*, a gently sloping zone where interlocking fields of the garrison's fire converged against any attacker brave enough to charge up it, crowned with a double row of *palisades* near the top. To an observer at the foot of the glacis, this alignment of low-lying, gradually-sloped structures presented an unbroken view-scape leading the eye to the trees lining the parapets and the very tops of the buildings within the town itself.

elin)—some distinguished a ravelin from the similarly-shaped demi-lune according to its location and function—a demi-lune was an outwork that protected the bastion directly behind it (contrasted with Ath's two counterguards, which were attached to the covered way), while a ravelin sheltered the curtain to its rear from view. Many contemporaries, however, ignored this distinction. Some maps of Ath published at the time show a hornwork between the Hainaut and Luxembourg bastions (primarily Allied, but also Nicolas de Fer's plan of the siege), but this was actually a ravelin with several smaller lunettes in front of it. Some maps also depict a hornwork on the north/northwest corner in front of the Flanders bastion. Instead, the counterguard extended along the *capital* (i.e. the central spine of the glacis) to its base and was flanked by a small number of inundated lunettes.

[11] Technically, the covered way and counterscarp were distinct, but contemporaries frequently used the two terms interchangeably.

From the outside the city walls were sheltered by the masses of earth and masonry in front of them, while defending troops on their firing steps of the curtain wall could follow their line of sight down into the ravelin and beyond to the base of the glacis. Traditional storm tactics were useless against a multi-layered defense of alternating ditches and walls defended by gunpowder weapons: instead of a straightforward attack, each concentric layer of defenses would have to be negotiated and taken in turn before its rear ditch could be filled and the next one overcome. To survive the concentration of defensive musket and cannonfire, besiegers needed to protect their approach to the covered way with trenches dug in the earth. Once they reached the counterscarp, it would have to be captured and secured, then the ditch in front of the ravelin, then the ravelin itself, then the main ditch behind the ravelin, and finally the main town wall itself. The author of the Goulon journal of the siege, with an engineer's eye for the ideal, would call Vauban's application of these principles at Ath a "perfect model of the Art, on whatever side you examine it" (p. 92). Vauban was now called upon to capture his own creation, a project he had frequently mulled over since it had been handed back to the Spanish in 1678.

With the siege line construction and other preparations wrapping up, Catinat and Boufflers reviewed the troops on the 22nd of May. At seven o'clock that same evening, the trenches were opened (Map 2.4).[12] The attack centered on the Brussels gate, though the first parallel would eventually envelop four bastions and more than a third of the town's circumference—from the Upper Dender River and the Burgundy bastion on the besieger's left to the Brabant bastion on the right. The trenchworks were divided into two approaches, the honor of commanding the right attack going to the senior Lieutenant-General René de Froullay comte de Tessé, and the left under the *maréchal de camp* Ferdinand comte de Marsin. A third, false, approach was led by the engineer/miner Mesgrigny and intended to stretch

---

[12] For a detailed exposition of the mechanics of Vaubanian siegecraft, consult his treaties: Rothrock (ed.), *A Manual*; and Sébastien le Prestre de Vauban, *De l'attaque et de la défense des places*, edited by Antoine-Marie Augoyat, 2 vols. (Paris, 1828), vol. 1, (hereafter Vauban, *Traité de l'attaque*). Duffy's modern *Fire and Stone* gives the best technical details and provides a wider chronological and geographical discussion. Excellent color illustrations from Vauban's works accompany selections of his text in Nicholas Faucherre and Philippe Prost, *Le triomphe de la méthode: le traité de l'attaque des places*, (Paris, 1992).

the garrison out even further. At each of the real approaches they would be assisted by one of Vauban's subordinates as an *ingénieur en chef* in command of a brigade of nine engineers. Three battalions guarded the posts at each attack while 800 soldiers from those regiments that would not serve guard duty in the trenches were divided into 50-man brigades and started turning dirt and placing thousands of fascines and gabions according to the engineers' directions. The trenches would eventually require thousands of man-hours to construct, with officers, engineers, carpenters, sappers and guards all playing their part. Those serving the first night silently filed off to their appointed positions and broke ground, with Vauban placing the first twenty fascines himself before moving to the next approach to oversee its trenchworks. This first night of digging went well: by Thursday morning the workers had advanced more than 2,500 feet towards the town, where the two approaches were connected by a single trench running parallel to the front of attack at 1,900 feet distant from the outworks. With this first parallel established, the besiegers managed to accomplish in one night what had been expected in two without suffering even a single casualty, thanks in part to rain showers which obscured their progress. Once within range of the garrison's cannon, they followed the prescribed rules for trench construction as written in Vauban's manual. The process for sapping began at the head of the trench, i.e. closest to fortress fire, where a lone sapper wheeled a wooden mantlet ahead of him as a portable shield while he interposed an empty gabion between himself and the fortress. He then scratched out the beginnings of a trench sprouting from the parallel and shoveled up to twelve cubic feet of excavated dirt into the wicker basket before creeping forward to repeat the process with another gabion. Behind him other workers would follow in his path, each widening and deepening the trench a little bit more and reinforcing the barrier with more earth and fascines until men could safely walk in the trenches completely concealed from the town's view. After an hour or two of such tiring and dangerous labor, those at the head of the trench would be relieved by fresh bodies in order to keep up a brisk pace. With the furrows taking shape, small groups from the regiments slated for guard duty could then take up their positions in the trenches, firing steps allowing them to peer over the rampart and return the defenders' fire. The trenches would eventually be widened to at least twelve feet so horse teams could haul artillery pieces to the siege batteries in relative safety. In the morning the 1600 workers would be relieved by 1600 others, and after every 24 hours a new lieutenant-general would serve as 'general of the day' under each approach commander's supervision, while the three guard battalions would be relieved in the evening by a similar number carrying fascines with them into the trenches. All these would be supported by an additional reserve of 250 fusiliers, 200 horse, and a company of miter-capped grenadiers, the shock troops of the early modern world. The siege of Ath would last long enough for each guard cohort to see a total of two days of trench duty.

Over the next two days the trenches crept closer, their approach saps zigzagging to protect the besiegers from enfilading fire—just before the point at which the garrison's cannon would be able to enfilade down the trench, the engineer would cut the trench back in the other direction, always being sure to offer only its cannon-proof breastworks as a target. The night of the 24th a second parallel capped the two approaches within 1,000 feet of the covered way. Four of the six battalions on trench duty now took up positions in these completed lines, with the remaining two guarding the approach trenches, the first parallel, and points in between. Tightening the noose around the fortress, the concentration of French troops forced garrison sentinels who had hidden themselves in the grass to retreat back into the covered way. At the same time the garrison increased its fire against the approaching besiegers, their thirteen active guns wounding ten French soldiers over the course of these twenty-four hours. The garrison's musketry increased in volume as they neared the works, but after five nights of trenchwork, the number of French wounded had increased to only fifty men of all ranks.

These losses were quite small given the fact that Vauban, against the taunts of the defenders and the wishes of his own troops, waited until the second parallel was complete before he sited the first artillery batteries. Thus, the furrows pushed forward these first five days succeeded without any visible assistance from their siege train or its brigadier of artillery Jean-Baptiste de Vigny. While the trenches snaked towards the covered way in this first phase of the siege, artillery was needed to prepare their goal for capture by smashing through the palisades and killing or maiming those on the counterscarp, as well as those in the ravelins to their rear who would offer them fire support. Once the guns necessary for the task had been manhandled into position behind twenty-foot thick cannon-proof parapets, the besiegers fired their first cannonade on the 27th (Map

2.5). The besiegers had seventy-three cannon in their artillery park to call upon, thirty-four of these specialized for breaching the thick fortress walls.[13] Forty-one smaller caliber pieces were also available to target the individual defenders, their guns, and the summit of the fortifications. Of this number, Vauban would use no more than forty-two cannon and twenty-two mortars at any one time. To the relief of those in the trenches and the consternation of the defenders, they finally opened fire *à ricochet* with thirty-six cannon in six batteries (A-F). The gunners were skeptical of the new technique of ricochet fire Vauban introduced, though they slowly came to appreciate the almost magical way in which the continuous barrage of solid shot bounced along the ground in front of the glacis, then ricocheted up over the crest of the covered way, shattering palisades and even razing the tops of the parapets, before plunging back down into the covered way itself to strike at cowering defenders. Wet weather hindered the first day's fire, making the gun platforms slippery and their powder damp, but with drier weather this innovative technique took a heavy toll on those sheltering unseen in the outworks. The fire from these thirty cannon was so effective in its ability to "seek them out in the most hidden of places" that the defenders soon refused to show themselves above their smashed parapet. Even more, the defenders' musketfire declined dramatically as the ricochet batteries kept up a continuous fire, and within a day of the first volleys all but two of the defending cannon had been silenced.[14] Inspecting the survivors after it was all over, the French would estimate that one hundred defenders had been killed each day by these rounds, while many other victims of ricochet fire lay in their hospital beds missing arms or legs.

A battery of twelve mortars was also constructed to aim at the Cambron and Brabançon ravelins (battery 1), while another battery of twelve mortars (battery 2) started bombarding the Recollets ravelin on the 28th. Three of the largest mortars in this second battery began firing soon after, and quickly shifted their fire in order to target the sluice gate retaining the waters of the Upper Dender. The heavy mortars made short work of the lock with their 540-pound

[13] Accounts of the exact numbers and calibers differ—some sources counted twenty 24-pounders and twelve 33-pounders.

[14] Quote from *Mercure galant*, Juin 1697, p. 224.

bombs. A direct hit on the 31st released within a few hours most of the water out into the river channel and flooded the town square for a number of hours. Through the frigid rain workmen took advantage of the ricochets to make communication ditches between the works while their zigzag saps snaked forward from the second parallel. The frontage of the attacks focused in on their targeted section of the Brabançon ravelin and the Namur and Limburg bastions to either side, and this narrowing of frontage allowed the besiegers to halve the manpower in the trenches. Only three hundred workers and three guard battalions would be necessary at each attack, thus requiring fewer days of trench duty for each cohort. In Vauban's theoretical treatises he called for a third parallel to be built 100 feet or so from the glacis in order to minimize the distance assaulting troops would have to travel on their charge up the covered way, but the garrison's dwindling resistance convinced him to forego this additional effort. He instead constructed small lodgments that branched laterally off of the saps, where small groups of grenadiers could watch over the workers. Although Vauban had a number of mounds (*cavaliers de tranchée*) built to peer down into the covered way, ricochet fire made their use unnecessary.

The result of Vauban's ricochet fire was evident when they captured the town's covered way on the night of the 29th. Through three days of heavy rain the zigzag saps had approached within a few dozen feet of the top of the glacis, yet the covered way itself had shown practically no resistance. To investigate further, a small reconnaissance party led by the comte de Marsin and the engineer Richerand snuck up to the crest of its salient angle and to their surprise discovered the counterscarp completely empty. Less than two days of ricochet fire from thirty-six cannon had forced the defending soldiers to abandon it entirely. The scouts quickly sent for other engineers and workers to bring up gabions and started making a lodgment without taking a single casualty. By daylight they had surrounded the angles at all three salients. Only in the *places d'armes* did they discover a few surprised defenders, who abandoned their posts immediately. All the next day French workmen shored up their lodgments along the newly-won ground and extended their saps to envelop more of the covered way to the left and the right, piling up dirt embankments on the inner lip of the covered way and turning it against the fortress. Heavy fire from the town ramparts had little effect: establishing these posts on the covered way cost the besiegers a paltry three dead and eight wounded, including two engineers. Almost every report from the siege camp over the next few days expressed two emotions: surprise at the ease with which the covered way was captured, and relief that Monsieur Vauban was not more seriously wounded by a sandbag-piercing musketball that bruised his collarbone.

Now that the covered way had been compromised, the lodgments could almost inevitably expand laterally along the crest of the covered way. Vauban, shrugging off medical attention for his wound, turned to implementing the next phase of his plan, capturing the Brabançon ravelin. First, at the same time as lodgments were being established on top of the covered way, French miners began 'descending the ditch' back at the base of the glacis, digging tunnels that sloped downward underneath both the glacis and covered way until they broke through the retaining wall at the bottom of the ditch in front of the ravelin. With the water level lowered to only three feet by the end of May, the besiegers would use these openings into the ditch as jumping off points for fascine bridges to cross the vanishing wet moat, under the supporting fire of their brothers in arms on the reversed covered way above them. While this subterranean project was in progress, the artillery too had a continued role to play (Map 2.6). To soften up the Brabançon ravelin and make an assault practical, the two batteries (1 and 2) of twenty-one 250-*livre* mortars and the single six-gun ricochet battery F that had been targeting the collateral Cambron and Recollets ravelins were supplemented by additional firepower. Two new ricochet batteries (G and H) were built on the far periphery of the attacks to enfilade the two collateral ravelins, while the existing central ricochet batteries (B–E) shifted their fire from the covered way, which was now crawling with their own men, to targeting either the Brabançon ravelin itself with ricochet fire (battery E) or firing on possible support positions with direct fire (B–D). With the covered way captured, four of the guns from ricochet battery A were moved forward to a new one that would be able to play a role in the final attack on the bastions themselves, while the two remaining guns kept an eye on any enemy efforts to repair the sluice gate. Enabling a forlorn hope of grenadiers to assault the central ravelin required two new breaching batteries (J and K, totaling five pieces) on the palisades to target both of its faces, while the interior and rear of the work was targeted by ricochet and mortar fire. Several miners worked to enlarge the breaches made by these guns, so that by the last day of May improvised bridges had reached the rubble that had accumulated at the bottom of the breach. With this talus obscuring the flanking bastions' view, grenadiers crossed the fascine bridge and established a lodgment on the work, repulsing several attempts to recapture it. The seventy-odd defenders that survived this struggle for possession of the outwork retreated back to the small brick-faced *réduit* that commanded the rest of the ravelin from behind its own moat. Their communication bridge connecting the réduit to the tenaille behind them had been destroyed by enemy bombs and there was little hope of retreat, as any defender crossing the practically-dry ditch was now visible to French musketeers in their lodgments on the periphery of the covered way opposite the points of the flanking bastions. With no hope of relief this last outpost on the ravelin surrendered at discretion on the third of June.

With the ravelin now cleared of enemy troops, the final objective was in sight: breaching the bastions themselves. Roeux could only watch as the French workmen enveloped the interior of the captured ravelin with trenchworks and established a base for a gallery which would lead to the tenaille and the Brussels gate beyond it (Map 2.7). The main targets, however, were the Namur and Limburg bastions. By the second of the month twelve breaching cannon in two batteries (M and N, located on the angles of the covered way just opposite the targeted bastions) fired their first volleys against the right face of the Namur bastion and the Limburg bastion's left face respectively. The fully-charged pieces fired in salvoes: first they targeted imaginary horizontal lines six feet above the waterline, then the gunners proceeded to create a vertical line upwards, working their way slowly up the wall to within six feet of the summit. The simultaneous impact of these precisely-placed cast iron rounds brought the masonry crumbling down into the ditch within a few days. Vauban also ordered a bombardment of the bastions' open-air interiors to prevent the polyglot defenders from shoring up the beleaguered walls. In addition to continued ricochet fire from battery F and indirect fire from the three heavy mortars in battery 2, the other twenty-one mortars were also moved forward and placed in a massive battery (3) in front of the now-secure ravelin. The adjacent flanks of these two bastions were also targeted by direct fire from two batteries of four cannon each (L and O)—these two gun emplacements also kept an eye on the tenaille, though the garrison had thus far failed to utilize it.

The musketfire from the town ramparts diminished daily as the garrison's casualties mounted and as the French continued to lay down a high volume of fire on the targeted bastions and their collateral flankers from an increasing number of angles. Breaching batteries continued to enlarge the gaping holes in the faces of the Namur and Limburg bastions, as French workmen started laying down fascine bridges to cross the final ditch opposite them. King William III had already abandoned a half-hearted attempt to lift the siege—it now accelerated towards its inevitable conclusion. The governor's troops, too few in number to begin with, were quickly losing the will to fight. From the French side, the situation was as bright as the Allies' was dark. Their biggest concern was not a relief army, but that they would be forced to abandon the siege due to a suspension of arms pursuant to the signing of peace. Barring such an event, Vauban was so certain of the inevitability of its capture that he boasted that even 10,000 defenders led by the Devil himself would be unable to prevent its capture:

> things are going marvelously; up till now the enemies have done nothing to further their cause and I see with great pleasure that I know their works better than they do. It is also true, and I can say this without being a braggart, that never has a place been attacked with as much art and speed at the same time.[15]

He was probably right, for the breaches in the faces of the two bastions were 'practicable' by the 5th. Workers laid down bridges of fascines and sandbags that almost completely spanned the 130 feet between the counterscarp and the base of each breach; a prefabricated *pont volant* promised an even quicker spanning of the tenaille ditch. Crossing over such passageways, assaulting troops would have a smooth ascent up the talus of rubble to the summit of the main wall, where twenty to thirty men abreast could pass through the opening into the interior of the fortress. They were likely to meet little resistance once at the top, since the plunging fire of mortar battery 3 and ricochet fire from batteries F and I had discouraged

---

[15] Rochas d'Aiglun (ed.), *Vauban*, vol. 2, p. 455.

the defenders from building ad hoc retrenchments behind the breach to serve as a last line of defense. With Vauban and Catinat scouting out the breaches in preparation for a general storm on the town and with twenty companies of grenadiers held ready for the assault, Roeux ordered the chamade drum to be beaten at the front of the attack and hung out a white flag to signal his willingness to negotiate surrender. Catinat postponed the impending attack. Hostages of equal rank were exchanged to ensure negotiation in good faith, and while these discussions were carried on a number of garrison officers descended the breach for a semi-friendly *tête-à-tête* with their peers in French service.

Terms were agreed upon that night, and the garrison was allowed to freely evacuate on the 7th through the breach in the Namur bastion, with the exception of five officers who were held back in reprisal for similar Allied behavior two years earlier at Namur. Overall the siege had cost the besiegers fourteen days of open trenches (twenty of investment) and the lives of fifty to seventy Frenchmen with 250 others wounded. Of the sixty-two engineers at the siege, two had been killed and seven others wounded seriously. Powder consumption had been less than half of what was stockpiled on site (266,400 *livres*, or almost 290,000 English pounds), while other munitions consumption was calculated at 34,000 *livres* of lead, 27,050 cannonballs (24,300 of these being for 24-pdr cannon), 3,400 mortar bombs, 950 grenades and 12,000 sandbags. The overall costs were calculated at 89,250 *livres*—in all a small price for such a large fortress.[16] After the comte de Tessé was made governor and his regiments received their orders to garrison the town, the trenches were filled in by 6,000 peasant workers and planning began on repairing the town's smashed fortifications. Looking forward, Louis had entertained thoughts of besieging Oudenaarde as well, but Vauban would not be ready for another siege for several months (many of his engineers had left to reinforce an attack on Barcelona), while the Allied army blocked an advance towards Brussels. Without an obvious target of opportunity, Boufflers was ordered to meet with William's plenipotentiary, Hans Willem Bentinck, first Earl of Portland and give new

[16] AG A^1 vol. 1401, #84. AG A^1 1401, #88, Estat des consummations de munitions qui ont esté faites au parc de l'artillerie devant Ath jusques au 4e Juin inclus. A^1 1400 #280 contains an earlier list dated 31 May.

impetus to the peace talks. The Treaty of Rijswijk was subsequently signed at the end of September, putting an end to the war, if not settling the issues for which it was fought.

Ath in Retrospect

Averaging more than a siege per year over forty years of service, Vauban brought his siege attack to perfection at Ath. His reliance on trench parallels was supplemented by the newly-perfected technique of ricochet fire. Unlike the easy conquests early in Louis' reign, here Vauban faced a strong fortress he himself had renovated—detractors could not belittle this accomplishment by pointing to the frailty of the fortifications as they might earlier captures. It was in many ways a model siege.[17]

That this siege was an exemplar and worthy of our attention is widely attested to, both at the time and afterward. Vauban's immodest appraisal of its unparalleled conduct was echoed by his peers. The Goulon journalist, proud of his own participation, declared bluntly: "No siege was ever carried on with so little loss or expence."[18] Particularly noteworthy to contemporaries was the first mature use of ricochet fire, publicized in the *Mercure galant* and highlighted by the famous scientist Bernard de Fontenelle in his 1707 eulogy of Vauban delivered to the *Académie des Sciences*. Participants recognized the siege as a *tour de force*, even a non-engineer like the chevalier de Quincy:

> Every single day I and my comrades went and viewed every detail of every part of the trenches. I was delighted to see and learn everything that occurred when attacking a fortress. There was much to learn since it was regarded as the most skillful, the most intelligent, and the most perfect attack ever conducted by the great Vauban. We went in such large numbers that we crowded the trenches, which led to a general order that forbade us from putting even a foot in them, under penalty of imprisonment. But despite this harsh measure, I found myself there from morning until evening.[19]

---

[17] On the other hand, the garrison's defense was excoriated by observers.
[18] "A Journal," p. 144.
[19] Lecestre (ed.), *Mémoires du chevalier de Quincy*, vol. 1, p. 45.

Engineers from other countries, even former enemies, traveled to Hainaut in hopes of seeing this siege conducted by the famous Monsieur de Vauban.[20] What is more, the attack retained its luster even when viewed with hindsight. Thirty years later the marquis de Quincy spent a dozen pages on the siege in his multi-volume history of Louis' wars, explaining to his audience that these details would allow one to "recognize Vauban's most perfect siege ever conducted."[21] The most famous engineer of the French Revolution and 'Organizer of Victory' Lazare Carnot proclaimed the attack on Ath Vauban's "chef-d'oeuvre."[22] Another turn-of-the-century engineer, Bousmard, summarized it thus: "never had a siege been so brief, cost the besiegers so little money, and proportionally been so deadly for the defenders."[23] The Napoleonic-era historian of the French engineering corps reiterated Vauban's opinion of its management:

> No siege has ever been conducted so methodically and with so little destruction. In no other has force played so small a role, nor art so large a role. Never have mistakes been avoided so skillfully, nor the faults of the enemy so well exploited. This was the triumph of Vauban. It was said that it was not so much a siege as an idealized version of one. It gave a great example to engineers.[24]

The mid-nineteenth century engineer/historian Antoine-Marie Augoyat indicated the persistence of this consensus decades later, repeating Vauban's claim that: "Never had a place been attacked with so much art and such speed, and with so little loss of life."[25] The recurrent identification of this combination of skill, speed and safety is particularly salient as it had special resonance for engineers. We turn now to a more detailed explanation of how and why this siege of Ath appealed so strongly to these technicians.

---

[20] With the Duke of Savoy no longer at war with France, he sent three of his engineers to observe the siege, though the town surrendered before they arrived. Rochas d'Aiglun (ed.), *Vauban*, vol. 2, p. 467 Barbesieux to Vauban, Marly, 22 July.

[21] *Histoire militaire du règne de Louis le Grand*, p. 307.

[22] Lazare Carnot, *Éloge de M. le Maréchal de Vauban*, (Paris, 1784), p. 16.

[23] Henri Jean Baptiste Bousmard, *Essai général de fortification et d'attaque et défense des places*, (Paris, 1797), vol. 1, p. 20.

[24] Allent, *Histoire du corps impérial du génie*, p. 351.

[25] Augoyat, *Aperçu historique*, vol. 1, p. 218. For more recent accounts, see Pujo, *Vauban*, pp. 217–218—"un modèle du genre"; and Blanchard, *Vauban*, pp. 349–350.

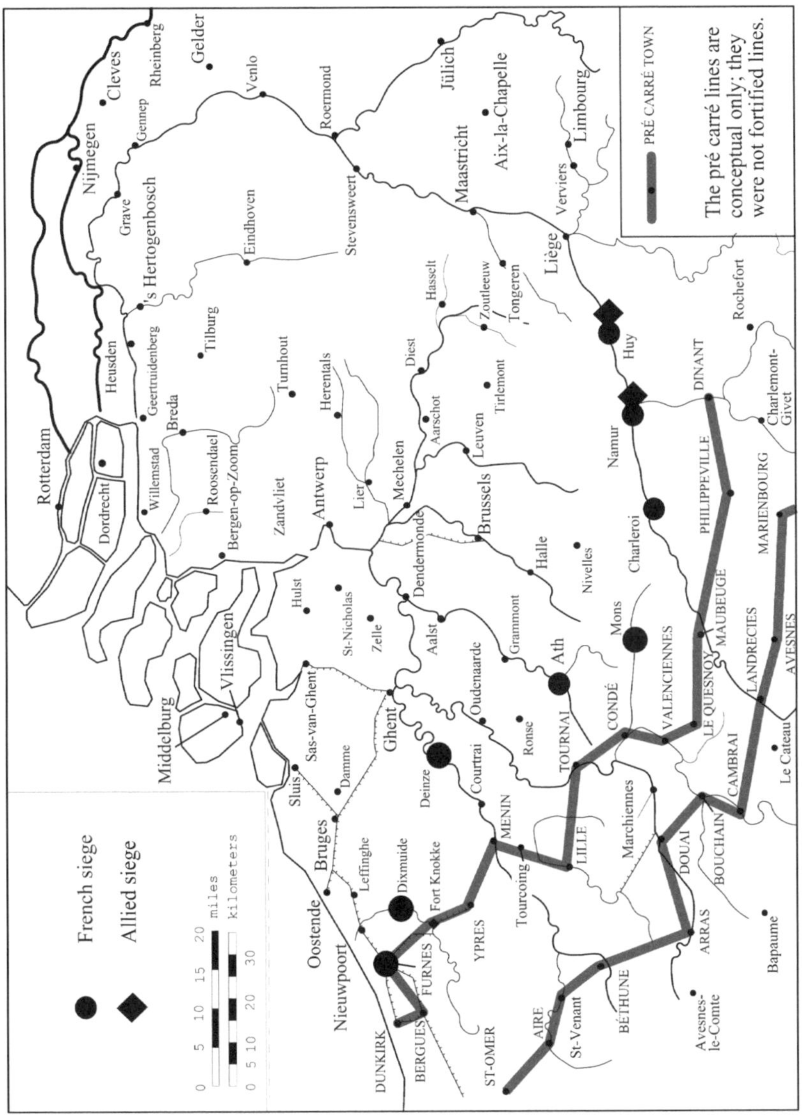

Map 2.1: The Low Countries Theater in the Nine Years War

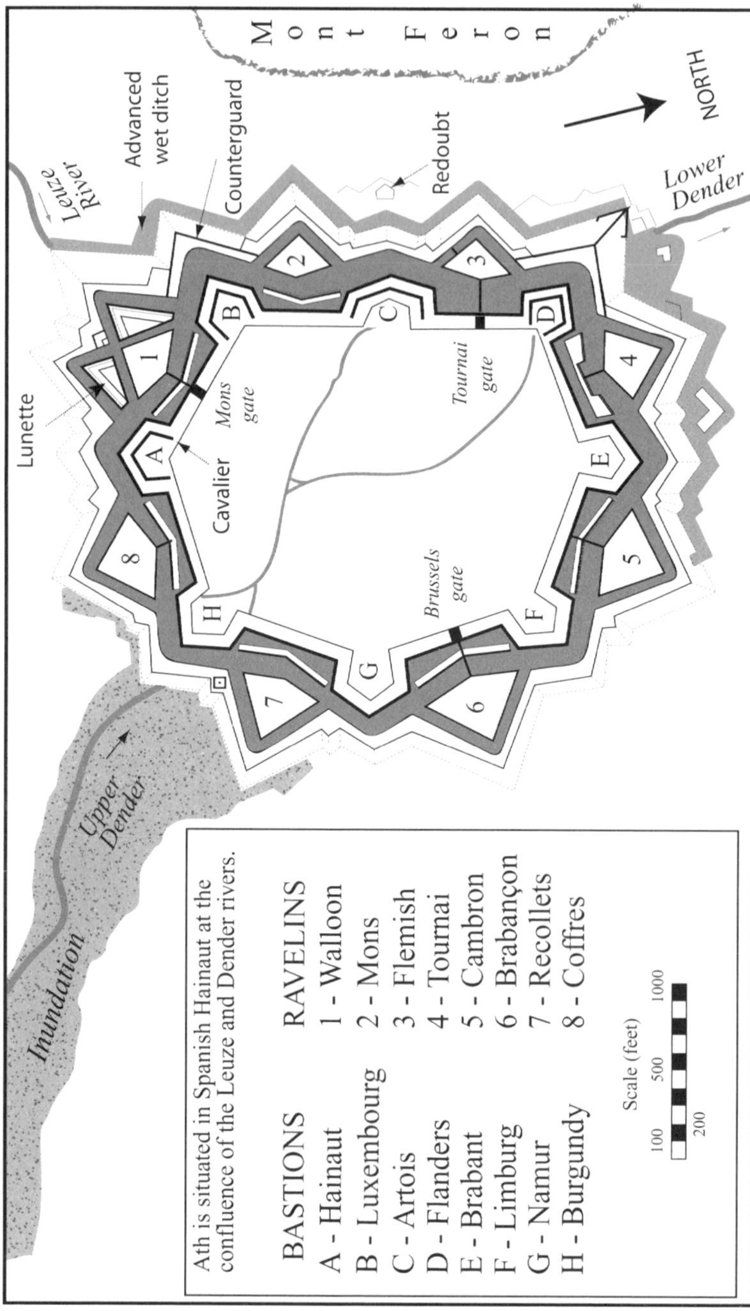

Map 2.2: The Fortress of Ath 1697

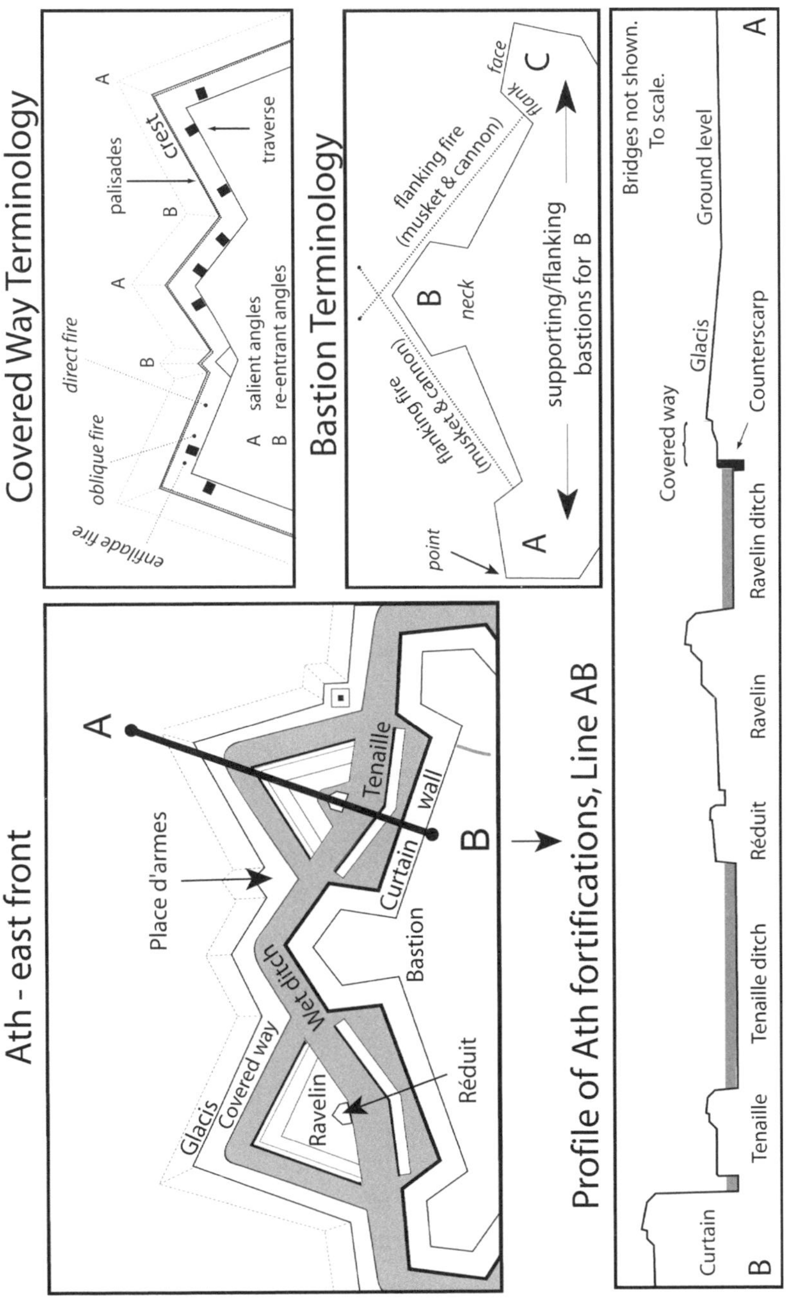

Map 2.3: Fortification Terms

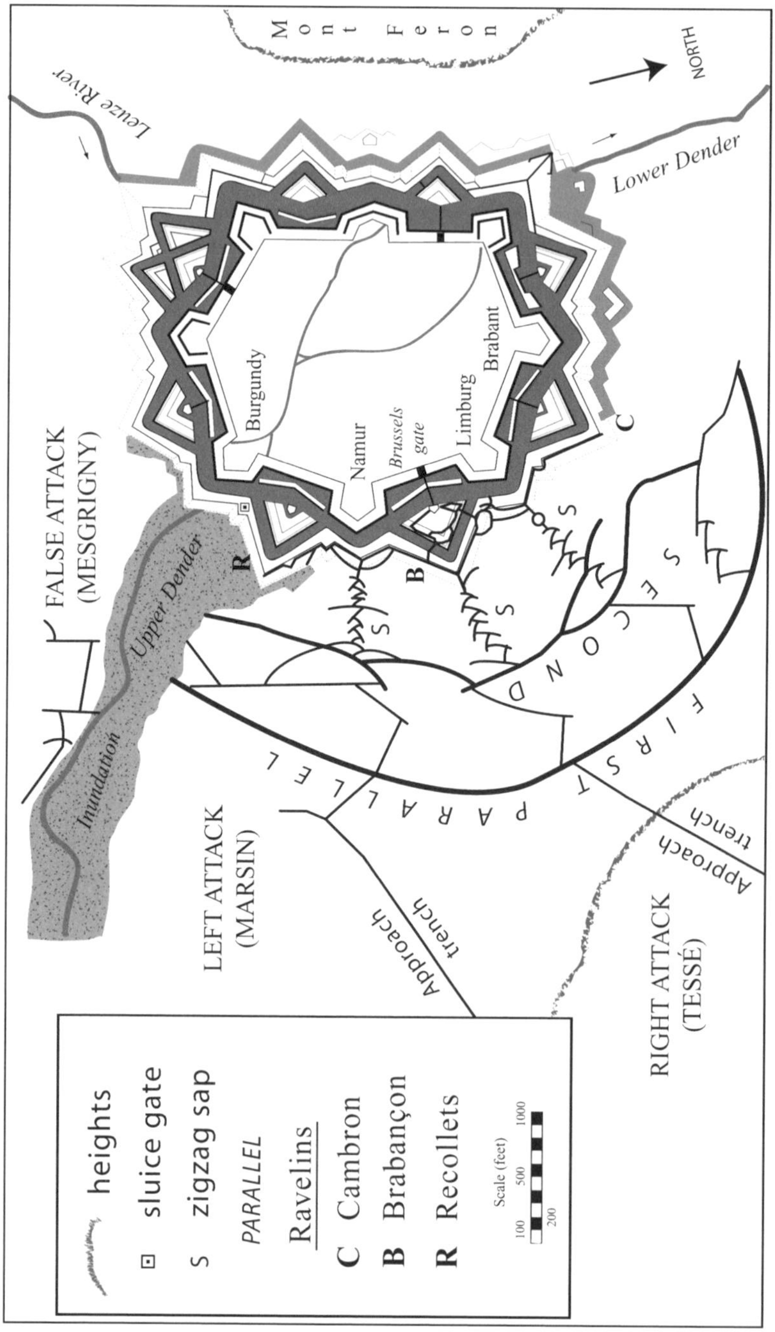

Map 2.4: The Siege of Ath 1697

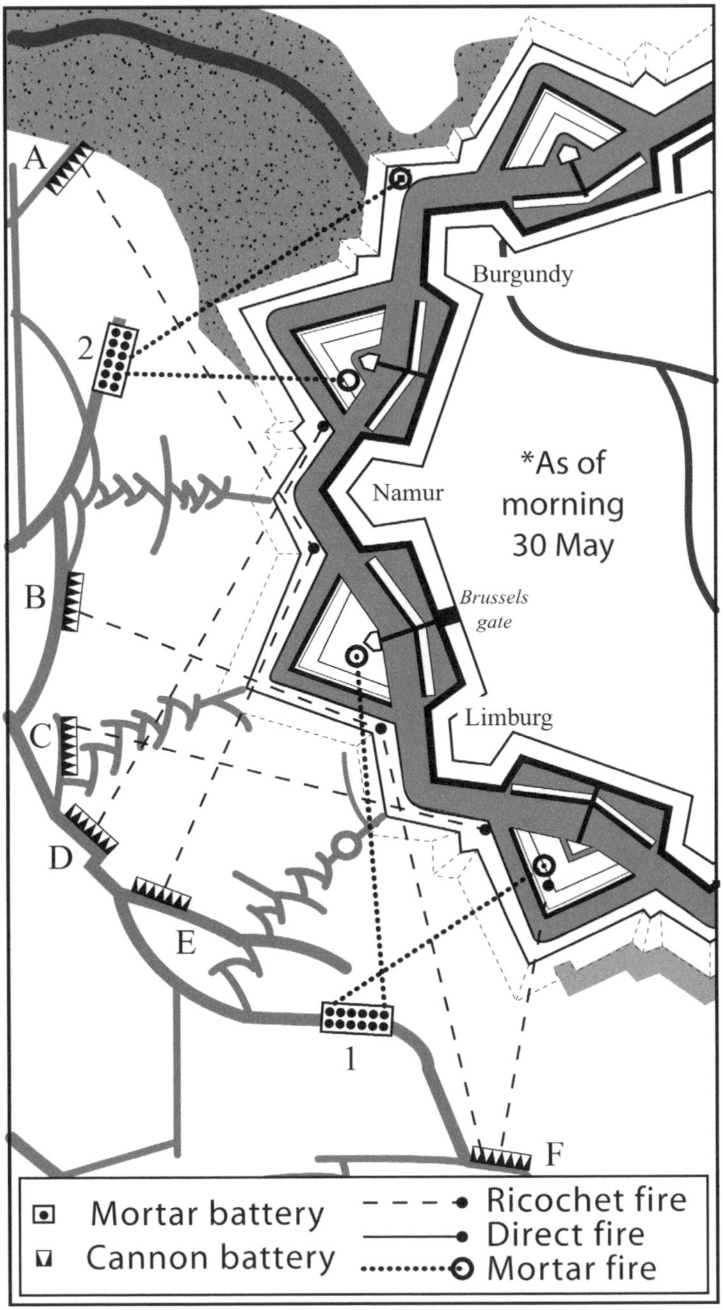

Map 2.5: Ath Phase I: Attacking the Covered Way

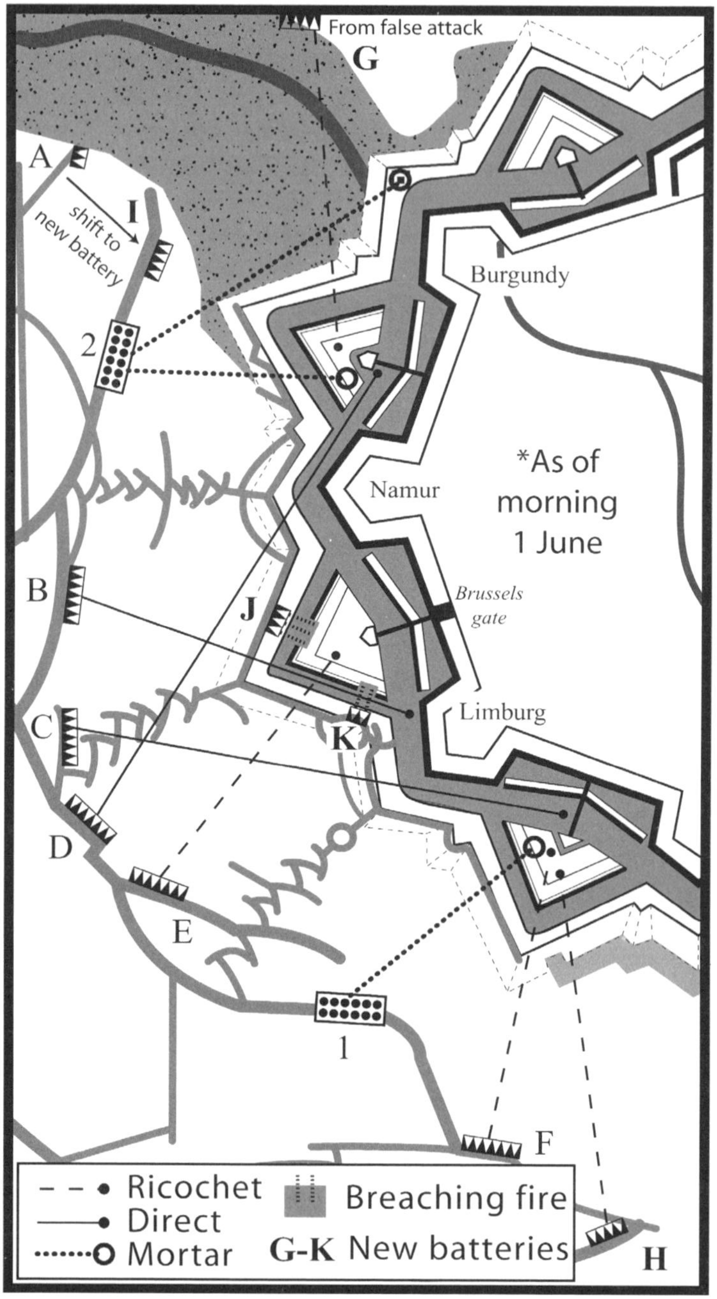

Map 2.6: Ath Phase II: Attacking the Ravelin

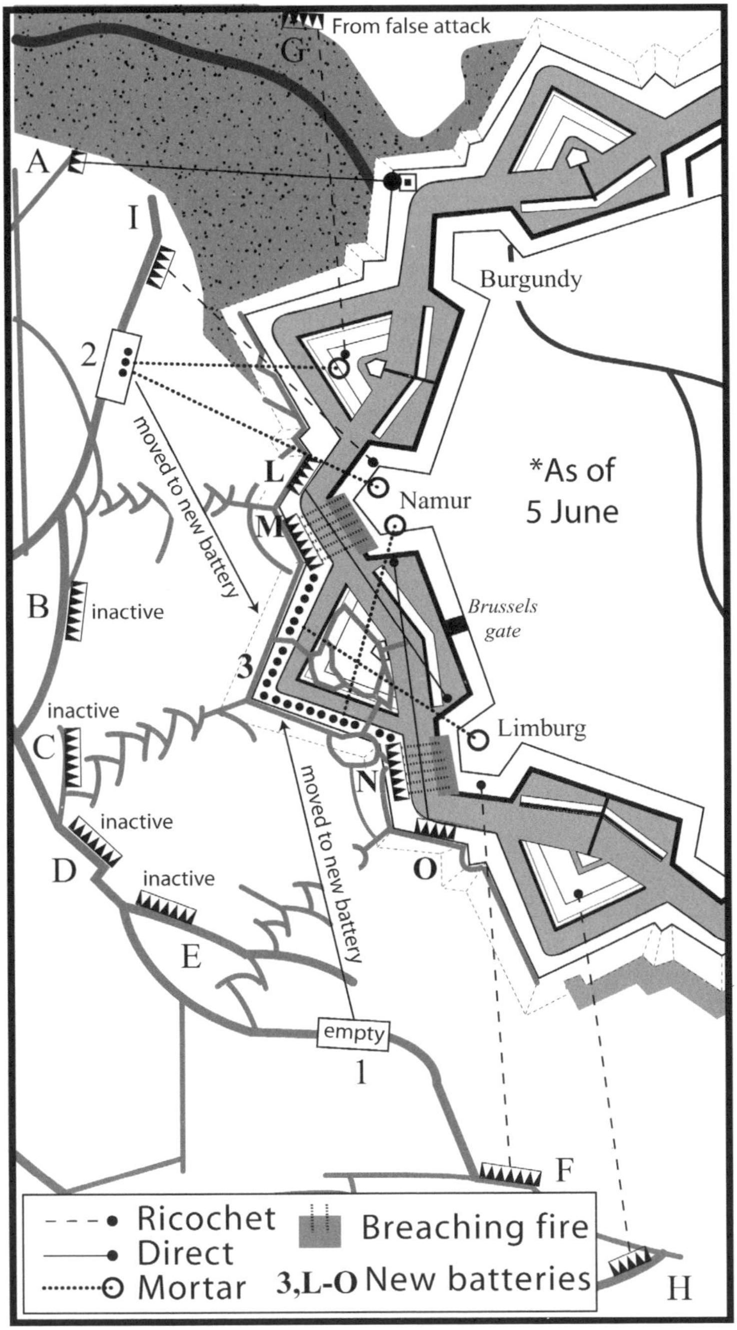

Map 2.7: Ath Phase III: Attacking the Bastions

CHAPTER THREE

EFFICIENCY AND THE PERFECT SIEGE

With over forty sieges to his credit, Vauban's followers could have chosen other sieges to emphasize. But several reasons converged to focus attention on the 1697 siege of Ath. First, given the importance of Vauban's legacy to the engineering community, a capstone of his career was only appropriate to show him at the zenith of his art. The last two sieges of his illustrious career (during the War of the Spanish Succession) were less heroic and less worthy of his true talents. His brief, unsuccessful attack on the forts surrounding Hulst in 1702 would hardly be a representative endpoint, as it was the only place to resist the *maréchal*'s efforts. In defense of Vauban's reputation, Louis had ordered him to besiege these forts (approachable only along narrow causeways) against the engineer's better judgment in order to divert the Allies' attention from their siege of Venlo. Short of men and supplies, Vauban could make little impression against either the forts or the Allies' focus on Venlo, so he was allowed to lift the siege after a week of lackluster effort. Nor would the 1703 campaign offer a proper tribute to this historical figure: his attack on the Rhenish town of Old Breisach was, like so many others, successful, but its conduct was contested by the commander, and this resistance to his ideas would become an increasingly common obstacle for his few remaining years of life. Ath, on the other hand, was a fitting tribute to chronicle the epitome of Vaubanian siegecraft in all its glory, but it was more than just that. This siege offered engineers the opportunity to illustrate a fundamental tenet of military engineering, the desire to constantly improve the efficiency of siegecraft.

1. *The Engineering Pursuit of Efficiency*

Vauban's improvement of the discipline of poliorcetics (attacking fortifications) resided in his detailed exposition of how to attack a fortress by managing the chaos of positional combat as much as in his successful practice of these skills. In this educational vein he was continuing a long engineering tradition, for since the birth of this profession its members have constantly sought to increase the productivity of labor through mechanical means. Before the development of discrete engineering specialties in the mid-eighteenth century, architects, military engineers, artillery engineers and civil engineers all learned their craft in a common artisanal fashion and all shared an emphasis on efficiency, defined at its simplest as a measure of output per unit of input.[1] This interest in improving productivity is seen most clearly in the literature on the design of fortifications. After decades of predominance the *trace italienne*-style fortresses began to fall prey to the growing power of siege armies. In response, engineers of the late 17th century sought to shift the balance back towards the defense, often justifying their proliferation of defensive designs in terms of economy: how wasteful, they reminded their readers, to expend so much money on a multi-bastioned stronghold only to see it surrender after a single breach in only one of its bastions![2] Later in the eighteenth century a representative of the increasingly-professionalized French military engineers, Fourcroy de Ramecourt, would take this to the logical extreme. He proposed a 'scientific' technique of determining optimal efficiencies by calculating a "moment of fortification" for each fortress, that is, a numeric ratio of the length of a fortress's defense during a theoretical siege relative to the expense

[1] On engineering (military and otherwise) in the early modern period, see Janis Langins' excellent *Conserving the Enlightenment*, which emphasizes the engineers' conservative pursuit of gradual, incremental improvements rather than radical, 'destructive' change. Other works on the engineers of the period include Simon Pepper, "Artisans, Architects and Aristocrats: Professionalism and Renaissance Military Engineering," in D. Trim (ed.), *The chivalric ethos and the development of military professionalism*, (Leiden, 2003); John R. Hale, *Renaissance Fortification: Art or Engineering?*, (London, 1977); on the artillery engineers Ken Alder, *Engineering the Revolution: Arms and Enlightenment in France, 1763–1815*, (Princeton, 1997); and more generally Hélène Vérin, *La gloire des ingénieurs: L'intelligence technique du XVIe au XVIIIe siècles*, (Paris, 1993); and Roger Hahn and René Taton, *Écoles techniques et militaires au XVIIIe siècle*, (Paris, 1986).

[2] For example, Blaise-François Pagan, *Les Fortifications de Monsieur le Comte de Pagan avec ses theorems sur la fortification*, (Brussels, 1668 edition of 1645 original), Preface. Later in the period, we could cite Jacob de La Vergne, *Nouvelle fortification imprenable par force d'armes* (Vienne, 1700), Dédicatoire; and, much later, Jean-Bernard Virgin, *La défense des places, mise en équilibre avec les attaques savantes et furieuses d'aujourd'hui*, (Stockholm, 1781), Dédicatoire.

of its construction.[3] Engineers similarly sought to constantly improve the conduct of the trenches. Here military engineers attempted to increase efficiency primarily in terms of economizing on wasted effort, which might include everything from superfluous trench and battery construction to excessive powder consumption. This pursuit of siege efficiency also encompassed a more general desire to improve existing techniques.[4] The journalist of the model siege of Ath noted Vauban's constant quest for ever-greater efficiency, elevating him as a model engineer whose "happy genius was always contriving new expedients to facilitate the approaches."[5] Towards the end of the eighteenth century the French military engineer Jean-Claude-Eléonore Le Michaud d'Arçon identified this as Vauban's guiding philosophy—and by extension the ideal for all engineers—striving to "obtain the most with the least."[6] The engineering craft revolved around a veritable cult of efficiency, one that became increasingly explicit as the eighteenth century progressed.

If engineers sought the most efficient siege possible, what exactly did Vauban's systematization of the attack actually entail? To answer this question historians use two complementary approaches. First, there is a brief discussion of the three tactical innovations Vauban introduced: trench parallels, *cavaliers de tranchée*, and ricochet fire.[7]

---

[3] Duffy, *The Fortress in the Age of Vauban and Frederick the Great*, p. 153; Vérin, *La gloire des ingénieurs*, pp. 368–378 and pp. 31–33 more generally; and Langins, *Conserving the Enlightenment*, pp. 337–338.

[4] Pagan, one of the more modest of the technicians, justified his attempts to meld experience with Geometry as a way to "add to the inventions of so many grand Captains and so many celebrated authors." Preface to his *Les Fortifications de Monsieur le Comte de Pagan*. Following this approach, Vauban's pupil Naudin pledged to his readers to "explain in depth those things that have only been treated superficially." M. Naudin, *L'ingénieur françois*, (Paris, 1695), p. ii.

[5] *A Journal of the Siege of Ath*, p. 112.

[6] Le Michaud d'Arçon, *Considerations sur l'influence du génie de Vauban dans la balance des forces de l'État*, (Strasbourg, 1786), p. 8.

[7] For sample modern discussions, see: Faucherre and Prost, *Le triomphe de la méthode*, pp. 52–53; and Lynn, *Giant*, pp. 569–571. Such accounts of Vauban's tactical innovations can be found in Fontenelle's 1707 eulogy, reproduced in Virol, *Les Oisivetés*, Annexe 2 esp. 480, and are rarely explained as clearly as N. Allard's brief 170-year old "Notice sur Vauban" in *Le Spectateur militaire*, 18 (1835), pp. 437–442. Paul Lazard's 1934 work already presented a slightly less nuanced approach: contrasted with the 23 pages spent discussing Vauban's three "systems" of fortifications, Lazard spends five on the attack, most of which quote Vauban. He does, however, admit that Vauban only perfected the procedures that had been haphazardly applied until then (394). More recent works pay even less attention to the attack.

This is usually followed by a recitation of the stages through which a Vaubanian siege would progress.[8] With this elaborate multi-stage process briefly sketched, Vauban is declared the engineer who systematized and perfected the early modern siege attack.[9] This hypothetical siege model does a good job of illustrating an idealized 'average' siege and its component parts, but it has a significant disadvantage for those who believe siegecraft was pivotal to the European way of war in the early modern world, since it assumes Vauban's style of siegecraft *a priori* rather than testing the extent of his influence. If sieges truly were central to the period's warfare, they deserve a far more detailed and empirical examination than they have received thus far. A more thorough investigation of Vauban's process of systematization is central to this endeavor.

To place Vauban back in his context is to resituate him within the larger context of engineering efficiency, and we will do so by relying largely on three prescriptive sources: Vauban's two most detailed and comprehensive presentations of his offensive ideas, and a particularly interesting account of his model siege of Ath. Vauban's original treatise on the siege attack, entitled *Mémoire pour servir d'instruction dans la conduite des siéges et dans la défense des places*, was drafted in 1669 at the request of the Secretary of War and director-general of fortifications François-Michel Le Tellier, marquis de Louvois, and was finished in 1672, in preparation for the upcoming war with the Dutch. A hasty work that included many undeveloped ideas, it would later be supplanted by a much longer work, his *Traité de l'attaque des places*.[10] This latter work was composed in 1704 at the king's behest

[8] For examples, see: Pernot, "Vauban, le siège devenu réglé," 256 and 258ff; Chandler, *Art of Warfare in the Age of Marlborough*, pp. 240ff.; Duffy, *Fire and Stone*; Childs, *The Nine Years' War*, pp. 92–96; and Guerlac, "Vauban," pp. 79–80.

[9] Guerlac briefly refers to Vauban's "spirit of critical appraisal, . . . love of logic, order and efficiency" in "Vauban," 77. Chandler briefly mentions his importance as systematizer in *The Art of Warfare in the Age of Marlborough*, p. 273; also John Lynn, "Vauban," *MHQ: The Quarterly Journal of Military History*, 1 (2) Winter 1989, p. 58; *Giant*, p. 571; and his "Vauban" entry in L. and M. Frey (eds.), *The Treaties of the War of the Spanish Succession: An historical and critical dictionary*, (Westport, CT, 1995), p. 459. Jean-François Pernot uses similar language in "Vauban, le siège devenu réglé ou l'économie des vies militaires," in A. Corvisier et al., *Les malheurs de la guerre I: De la guerre à l'ancienne à la guerre réglée*, (Paris, 1996), p. 255.

[10] For summaries of these treatises, see Blanchard, *Vauban*, pp. 167ff.; and Virol, *Les Oisivetés*, pp. 27–40 (1669/1672) and pp. 41–53 (1704/1705). It should be noted that Rothrock's readily-accessible English translation is of the earlier of these two works.

for the royal grandson Louis Bourbon, Duke of Burgundy, and it provides a detailed illustration of Vauban's reflections over his lifetime of service. These two works offer useful bookends to his career, allowing us to see the progression of his ideas as well as his overriding concern with efficiency, which serves as the thread of continuity throughout his career as well as the connection between him and his engineering peers. Our anonymous engineer's account of the 1697 siege of Ath provides a complementary view of the great technician in practice, locating Vauban explicitly within this broader engineering cult of efficiency. While most siege journals were content to describe events at a siege in an objective tone, this narrative emphasized how efficiency was about improving *every* aspect of the attack, explaining why Ath was, for engineers, the perfect siege conducted by the model engineer.[11] Vauban's quest for the most efficient siege possible, rather than just his tactical innovations or codification of a set number of siege stages, embodies the systematizing legacy he sought to pass on to his students, and while he brought siegecraft to a higher level, he was building upon the ambitions and innovations of the many engineers who had come before him.

2. *Minimizing Delays, Casualties, and Costs*

Beyond capturing a place, the engineers' goal was to conduct the attack "by the shortest, most reasonable and least bloody route possible."[12] Here Vauban identified the engineers' three most important measures of a siege's efficiency: its length, its financial costs, and its cost in human lives. Though each of these measures was to be minimized, Vauban focused most consistently on the last of these. Just as he saw the wealth of a nation residing in its people, so too did the strength of an army depend on its soldiers.[13] Thus the desire, reflected in many of his writings, to improve their composition, morale, pay, training, and conditions of service. Vauban's humanitarian spirit recoiled therefore at the prospect of needlessly-spilt blood:

[11] We can contrast the overtly pedagogical nature of the Ath text with the straightforward reporting of another Vaubanian siege found in a treatise published by another of his subordinates, Naudin. See his *L'ingénieur françois*, pp. 288–290.

[12] Vauban, *Traité*, pp. 201–202.

[13] Virol, *Les Oisivetés*, p. 295.

> If this were absolutely necessary, and if there were no other way of advancing the works, it would be necessary to accept it patiently. But since it is possible to push the trenches forward just as quickly without exposing anyone, is it not a horrible cruelty that we expose 400 or 500 men to enemy fire every night, with not a single one of them able to fire back?. . . . The result is that our workers and those who guard them, being protected by only the darkness, are miserably killed by an enemy who is able to pick them off at leisure.[14]

For Vauban, casualties were a critical factor in judging a siege—an attack should only be considered well-conducted if the besiegers suffered fewer losses than the garrison.[15] The latest reports from the 3 June 1697 edition of the *Gazette d'Amsterdam* had already noted his concern over the casualty rate during the siege of Ath, writing that the siege was going slower than expected "because M. de Vauban . . . judged it reasonable to conduct his attacks a little more slowly in order to succeed with less risk and fewer casualties." In his primer for the Duke of Burgundy, Vauban elevated its seventh rule to axiomatic status and indicated the compatibility of both success and saving lives: "Use the sap as soon as the open trench becomes dangerous, and never expose yourselves or use force where labor [*industrie*] will suffice, since *industrie* is certain, whereas force is much less certain and usually risks much."[16] Vauban shared this objective with other engineers, the Ath journalist boasting how the master had been able to balance both progress and safety, with 1500 toises (one toise equaling 1.95 meters or 6.4 feet) of trenches dug the first night "without the Loss of a single Man" (p. 98). Jacob de La Vergne, a contemporary engineer in Austrian service, similarly promoted efficiency not only in terms of minimizing unnecessary trenchworks, but of saving lives as well.[17] This was in fact a common concern in the

---

[14] Vauban, *Mémoire pour servir*, p. 15. See also his condemnation a year later at Maastricht of the unnecessary deaths of more than one hundred men due to the "negligence or vanity" of several unnamed officers. Rochas d'Aiglun (ed.), *Vauban*, vol. 2, pp. 94–95. Fontenelle's eulogy of Vauban contended that the deceased would have gladly accepted delays if he could have suffered fewer losses as a result. See also: Guerlac, "Vauban," p. 79; Blanchard, *Vauban*, for Vauban's complaints at Maastricht 1673 reported on p. 183, and Blanchard's description of his approach as founded on "méthode, rapidité, souci d'éviter les pertes inutiles" on p. 320; Pernot, "Vauban, le siège devenu réglé," 255; and Virol, *Les Oisivetés*, p. 253.

[15] Vauban, *Traité*, p. 56. In large part this was because those in command deemed their soldiers' lives of only secondary importance, as we shall see in a later chapter.

[16] Vauban, *Traité*, p. 261; also p. 72.

[17] La Vergne, *Nouveau exercice du gabion, et de la fascine,* (Vienne, 1698), pp. 23 and

seventeenth and earlier centuries, when governments could barely keep their units in the field up to strength.[18] De Ville's worry over the lives of the soldiers illustrates this long-standing concern and stands as a mission statement for military engineers:

> We must search for every possible invention to conserve our soldiers, who are as important to the commanders as their own limbs. The spirit, science and experience of commanders can only be seen in the courage of the soldiers: generals have the theory of war, the soldiers the practice. They are the means and the instrument by which Princes conserve, expand and conquer States; this is why if they love their honor and their well-being, they must value their troops and protect them as if their own lives depended on it.[19]

In a sense, this humanitarianism, however laudable, was beside the point. Nor for that matter, is casualty avoidance a technique in itself. Vauban could have a low tolerance for casualties and even inculcate that priority in others, but he needed specific tactics to reconcile this objective with the overarching need to capture a fortress, to find the optimal intersection that would minimize delays, casualties and costs all at the same time. He needed to improve upon the techniques of past technicians.

Vauban believed, like his predecessors, that in order to reconcile avoiding unnecessary costs and bloodshed, one must apply a mixture of Reason (in the form of geometrical axioms) and practical construction techniques to this problem set. The results were the combination of specific tactical techniques that coalesced into the ideal siege, an aspiration that all engineers should attempt to implement in reality. Many scholars have cited this rational, systematizing approach of his, but see its manifestation as either his three tactics, or a rigid, 'scientific' progression of the siege stages.[20] Even

---

56–57. La Vergne's personal history is uncertain; he was in Austrian service since the 1680s, making it possible, given his name, that he was an exiled Huguenot who may have served under a young Vauban. His treatises made no mention of Vauban's specific tactical innovations, and instead he proposed institutional solutions to improve the efficiency of Austrian siegecraft.

[18] Mark Fissel contends that the concern for limiting casualties is a universal trait of military commanders. Fissel, *English Warfare, 1511–1641*, p. 181.

[19] Anthoine de Ville, *Les fortifications du chevalier Antoine de Ville*, (Lyon, 1628), pp. 314–315. This was largely the perspective of the 4th century Roman reformer Vegetius, who emphasized how the soldiers deserved the respect of their commanders.

[20] See for example claims that Vauban insisted on rigid 'systems' in Lynn, "Vauban," *MHQ*, p. 58; and the brief comments of Pollak in *Military Architecture*, p. xxxiv.

the most nuanced discussions referring to perfecting and systematizing pre-existing techniques are, unfortunately, exceedingly vague.[21] Vauban, however, had identified at the beginning of his career how all-encompassing the fundamental flaw in the French tradition of siegecraft was:

> The confused conduct of the attacks that we have followed up to the present in our sieges, following neither plan nor design, is one of the most pernicious defects which can be imagined; we work all day without ever knowing what will need to be done two hours later. Thus everything is in disorder and without reason, so that the trenches are always poorly positioned and poorly constructed. The batteries and assembly areas are never where they should be; measures are never taken to establish firing positions properly; we are never in a position to repulse a sortie; and never, or at least very rarely, does the trench not extend one half or one third beyond where it should, which results in it being enfiladed or ending in a dangerous escarpment. It follows then as a necessary consequence that many men are lost, much money is wasted, little damage is done to the enemy, little progress is made, little effect is seen from the artillery fire, little support is offered for the lodgments, while we operate under a perpetual risk of being defeated by the first sortie conducted with the least bit of vigor. We will always face such dire straits until we change our techniques.[22]

No single tactic (or even the famous three tactics) could solve all of these problems—the problem was much larger. As his enumeration of faults and dangers suggests, Vauban's challenge was to systematize the *entire* process of the siege in order to make it more efficient. Specific tactical innovations played a large role, but they could only work if they were combined with the engineer's fundamental commitment to constant improvement.

3. *Eliminating the Attack's Weaknesses*

Declaring Vauban a systematizer of past techniques requires that we at least make an effort to determine what exactly he was attempting

[21] Childs, *Nine Years War*, p. 93; Lynn, *Giant*, pp. 568, 571. In general, Blomfield's biography provides the broadest chronological view of Vauban's work, though he too limits his contextualizing discussion almost solely to military architecture rather than the siege attack.

[22] Vauban, *Mémoire pour servir*, p. 14.

to systematize. A perusal of earlier 17th century works provides a strong basis for appreciating how heavily Vauban relied on prior practice. Unfortunately, we are poorly served by the existing historiography. The current scholarship on 16th and 17th century warfare does not address the issue of the siege attack in any detail; the *trace italienne*'s defensive features and extant architectures monopolize discussion just as they do in Vauban's age.[23] The gallocentric focus of the Louisquatorzian scholarship (and of Vauban's own writings, it should be said) also skews our perception of the *maréchal*'s originality, making it easy to forget that France, divided by civil wars for fifty years, lagged far behind the siege expertise of other powers, who had been slowly improving their techniques during decades of positional warfare funded by fiscal-military states.[24] With our knowledge of other aspects of Renaissance warfare, we would also expect significant experimentation long before the age of Vauban. Given the crushing burdens early modern states faced supporting their war efforts, there must have been intense pressure to develop an alternative to the lengthy, bloody assaults on new-style fortresses and the glacial pace of plodding blockades.[25] There were plenty of opportunities to implement such theories as well: the hundreds of sieges conducted in the Valois-Habsburg wars, over the course of the Dutch Revolt, and during the waging of the Thirty Years' War offered testing grounds where such techniques must have been developed and refined.[26] New weapons and increasing army sizes would require

[23] The single noteworthy exception is Pepper and Adams' *Firearms and Fortifications*. For the only recent overview of pre-Vaubanian engineers and their defensive works, see David Buisseret, *Ingénieurs et fortifications avant Vauban: L'organisation d'un service royal aux XVIe–XVIIe siècles*, (Paris, 2002). Jean-François Pernot focuses on a single engineer in "La Guerre et l'infrastructure de l'État moderne: Antoine DeVille, ingénieur du Roi (1596?–1656?), la pensée d'un technicien au service de la mobilisation totale du royaume," *Revue d'histoire moderne et contemporaire*, 34 (1987): 404–425.

[24] Lynn mentions this point briefly in *Giant of the Grand Siècle*, p. 568.

[25] Parker discusses Philip II's ruinously-expensive attempts to bring Dutch rebels to heel and James Tracy chronicles the father's struggle to pay for his own wars against French, Ottoman, and German Protestant forces. James Wood, John Lynn, David Parrott and Guy Rowlands offer recent accounts of how the French state attempted (often with limited success) to support 250 years of seemingly-perpetual internal and external wars.

[26] For an excellent case study of how wartime experience could accelerate the dissemination of new military ideas and encourage experimentation, see John R. Hale, "Tudor Fortifications: The Defence of the Realm, 1485–1558," reprinted in Hale (ed.), *Renaissance War Studies*, (London, 1983), pp. 79ff.

developing technological, administrative and logistical infrastructures (i.e. social technologies), but the use of these weapons and men in a siege were more constrained and therefore largely up to the individual commander or engineer, allowing much more flexible experimentation. Furthermore, an innovative mindset was clearly present in the 16th century. Renaissance historians are familiar with thinkers such as Maurits of Nassau and Gustavus Adolphus, who used the geometrical and arithmetical postulates of Classical authorities to develop battlefield tactics like the countermarch, in spite of the extreme difficulty of implementing such complex maneuvers in the fluid condition of battle.[27] It would have been just as easy for engineers before Vauban to improve the more static siege attack, especially since these same specialists were already applying geometric principles to the design and construction of *trace italienne* fortresses.

Beyond such hypotheticals, there is more concrete evidence of earlier advances. Though France may have lacked experience with the latest advances of siegecraft, its neighbors to the north and south had decades of recent experience conquering and defending some of the strongest fortresses in Europe. Duffy argues that in this school of hard knocks Dutch, Spanish and even a few French engineers in foreign service quickly caught up with and surpassed the cutting-edge Italian engineers.[28] If contemporary accounts are to be believed, nameless engineers had overcome many of the advantages of the *trace italienne* well before Vauban arrived in the 1660s. The French marshal Gaspard de Saulx, seigneur de Tavannes described circa 1620 the innumerable techniques foreigners had perfected in order to capture strongholds. He contended that military engineers knew all of the weaknesses of the latest fortification systems, and implored these same technicians to develop new defensive works that would return the advantage to the defensive:

> Thirty years ago fortresses were so well defended by the ignorance of the age that many were judged impregnable, and even those that were

[27] For an overview, see Parker, *The Military Revolution*, pp. 16–24. See also the articles by Patricia Cahill and Timothy J. Reiss in D. Glimp and M. Warren (eds.), *Arts of Calculation: Quantifying Thought in Early Modern Europe*, (New York, 2004).

[28] *Siege Warfare*, p. 54. Reginald Blomfield made a similar argument forty years earlier, in *Sebastien le Prestre de Vauban*, pp. 27ff. See also Evans' discussion in his Introduction to *The Works of Sir Roger Williams*, pp. cxxviii–cxxxii. For De Ville's foreign tutelage, see Pernot, "La Guerre et l'infrastructure de l'État moderne," p. 405, and more generally pp. 407–413.

> very weak were quite difficult to reduce.... Nowadays the assailants have gained the upper hand, and the defense of towns is so weakened by the experience [of the besiegers] that one can say that they cannot resist without the outside assistance of an entire army.... Now Spanish and Dutch captains have made the capture of towns an art, and they can predict the number of days that a fortress can resist, regardless of its strength.[29]

To cite another witness, one of Vauban's predecessors, Blaise-François comte de Pagan, claimed around 1640 that "All of Europe is shocked by the weak resistance from even the strongest of today's fortresses, which cannot resist more than six weeks; even the best can fend off collapse only with the help of a relief force."[30] Whether these impressions are supported by a representative dataset of real sieges is a question for future research, but these examples suggest that earlier military thinkers had already developed many of the tools Vauban would build upon in Louis XIV's reign. Certainly the age-old techniques of surprise, escalade and stratagem (see chapter 7) were not new, and it is not surprising therefore that Vauban was the first engineer to devote two of his treatises *solely* to the craft of capturing a fortress by siege, rather than distract his attentions with coverage of fortification design, artillery specifications and non-siege techniques.[31] He thus devoted several pages to each tactic, elaborating on the brief discussions found in previous published manuals

---

[29] Gaspard de Saulx, seigneur de Tavannes, *Mémoires de Très-Noble et Très-Illustre Gaspard de Saulx, seigneur de Tavannes*, (Paris, 1838), pp. 177–178. The comments of Jacques François de Chastenet, son of the previously mentioned and less-well-known memoirist Puységur, made a similar claim. Jacques François de Chastenet, marquis de Puységur, *Art de la Guerre, par Principes et par Régles*, (Paris, 1749), vol. 1, p. 37. Tavannes even foreshadowed a claim heard in the Vauban era when he cited the early 17th century belief that with "the counterscarp taken, a town was half captured." The Frenchman René Le Normant claimed that Maurits of Orange could also predict a fortress's fall within days. *Discours pour le restablissement de la milice de France*, (Rouen, 1632), p. 225.

[30] Pagan, *Les Fortifications de Monsieur le Comte de Pagan avec ses theorems sur la fortification*, (Brussels, 1668), Preface, and pp. 29–30. Antoine De Ville, writing in the 1620s, also noted that sieges had become much less bloody in recent years. *Les fortifications*, p. 281.

[31] Vauban's *oeuvre* was much broader than just the attack and, of course, went far beyond military tactics and strategy. At the end of his life he would pen a treatise on how to defend fortresses as well, but unlike all of his contemporaries and predecessors, Vauban never composed an exposition on the art of fortifying, a task which had taken pride of place from the beginning of Renaissance engineering. He also abandoned the rhetorical reliance on Classical precedents so popular with his predecessors.

which usually amounted to only a line or a single paragraph. Spending hundreds of pages on siegecraft rather than a few dozen, Vauban was attempting to bring French practice up to the expertise of siegecraft as it was practiced in the Low Countries. In the process, over the course of his career he elevated the level of siege efficiency far beyond the achievements of his peers.

As Vauban's complaints about confused attacks indicate, one of the biggest challenges was the threat of garrison sorties. Up to several hundred men strong, garrison infantry supported by cavalry would sally out to attack the heads of the trenches, forcing workmen to flee and razing the abandoned trenchworks.[32] Earlier military engineers had spent much time fretting over how to prevent these disruptions, though their recommendations usually focused on the construction of redoubts within the lines, and ensuring that the trench guards were properly-armed, well-disciplined and well-rested.[33] Vauban repeated these and other recommendations, but his distinctive contribution in this regard was his recognition that trench parallels were the most efficient counter-measure of all.[34] A series of (ideally three) trenches, each dug parallel to the front of attack and progressively closer to the fortress, these lines would envelop the defenders' front and shelter the besieger's early batteries and zigzag approaches. According to the historiography, his first use of parallel trenches was at the siege of Maastricht in 1673, where their utility became immediately evident. In his 1704 treatise he concluded his discussion on parallels by noting that they were singularly effective

---

[32] This is an important aspect of the defense that has received little discussion in the Military Revolution and siege warfare literature, likely because the scholarship has followed military engineers' emphasis on the form of fortifications, particularly the shape of bastion's salient angles. On the conveniency of seeing a fortress's form as the proverbial 'nail' for the geometer's 'hammer,' see Langins, *Conserving the Enlightenment*, pp. 27–30.

[33] See for example De Ville, *Les fortifications*, pp. 321–322 and 399–403; as well as his praise of sorties for the garrison commander in *De la charge des Gouverneurs des Places*, (Paris, 1640), pp. 476ff.

[34] This is the most frequently mentioned of his innovations. Blomfield, *Sebastien Le Prestre de Vauban*, pp. 61–62; Rothrock (ed.), *A Manual*, p. viii; Pujo, *Vauban*, pp. 67–68; Blanchard, *Vauban*, p. 164; Lynn, *Giant*, pp. 569–570; Lynn, *Wars of Louis XIV*, pp. 76–77. Confusingly, Vauban called them either *places d'armes* or simply lines. Neither term provides the specificity of 'parallels' and both terms were already used to refer to other parts of siegecraft—the *place d'armes* on the covered way, and the siege lines of circumvallation and contravallation that surrounded an invested fortress.

in preventing and neutralizing enemy sallies. Their gradual development is the most observable example of not only Vauban's pursuit of ever-increasing efficiency, but of the larger engineering interest in improving efficiency over the long-term.

Most scholars mention that Vauban's trench parallels were spurred by his second-hand knowledge of the Turkish trenches at the siege of Candia (Crete)—Louis had sent an unsuccessful French expedition to help rescue the Mediterranean island from Ottoman depredations.[35] The marquis de Puységur (in French service from 1617 through the 1650s) had already referred to the use of parallels as the *maxime des Turcs*, possibly suggesting a more widespread familiarity with the technique than currently recognized.[36] The parallels as they appeared at Candia, however, were far from how Vauban implemented them at Maastricht just a few years later: the excessive number of Turkish branches that sprouted from the approach trenches signaled an inefficient use of labor and did not even connect the saps together.[37] It would take someone attuned to the primacy of efficiency to see within this honeycomb of trenchworks the simplified three-parallel approach Vauban made famous. But the Ottomans were not the only ones in need of greater efficiency. Indicative of Vauban's own unsettled ideas, a wide variety of transitional styles can be found in his 1672 treatise, and while none incorporated the Turkish proliferation of works, only some showed the parallel branches making contact with their neighbors, while

---

[35] Vauban's secretary Pellison claimed that Vauban had told him that Candia was his model. Georges Michel wrote that Vauban deserved the credit nonetheless, because "in war, the art consists as much in the application as the invention." *Histoire de Vauban*, (Paris, 1879), p. 81. Michel traces the transmission via an engineer named Paul, while F.J. Hebbert implies that the Huguenot Charles Goulon was a possible vector. F.J. Hebbert, "The Memoirs of Monsieur Goulon," *The Journal of the Society for Army Historical Research*, 69 (279) (1991), p. 161. A mid-eighteenth century author claimed that Vauban had learned the technique from an Italian engineer in Turkish service. Tileman van der Horst, *Essai sur la fortification*, (La Haye, 1755), p. 30.

[36] Puységur, *Les mémoires de messire Jacques de Chastenet, chevalier, seigneur de Puységur, colonel du régiment de Piedmont, et lieutenant général des armées du roy. Sous les règnes de Louis XIII et de Louis XIV*, 2 vols. (Amsterdam, 1690), vol. 2, p. 505. This author was the father of the better-known Jacques François de Chastenet, author of the 1749 treatise *Art de la guerre par Principes et par Régles*.

[37] A reproduction of these trenchworks can be seen in Duffy, *The Fortress in the Age of Vauban*, p. 220. It is conceivable that their huge numbers encouraged further thought on their utility, and how they might be improved.

several others included mistakes he would later warn against, such as parallels being enveloped by the front under attack.[38] By 1683 the Turks threatening Vienna knew to join their approaches with parallels, but their endless multiplication suggests that Vauban's concern for efficiency had still not found purchase with Kara Mustafa's army.[39]

There were more economical examples of trench parallels, both real and theoretical, much closer to Vauban. This is hardly surprising as it would be difficult *not* to dig trenches parallel to the works under attack considering the dictates of geometry and the hundreds of sieges conducted during the period.[40] A few isolated examples will have to suffice until further research can indicate how exactly they differed from Vauban's tactic, though they likely differed only in the increased efficiencies Vauban was able to achieve. At Amiens in 1597 the French king's approaches included several parallels and quasi-parallels.[41] The royal army at the siege of La Rochelle in 1573 utilized one large parallel trench to envelope the entire front under attack and another smaller one on the counterscarp.[42] Illustrations of the sieges of Groningen 1594, Grave 1602 and Jülich 1610 in Wijn's study of Dutch Revolt siegecraft also portray trenches dug parallel to the attacked fronts.[43] Duffy's reproduction of the attack

[38] Compare, for example, the easily-accessible reproductions in Rothrock (ed.), *A Manual*, Plate 8 (p. 64) with Plate 9 (p. 66). Consult also the variety of other parallels on p. 72 (Plate 11), 82 (Plate 16), 98 (Plate 20), 104 (Plate 22), and 124 (Plate 25). This last is closest to the ideal, but with only the second and third parallels. A plan of Maastricht's not-quite-ideal trenches can be found in Parent and Verroust, *Vauban*, p. 113. For a comparison with the mature version, see Plate 14 of Vauban, *Traité*.

[39] Depictions of Vienna's labyrinthine trenches can be found in many works, including Thomas Barker, *Double eagle and crescent; Vienna's second Turkish siege and its historical setting*, (Albany, NY, 1967), pp. 250–251; John Stoye, *The Siege of Vienna*, (Edinburgh, 2000), Plate VI; Duffy, *The Fortress in the Age of Vauban*, p. 230; and Childs, *Warfare in the Seventeenth Century*, pp. 130–131.

[40] Later in the eighteenth century the chevalier de Folard would argue that the Ancients had also used trench parallels (among many other techniques that deserved to be resurrected). Jean Charles, chevalier de Folard, *Histoire de Polybe, nouvellement traduite du Grec . . . avec un commentaire ou un corps de science militaire . . .*, (Amsterdam, 1753), vol. 2, pp. 161ff., esp. 168.

[41] Olivia Carpi-Mailly, "Amiens au XVIe siècle: Le destin d'une ville frontière," in P. Nivet (ed.), *Picardie, terre de frontière: actes du colloque, Amiens, 26 avril 1997*, (Amiens, 1998), Plate 9.

[42] See the map in Wood, *The King's Army*, pp. 256–257.

[43] Wijn, *Het krijgswezen in den tijd van Prins Mauritz*, p. 296 (Plate VII), p. 289, and p. 287. On 289 Wijn summarily stated that besiegers did not use parallels, but it

on Montauban in 1621 similarly shows the town being attacked on three sides by cannon behind entrenchments roughly parallel to their targets, although with cosmetic angled redans.[44] A plan of the 1636 siege of Dôle conducted by Condé depicts two parallel trenches attacking the Besançon gate.[45] French maps of the Spanish sieges of Dixmuide in 1647, Ypres in 1647 and Arras in 1654 also show parallels and zigzag approach trenches among the siegeworks.[46] Vauban was present in the relief army that lifted the siege of Arras, offering at least the remote possibility of an alternate origin of his parallels. In any case, these West European versions are certainly closer to Vauban's efficient ideal than the profligate Turkish examples.[47]

Earlier theoreticians also reproduced trench parallels in their own works. In Papillon's *Practicall Abstract of the Arts of Fortification and Assailing* published in 1645, we find a trench at the foot of the glacis that looks much like one of Vauban's third parallels as he illustrated them in his 1672 manual.[48] The Englishman also depicts an "engirting circulatory trench" that surrounds the fortress near the glacis, with outlets for troops to storm the counterscarp—in other words, a contravallation line abutting the counterscarp which essentially served as an overextended version of a third parallel. That Vauban's parallel eliminated the need for contravallation lines where the attacks were located suggests a logical genesis: the first parallel may have been an atrophied contravallation line—at Ath 1697 the journalist noted that "the first parallel was designed merely as a check upon

is difficult to imagine what else these trenches dug parallel to the front of attack could be. Presumably Wijn was referring to the systematic, fully developed series of three parallels Vauban is famous for?

[44] *Siege Warfare*, p. 119. This engraving is surprisingly similar to the more generalized plan of Venlo 1702 in Nicolas de Fer's *Les forces de l'Europe ou description des principales villes avec leurs fortifications*, (Paris, 1705), Plate 24. De Fer's maps of both sieges and fortifications, however, are often extremely reliable

[45] Pierre Bertin, "Guerre de Trente Ans: Le siège de Dôle en 1636," in *Revue historique de l'armée*, (1970), fourth plate after 8 (unpaginated).

[46] *Le cabinet du Roi: recueil d'estampes de differents auteurs concernant les bâtimens, les tapisseries, tableaux, conquêtes et autres sujets qu'on trouve dans les maisons royales*, (Paris, 1679–1743), VII.

[47] Childs is exceptional in his contention that rather than Candia, "Vauban probably drew on the less methodical work of his European predecessors." *The Nine Years War*, p. 93. See also the brief mention of pre-Vauban parallels in Jean-Marie Goënaga's entry on "Sièges" in F. Bluche (ed.), *Dictionnaire du Grand Siècle*, (Paris, 1990), p. 1449.

[48] Plate 24, discussed on p. 112.

the first sorties from the town, and served for a kind of circumvallation" (p. 102). Papillon encourages such a theory by depicting the contravallation trench doubling as a first parallel, with the approach trenches zigzagging forward from it.

Admittedly, the trenchworks seen in earlier sieges and treatises do not yet resemble the fully-developed attacks portrayed in Vauban's final manual, though we should note that these are much closer to his model works at Ath than those Turkish trenches said to be Vauban's inspiration. More important, however, is the fact that Vauban's full appreciation of the functions of trench parallels, and even their proper application, only developed as he sought to perfect the siege attack over time. In his early treatise he focused on their use as an assembly area for storming the covered way, a function only appropriate to the third parallel on or near the glacis, and many Vauban scholars confine their discussion to this aspect.[49] A large part of this functionality came from the additional room besiegers had to maneuver around one another in Vauban's spacious trenches, and in this he was only trying to systematize best practices proposed far earlier by engineers like Errard Bar-le-Duc and De Ville.[50] The success of the parallels at Maastricht in 1673 demonstrated their utility as a counter-measure against garrison sallies, and in 1687 Vauban assured Louvois that their trenches would always be safe from garrison sorties "when we follow the rules established since the siege of Maastricht."[51] The fact that the application of these parallels was often less than perfect forced him, in his twilight years, to further systematize the technique:

[49] On their assembly function: Vauban, *Mémoire pour servir*, pp. 98–103, where he referred specifically to the use of the third parallel as an assembly area. See also p. 84 for hypothetical language that suggests he had not yet tried such a tactic. Examples of the modern focus on assembly areas can be found in Blomfield, *Sebastien Le Prestre de Vauban*, p. 62; Lynn, *Wars of Louis XIV*, p. 76; also in *Giant*, p. 570.

[50] Jean Errard Bar-le-Duc, *La fortification demonstrée et réduicte en art par feu I. Errard de Bar le Duc Ingénieur du Treschrestienne Roy de France et de Navarre*, (Paris, 1622), Premier livre, chapitre X, wherein he also stresses the efficiency advantages of a *limited* number of spacious trenches; and De Ville, *Les fortifications*, p. 299. We find mention elsewhere of siege trenches varying in width according to the commander's preference, e.g. Papillon, *A Practicall Abstract of the Arts of Fortification and Assailing . . .*, (London, 1645), p. 111; and Claude Flamand, *La guide des fortifications et conduite militaires pour bien se fortifier et deffendre*, (Montbéliard, 1611), p. 201.

[51] Rochas d'Aiglun (ed.), *Vauban*, vol. 2, p. 283. Also Vauban, *Traité*, pp. 82–84 and 91–95. For a relatively rare modern appreciation of their utility in this respect, see Duffy, *The Fortress in the Age of Vauban*, pp. 78–79.

> As no certain rules have been given until now [1704] for the practice and siting of parallels, this has always engendered some confusion, and they are not always well executed. I can even say that since the siege of Maastricht [1673], where I first used them, they have been properly implemented only at the siege of Ath [in 1697].[52]

Notice that systematization meant maximizing efficiency, not simply following the rules rigidly. After all, Vauban chose to forego a third parallel at Ath because it would have decreased the efficiency of the siege. What made parallels truly appealing to the engineers was less that they came in threes and more that they performed several functions simultaneously with surprisingly little effort. As Vauban explained, when implemented wisely they protected both the flanks of the trench and the zigzagging approach trenches that moved forward from them, while they also supported the batteries, provided a post for the guard battalions so as to not impede the workers in the approach trenches, allowed a wide frontage for maximum concentration of firepower, served as communication between the various approach trenches, and also served as a line of contravallation against the garrison.[53] A few of these functions had undoubtedly been fulfilled by earlier trenches dug parallel to besieged fortresses, but what differentiated Vauban's trench parallels was that they were intended as an efficient *system*; these works were intentionally designed to serve multiple, mutually-reinforcing purposes in the most economical way possible. This was the quintessence of efficiency: minimizing input, i.e. the labor used to plan and construct the trenches, while maximizing output, the advantages gained from these works.

A second aspect of the siege attack that Vauban sought to make more efficient was the targeting of troops in the covered way. Whether shielded by the crest of the glacis and palisades or by traverses segmenting the covered way itself, defending soldiers were effectively sheltered from most of the besieger's fire behind such works and thus more ready to resist a storm of the counterscarp. Vauban's second "innovation," *cavaliers de tranchées* or trench cavaliers, triumphed over this challenge, as it allowed besieging musketeers to fire down into the exposed covered way from raised firing platforms on or near the glacis. This tactic too illustrates the gradualistic improvement of

[52] Vauban, *Traité*, p. 83.

[53] They also decreased the reliance on cavalry, a group whose archly-aristocratic members were traditionally one of the fiercest opponents of low-born engineers.

the siege attack: earlier sources indicate a widespread use of such techniques well before Vauban's first adoption in 1684, a fact rarely-remarked upon in the biographical literature.[54] From prehistory onward fortified communities and their attackers recognized the axiom that all things being equal, higher positions command lower ones. The Roman soldier, used to digging a fortified field camp each night, as a last resort would exert Herculean efforts building long siege ramps up to the top of a tall city's walls. More sophisticated siege towers and belfries enabled medieval besiegers to 'overtop' towering enemy walls, a practice that continued into the 15th century.[55] As Vauban would do at Ath, fortress designers had been building *cavaliers* in the middle of the new angled bastions and along the curtain walls since the sixteenth century in order to expose attacking troops in their trenches, hence the later addition of *de tranchée* to *cavalier.* Outside the city walls, Renaissance attackers replaced the awkward and now, thanks to defenders armed with gunpowder weapons, dangerously-vulnerable siege towers with smaller mounds of earth for batteries, using 'mounts' or trench cavaliers to fire down onto garrison troops defending a breach.[56] The mid-seventeenth century English writer Papillon, keeping his compatriots abreast of Continental siege tactics, discussed cavalier platforms raised beyond the covered way to breach the bastions hidden behind the covered way, "four

[54] For brief secondary discussions of *cavaliers de tranchée*, see: Michel, *Histoire de Vauban*, pp. 187–188; Blomfield, *Sebastien Le Prestre de Vauban*, pp. 91 and 110; Chandler, *The Art of Warfare in the Age of Marlborough*, p. 264; Pujo, *Vauban*, p. 121; Lynn, *Giant*, pp. 570–571; and most strongly of all, Faucherre and Prost, *Le triomphe de la méthode*, p. 53. For Vauban's early treatment of trench cavaliers, see *Mémoire pour servir*, p. 104.

[55] Salamagne, "L'attaque des places-fortes au XVe siècle" on *bastilles* and *bastides*, esp. p. 106. Folard also notes the use of cavaliers by the Romans, in *Histoire de Polybe*, vol. 2, pp. 171ff., and 254–255, explaining it with an early version of 'Inherent Military Probability'.

[56] Wauwermans, "L'architecture militaire flamande et italienne," 166; Evans (ed.), *The Works of Sir Roger Williams*, p. cxxx; Duffy, *Siege Warfare*, p. 95, and see the illustration on p. 83 for a clear example of cannon-laden trench cavaliers in the 1616 illustration of the 1592 siege of Coevorden; Eltis, *The Military Revolution in Sixteenth-Century Europe*, p. 88; and Olaf van Nimwegen, "Maurits van Nassau and siege warfare (1590–1597)," in M. van der Hoeven (ed.), *Exercise of Arms: Warfare in the Netherlands, 1568–1649*, (Leiden, 1997), pp. 127–128. The English editor of Goulon's memoirs mentions on p. 22 that the Turks at Candia also used a cavalier in their attack—this is the same siege where trench parallels were used. Duffy treats the siege's main events in *The Fortress in the Age of Vauban*, pp. 218–221.

yards high, and sometimes more."[57] The leap to artificially-constructed heights was not a large one given the topographical variations inevitable outside many fortresses, not to mention when gunpowder artillery forced fortress designers to significantly decrease the height of their walls, making trench cavaliers more practical than the lumbering medieval siege towers. At most, Vauban might lay claim to publicizing the use of trench cavaliers specifically against the defenders in the covered way. Renaissance commanders had already used trench cavaliers to overcome *their* greatest challenge, breaching the thick walls and exposing the defenders behind a retrenched breach. Vauban used this same technique against the garrison troops themselves, since capturing the covered way had by this point replaced storming a breach as the toughest task for besiegers. Here too, the most we can claim is that Vauban only incrementally extended techniques already introduced by past engineers.

Trench cavaliers were not enough for a profession intent on constant improvement, however, for further improvements would be found towards the end of Vauban's life. His last tactical innovation was ricochet fire and it too offered increased efficiencies for attacking the covered way. Ricochet rounds could smash the summit of the enemy ramparts and its palisades and even bounce over obstacles to strike at the men sheltering behind them.[58] Indicative of the long-term process of research and development, Vauban first implemented the technique at the siege of Philippsbourg in 1688, and later revisited it and perfected its implementation at Ath nearly a decade later, where the proper disposition of these batteries were considered fundamental to the attack's finesse.[59] Again however, Vauban was simply extending and perfecting previous experiments, for it is quite likely that ricochet fire was 'discovered' far earlier, when an absent-minded gunner (or one rationing a dwindling pow-

[57] Papillon, *A Practicall Abstract*, p. 113. On the author and his work, see Pollak, *Military Architecture, Cartography and the Representation of the Early Modern European City*, p. 79. See also De Ville's explicit discussion of trench cavaliers in *Les fortifications*, pp. 306–307, both as breaching batteries *and* (p. 313) targeting outworks.

[58] Vauban, *Traité de l'attaque*, vol. 1, pp. 111ff. For modern discussions, see: Lazard, *Vauban*, pp. 276–277, p. 475; Blomfield, *Sebastien Le Prestre de Vauban*, pp. 110, 140; Duffy, *Fire and Stone*, p. 118; Chandler, *The Art of Warfare in the Age of Marlborough*, p. 254; Pujo, *Vauban*, pp. 217–218; Faucherre and Prost, *Le triomphe de la méthode*, pp. 53 and 64, where they deem it the most effective ("*la plus performante*") of Vauban's innovations; and Lynn, *Giant*, p. 570.

[59] *Relation du siège d'Ath*, p. 20.

der supply, or one pressed for time . . .) accidentally charged a cannon with too little powder, resulting in the lower velocity shot that would bounce along the hard ground.[60] We even see the germ of the idea in Vauban's 1672 treatise, where he warned that a garrison's enfilading fire was particularly dangerous from far away, as a spent cannonball could quickly plunge onto troops otherwise shielded behind a breastwork.[61] From this realization it is not an impossible step for a bright mind to improve on this accidental discovery by artificially decelerating the round's velocity with a smaller powder charge. Nonetheless, until new scholarship examines the early 17th century siege attack in greater detail, Vauban must remain the popularizer of this tactic.

Though it only proved its utility at the end of Vauban's career, ricochet's significance for engineers derived from the advantages it offered over existing techniques. Clearing defenders from behind their traverses and ramparts was, after all, a task shared by several other better-known tactics. *Cavaliers de tranchée* similarly exposed the besieged in their outworks—though they required much more labor, while mortars firing exploding bombs and *pierriers* lobbing stones served the same purpose. Before Vauban had discovered ricochet fire he had praised this type of indirect fire, applauding its increased efficiency over cannon at the siege of Luxembourg in 1684 by noting that "fifteen mortars shatter [a garrison's] morale far better than sixty well-served cannon."[62] Once ricochet's utility became evident, however, he scaled back his reliance on mortar fire. Like parallels, this too was the perfect engineering technique because it was not only effective, but it was incredibly efficient as well. As Vauban beamed, "it is the best and most excellent manner of usefully employing cannon in sieges that has ever been used." He then backed up this claim by enumerating eight advantages it had over conventional cannonfire. First, it would quickly dismount the garrison's batteries; second, it would chase away enemy soldiers at the point of attack; third, it would destroy the bridges which allowed communication

[60] See the brief mention of 16th century ricochet fire in Wijn, *Het krijgswesen in den tijd van Prins Mauritz*, (Utrecht, 1934), p. 262. See as well Louis Susanne, *Histoire de l'artillerie française*, (Paris, 1874), p. 153; and Jürgen Luh, *Ancien Regime Warfare and the Military Revolution*, (Groningen, 2000), p. 128 note 9.

[61] Vauban, *Mémoire pour servir*, p. 24.

[62] Quoted in Lazard, *Vauban*, p. 473.

between the covered way and the inner works; and fourth, it would smash the palisades bristling from the covered way and chase enemy soldiers from the counterscarp. Several other advantages had particular appeal for engineers. By merely traversing the guns a few degrees these same batteries that had facilitated capture of the covered way could now strike the men in the outworks and assist in their capture and even reach those behind the curtain wall as well (see Map 3 for an illustration of this at Ath 1697). This was of "great economy, in that these can used throughout the entire length of the siege without having to construct new batteries." Waiting to place the ricochet batteries at Ath in the second parallel meant that only three new batteries of cannon had to be constructed before the covered way was captured. Further, ricochet cannon "consume seven or eight times less powder than the other batteries and their fire is never wasted."[63] Vauban summed up ricochet fire's appeal to the cult of efficiency by noting that it would allow the besiegers to fire "more accurately, more quickly, and much more efficiently [*efficacement*] than the other methods."[64] To these benefits the Ath journalist added two additional perks: when on the periphery of the main trenches (as batteries A, F, G and H were), the gunners did not have to worry about their misfires hitting their own troops, while the garrison would focus its firepower on these ricochet batteries and "leave the Trenches at quiet, which by this means are carried on betwixt the two Fires in great security" (p. 105). The results were clearly superior to the older techniques:

> From the time the Ricochett Batteries were establish'd, the Enemy were obliged to keep off their Defences, which they quitted as soon as ever those Batteries begun to fire, which was more than Cannon-Batteries in the Front, or Bomb-Batteries would have been able to have effected in four or five days time (p. 127).

---

[63] Notice also the overriding importance of efficiency—the ricochet batteries B–E were critical to the capture of the covered way but were not necessary in the final phase of the siege. Therefore their gun crews were sent to assist the other batteries. To attempt to relocate them would have been an inefficient waste of both powder and labor.

[64] Vauban, *Traité*, pp. 114–115. The guns could fire more quickly because they did not have to be repositioned after each shot, since they did not recoil with ricochet's smaller charges of powder. This is also noted by the *Mercure galant* in its account of Ath 1697, p. 225.

Ricochet fire's ability to serve several functions throughout the entire length of a siege epitomized the long-standing engineering goal of minimizing effort while maximizing productivity.[65]

4. *Improvement through Management*

Vauban's quest for the ideal siege went far beyond his tactical innovations. He surpassed his predecessors with the lifetime of sustained effort he dedicated to systematizing the siege, to making each attack even more efficient than the last. The difference between theoretical and real efficiency was, for the engineers, largely a matter of supervision, reflected in Vauban's frequent mention of the need to "conduct" (*conduire*), "manage" (*ménager*) and "regulate" (*régler*) the attacks, as well as his emphasis on the need for *industrie*, diligence and economy.[66] Such language reminds the engineers to keep an eye on the constant adjustments necessary to conduct an efficient siege towards its goal, to match the ideal siege with the reality as it was encountered on the ground. Vauban's model behavior as siege manager manifested itself in three general areas: in the attention he paid to siege preparations, in the emphasis he placed on the intelligent implementation of the ideal attack, and in his tireless efforts to evaluate the effectiveness of past practices. Vauban exhorted engineers-in-training to internalize these virtues of preparation, application and evaluation, for this was the only way to manage the siege properly. The laudatory journal of his Ath attack furthered this effort by illustrating how this was done in a model siege.

Planning is critical to war. For centuries military treatises had encouraged peacetime 'war-gaming' in order to improve practice in

[65] For other samples of techniques predicting Vauban's tactics, see Blomfield, *Sebastien Le Prestre de Vauban*, p. 22; and van Nimwegen, "Maurits van Nassau and Siege Warfare," pp. 137–138.

[66] Guignard noted the progressive refinement of siege techniques over the past two millennia, suggesting that the dialectic between attack and defense was eternal. *L'Ecole de mars, ou mémoires instructifs sur toutes les parties qui composent le Corps Militaire en France*, (Paris, 1725), vol. 2, p. 425. Langins discusses the important administrative and managerial aspects of engineering in *Conserving the Enlightenment*, especially chapter 5, 'Desk Jobs.' Erik Lund discusses a variety of managerials functions generals fulfilled in *War for the Every Day: Generals, Knowledge and Warfare in Early Modern Europe, 1680–1740*, (Westport, CT, 1999).

wartime.[67] More broadly, whether preparing for the next siege, an upcoming campaign or a future war, such preliminary attentions made it is less likely that wartime reactions would be slow, options limited and outcomes unforeseen.[68] Military engineers took this practice more seriously than most, especially in their *métier* of siegecraft. Among other things they would be required to request the supplies for a siege far in advance, and would also be expected to plan out the trenchworks before turning over the first shovel of dirt. Engineers before Vauban had, not surprisingly, preached the importance of preparation, both in general and specific terms.[69] Drawing on this engineering tradition, Vauban's welter of concrete details brought the concern to a new plateau.

The only way to systematize an attack was to plan its course in advance according to rational rules of siegecraft. Always with an eye to efficiency, Vauban recognized that it was the engineer's duty to study every possible fortification, as they all offered opportunities for improvement. This preparation was practical as well as philosophical, for fortresses passed easily from one sovereign to another as military and diplomatic fortunes shifted back and forth. Thus what was being defended one year might need to be besieged the next year. For this reason Vauban had refused to follow Louvois' short-sighted order (based on security concerns) to destroy the plans of the towns they had just won at the Peace of Nijmegen (1678).[70] The future

[67] The universally-lauded Vegetius was only one of many to call for war preparations and drill to improve an army's performance. See Alexander Murray, *Reason and Society in the Middle Ages*, (Oxford, 1978), pp. 127ff. for this long-standing interest in prudence. Machiavelli was only one of the many early moderns to encourage his readers to constantly war-game while out riding, and the chevalier de Folard would not be the last. More generally, see Roger Manning, "Poaching as a Symbolic Substitute for war in Tudor and early Stuart England," *Journal of Medieval and Renaissance Studies*, 22 (1992), 185–187.

[68] The military interest in planning and preparation is indicated by several popular aphorisms. Hence the military adage that "the plan is nothing, the planning is everything," often attributed to either Dwight D. Eisenhower or Winston Churchill. We could also cite its corollary: "no plan survives first contact with the enemy." One of the oft-mentioned dangers: preparing for the *last* war rather than the next one.

[69] For example, the 'Preface à la Noblesse françoise' in Errard Bar-le-duc's *La fortification demonstrée et réduicte en art* compares peacetime preparations for war to the rhythms of civilian life, justifying the King's need to prepare the 'ship of state' in winter for summer sailing. La Vergne's *Nouveau exercice du gabion* reminded his readers that practicing his procedure would save both time and men, "it being better to fatigue the soldier in drill than to have them killed needlessly" (p. 26).

[70] Virol, *Les Oisivetés*, p. 233.

was uncertain, both on a tactical and strategic level, thus the need to anticipate future contingencies.

With a potential target identified, the chief engineer had to carefully manage the logistical demands of siegecraft, starting long before the siege began.[71] Although engineers had little control over whether the supplies reached the besiegers—this was the purview of provincial *intendants* in the French case and of the Dutch *Raad van Staat* field deputy for the Allies in the Low Countries—the chief engineer was responsible for providing administrators with a list of the required material. Vauban gave his readers an indication of the types of supplies needed for a siege, what the magazines should contain as well as suggested prices for labor and materials (everything from lumber to rope to gunpowder to wheat to cheese to tobacco), something we find in the planning documents of specific sieges. The engineer also focused on how exactly the town would be attacked. Engineers naturally pronounced on many of the quotidian details of siegecraft: the proper construction of trenches and batteries, fascines and gabions, inundations and drainage works.[72] The chief engineer would already be mulling over the task even before he received orders to besiege it. In the most extreme case, Vauban had looked forward to recapturing Ath since at least 1691, and likely since it was turned back to the Spanish in 1678.[73] Our journal of the siege emphasized his exemplary behavior:

> The truth is, as [Ath] was a regular fortification, the rules and proportions of which are well known, the attack was so much the easier, requiring no farther study than the knowledge of the ordinary rules, which are inseparable from that Art; which is not the case in places of an irregular or bizarre construction, which oblige the engineer to

---

[71] See Vauban, *Traité*, pp. 14–16. In his 1672 work, he stressed the need for advanced planning so as to be able to surprise the enemy with an uncontested investment. Vauban, *Mémoire pour servir*, p. 45.

[72] Erik Lund discusses several of these details in *War for the Every Day*, pp. 84–91. In these every-day tasks engineers served as managers and quality-control to the officers who directly supervised the soldiers. The engineers' many complaints of the existing practices of officers and men in sieges, as well as their frequent repetition of precise instructions on how to construct siege works and their tools, indicate that the agricultural/managerial expertise derived from Lund's 'economy of knowledge' (i.e. the practical knowledge ubiquitous in an agricultural age) was deemed inadequate by efficiency-conscious engineers.

[73] Rochas d'Aiglun (ed.), *Vauban*, vol. 2, p. 327, Vauban to Louvois, Mons, 17 July 1691. See also Vauban quoted in Augoyat, *Aperçu historique*, vol. 1, pp. 218–219.

> search for expedients in his own imagination, whose purposes should be answerable to the irregularity of the place, in which respects they are often preferable to the regular ones. Nevertheless, as this regularity was accompanied with great skill and judgment, as the movements of the waters were well contrived and disposed; and, in a word, as Monsieur de Vauban had piqued himself on this work, a particular application was necessary, and a plan of attack to be formed beforehand, which in its way should be as perfect as the fortification; and it is not easy to say which has the advantage, as appears by the plan.[74]

The not-so-subtle lesson: even the great master, when attacking a regular fortress that approached the theoretical ideal elaborated in siege manuals and one that he himself had constructed, even here Vauban was unwilling to risk accidents and labored to perfect his planned attack even before he arrived on site.[75] Only with such prescience could one hope to overcome the many advantages of a well-designed fortress.

Encamped before the city, engineers were to leave nothing to chance, excepting their own personal safety. No better model could be imagined than Vauban's service at Ath: a 64-year old man spending most of the day and night in the trenches in order to identify and make use of every advantage possible, all the while fighting off colds, an infected tooth and a severely bruised shoulder (injured by a musketball that managed to pass through a sandbag). They were to follow Vauban's lead by reconnoitering the ground around the place many times in order to assure themselves that neither the terrain nor the fortifications deviated from the assumptions made in their project.[76] With his frequent surveying of Ath Vauban was taking to an extreme the advice of earlier engineers who stressed the importance of scoping out the target of attack.[77] An increasingly

[74] *A Journal of the Siege of Ath*, p. 92. The French original also notes that he took great care to consider all the advantages and disadvantages of each potential approach before making his choice. *Relation du siège d'Ath*, p. 19.

[75] De Ville had described the irregular fortress as the best test of an engineer's ability to implement theory in the real world. Pollak, *Military Architecture*, p. 22.

[76] Rochas d'Aiglun (ed.), *Vauban*, vol. 2, p. 452, Vauban to Louis XIV, Ath, 21 May 1697; and *A Journal of the Siege of Ath*, p. 96. In a much earlier discussion, Claude Flamand, an engineer in the service of the Duke of Württemberg at the turn of the seventeenth century, repeated the "well-known proverb" that "a place well-reconnoitered is half-taken." *La guide des fortifications*, p. 172.

[77] For example, Flamand spent almost twenty pages describing how to measure and draw a plan, while De Ville spent seven pages on the need to reconnoiter a fortress before attacking it, several of which focused on how to map the fortifications.

important part of such reconnaissance was externalizing the attack in a visual, easily-disseminable format. Just as plans of fortifications were drawn before they were constructed, so too were engineers to commit to paper their intended plan of attack.[78] A necessary and early part of any nobleman's education included drafting lessons; it was also the medium by which engineers communicated both with each other and with their superiors.[79] Once again serving as a model of best practices, Vauban's 1672 manual provided two dozen plans, i.e. aerial-view maps. A quickly-sketched, symbolic representation of the approaches allowed them to test their ideas, to have something to show to superiors for approval, while copies were made available to those directly in charge of overseeing the digging.[80] This more generalized visualization took advantage of the shift over the period in the representation of sieges, from a horizontal view (an elevation) to a more abstract and 'objective' aerial view (an ichnographic plan).[81]

Flamand, *La guide des fortifications*, pp. 172–190; and De Ville, *Les fortifications*, pp. 282–288. Compare this with Vauban, *Traité*, p. 20; pp. 39–40 on the need to make as many reconnaissance trips as are needed in order to acquire a full understanding of the fortification to be attacked: "You must neglect nothing in this matter, because you will have a great advantage over a well-reconnoitered place."

[78] Virol, *Les Oisivetés*, pp. 34–36. See Vérin, *La gloire des ingénieurs*, pp. 160–166 for the engineers' early use of plans, and more generally, John R. Hale, "A humanistic visual aid. The military diagram in the Renaissance," *Renaissance Studies*, vol. 2, 1988. Simon Pepper notes the early reliance on models, especially for cartographically-ignorant patrons, in "Artisans, Architects and Aristocrats," pp. 120, 128–130. For maps in early modern Europe generally, consult the collection of articles, especially David Buisseret's "Monarchs, Ministers and Maps in France before the Accession of Louis XIV," in D. Buisseret (ed.), *Monarchs, Ministers and Maps: The Emergence of Cartography as a Tool of Governance in Early Modern Europe* (Chicago, 1992).

[79] Manuals were published on the methods of drafting, and there is at least one letter where the Secretary of War's office attempted to enforce professional standards by sending recommendations to an engineer at the siege of Douai on how to improve his plans by coloring each day's trenches a different color and adding a scale. AG A[1] volume 2382, #51, 19 August 1712. On the advances in mid-eighteenth century engineering visualization, see Langins, *Conserving the Enlightenment*, pp. 234–256.

[80] Unfortunately, most of these maps and plans have been separated from the letters they were enclosed with and apparently lost. For a surviving example, see the French engineer Charles-Guy Valory's map of the attacks at Menin in AG A[1] 1939, #208. It is said that Vauban had a private plan of his attacks on Ath that he let no one see, though Louis' Court requested he send plans of the attacks as frequently as time allowed.

[81] Martha Pollak discusses this shift in "Representations of the city in siege views of the seventeenth century: The war of military images and their production," in J. Tracy (ed.), *City Walls*, pp. 605–646. For a discussion of the artistic perspectives used in portraying military architecture, see Pollak, "Military Architecture and

With scale enforced, this new tool allowed a more accurate representation of the works to be made, as well as providing a more geometric and (at least on paper) controlled portrayal of the siege. Again exemplifying the engineering principle of economy, Vauban judged the quickly-sketched plan by a reconnoitering engineer worth far more than a precise map that took several days to draw, here following in the footsteps of previous engineers who had refused overly- (or spuriously-) precise measurements in the field.[82] Emphasizing the importance of constant improvement, he included a plan of both the original attack on Lille 1667 and an improved version which corrected the aforementioned mistakes (Plates 28 and 29 in the Rothrock edition), allowing his readers to discover through direct comparison the many potentially-fatal mistakes that had been narrowly avoided.

Preparation did not end once the trenches had been opened, for the engineer was expected to be always looking ahead, working to eliminate the next defending obstacle before it even arose as a threat. Claude Flamand was only one pre-Vaubanian technician to note the need to exercise "good judgment and to foresee everything that might happen in order to not commit any errors or faults if possible, because it is not enough to besiege rashly without considering the consequences."[83] The journalist of Ath marveled at how highly the art had advanced under Vauban's hand: he enthused not only at the skill with which the route of the zigzag approaches was traced, but also the exactitude with which Vauban insisted that each toise of trench match the plan. As a result, these works "were conducted so

Cartography in the Design of the Early Modern City," in D. Buisseret (ed.), *Envisioning the City*, (Chicago, 1998), pp. 109–124; and for the Renaissance addition of perspective, Langins, *Conserving the Enlightenment*, pp. 22–26. George Satterfield also discusses briefly the 17th century shift of military cartography from focusing on the fortresses proper to inclusion of the surrounding countryside as well. *Princes, Posts and Partisans: The Army of Louis XIV and Partisan Warfare in the Netherlands (1673–1678)*, (Leiden, 2003), pp. 273–274.

[82] Vauban, *Mémoire pour servir*, p. 25. Vauban also considered the plans-reliefs an inefficient use of money—they were far too expensive. In his later work he stressed that engineers should study whatever existing maps were available before the siege so as to have all the better an indication of the fortifications to be attacked. Vauban, *Traité*, p. 38. On earlier engineers' rejection of overly-accurate maps, see Virol, *Les Oisivetés*, pp. 34–36. A similar rejection was made for gunnery range tables that were far more detailed than what gunners in the heat of battle could use. A.R. Hall, "Gunnery, Science, and the Royal Society," in J. Burke (ed.), *The Uses of Science in the Age of Newton* (Berkeley, 1983), p. 134.

[83] Flamand, *La guide des fortifications*, p. 191. He then expands his discussion to a broader one of gathering intelligence. He reemphasizes the point on p. 198.

methodically . . . that one could have imagined this the model of a siege, such as is shown to students at the Academy at Grenelle, rather than a real siege."[84] More importantly, however, was the overall design of the attack, and this is where Vauban truly surpassed his peers and predecessors. He planned the entirety of the trenches and batteries as a mutually-reinforcing system, with every part geared toward preparing the fortress for capture. Rather than seizing the covered way and other outworks by the less efficient techniques of *vive force* or mining, he sought to diminish the resistance besieging troops would greet long before they reached the covered way. On a small scale, this planning meant that approach trenches would need to zag right before the garrison's fire could enfilade them, and cannon and mortar fire would need to be directed onto positions that were not yet a threat and would need to continue targeting them even after they had been neutralized, so that they would not be repaired or reestablished. Even the pace of the trenches' advance might need to be managed so as to synchronize their arrival at the covered way, lest an impatient commander waste time waiting for a slower attack to catch up with the faster one. The account of Ath illustrates several other examples. Detachments might need to be positioned so as to preempt the trenches from being enfiladed (p. 101); the retaining sluice had to be destroyed in order to prevent the garrison from flooding the ditch when they would later attempt to pass it and two guns from battery A would be trained on the gate in case the garrison tried to repair the damage (pp. 114 and 133). Materials for crossing the ditch also needed to be stockpiled in the trenches and ready for use (p. 134). Vauban even coordinated his destruction of the fortifications with the advance of the saps by preserving the salient angle of the ravelin as a shield and thereby minimizing the besiegers' need to construct breastworks for the trenches (p. 139). As we saw with ricochet fire, he was similarly interested in minimizing the repositioning of gun batteries—the fewer the batteries and the more targets each battery could hit, the greater the savings in time, money and labor.[85] The efficient coordination

[84] *Relation du siège d'Ath*, p. 27.

[85] In addition to the efficiencies of ricochet fire mentioned earlier, we could also cite *A Journal of the siege of Ath*, p. 132. As Vauban later explained, this is "a significant advantage and a considerable economy [*ménage considérable*]." *Traité*, p. 108.

of these many tasks required expert management and detailed planning, with the chief engineer at the center of this process.

It was not enough to comprehend these more efficient tactics, however, because one could only realize their greater efficiencies if they were applied intelligently at an actual siege. It was as important to know *when* to use each technique as it was to know *how* to use each technique. Vauban illustrated the need for constant attention at the same time as he excused himself from not reporting to his superior more frequently regarding Ath. Even though the garrison's performance was sub par, he explained his lack of reports on the siege's progress to Court as follows: "one could write a book on everything that has happened at our attacks since my last report, but I must focus my attentions in order to profit from the mistakes of the enemy, and to take all measures necessary to prevent any reverses that could result from the many advantages that this place has."[86] Here too Vauban reflected the conventional engineering mindset, one which emphasized the importance of application over theory.[87]

While the relative importance of lives, time and money depended on the strategic situation, engineers believed that in general they should try to economize on all three. This explains why Vauban's perfectionist tendencies did not at first lead him to demand a strict adherence to all of his many siege techniques. He recognized that a commander who focused on only two of its objectives, capture and casualty-avoidance for example, could still conduct a far from efficient siege, since it might well be to the detriment of the other outcome, length. This is the real crux (and paradox) of Vauban's work. 'Better safe than sorry,' although preferable to its alternative, was not Vauban's mantra. Instead he expected skilled, experienced engineers to first assess the situation and determine the most efficient way to attack the town, then to carry out that plan. In this respect he mirrored the traditional engineering drive to find the application of theory approaching closest to the ideal.[88] There were too many enemy moves for each one to be guessed in advance, but drawing on his vast experience, Vauban could predict many of them in a given situation. He then codified the techniques in his manuals that would

[86] Rochas d'Aiglun (ed.), *Vauban*, vol. 2, p. 454, to Barbesieux, 2 Juin 1697.
[87] For example, *Les fortifications de Monsieur le comte de Pagan*, Preface and p. 9.
[88] Vérin, *La gloire des ingénieurs*, pp. 327–333.

either preempt most of them or at least minimize their disruptiveness. His early works in particular must be seen in the context of his larger message: these 'rules' for the attack were contingent on a whole range of factors, such as the garrison's size and morale, the strength of the fortifications, the size of the besieging force and the terrain, to name but a few. His writings on the attack go far beyond strict laws as he assisted his readers in applying the abstract axioms to each unique situation.[89] To take one example, when deciding whether to accelerate an attack or not he elaborated on the variety of issues to consider:

> if the defenders are under strength or if the town is un-garrisoned, if the defenses are poorly maintained, if the fortifications are covered, if the outworks are exposed and poorly supported (as hornworks and crownworks with dry ditches often are), if there are damaged or imperfect works which you can lodge in, if the glacis does not mask the curtain wall, if there is a ditch hidden from the town's view, if the glacis is properly palisaded, and if there are houses, cottages, cellars, gardens, ravines, etc. near their advanced works that could facilitate communication between the detached lodgments. These are the most essential observations which will determine whether you should accelerate the attack or not. There is yet another circumstance in which I would not hesitate. For example . . .[90]

So many factors required expert synthesis. Faced with a new or different situation, you must always return to first principles and judge each maxim's relevance to the case at hand.

His detailed discussions and many conditional statements are a direct result of the infinity of variations an army might meet: rather than leaving it to the 'imagination' of the engineer, Vauban helps the reader by providing a list of guiding principles and a plethora of specific options to be adapted to the situation at hand. Vauban's techniques and tactics were not to be followed blindly in lock step, but only with a full understanding of why they were needed:

> As for other places which are less formidable and defended by mediocre garrisons, it would be useless to follow point by point the rules that I have established here; they are only mandatory when the situation

[89] Once again, he was only following the conventional engineering advice to not, as Pagan put it, "follow the rules blindly." *Les fortifications de Monsieur le comte de Pagan*, p. 28.

[90] Vauban, *Mémoire pour servir*, pp. 133–134.

> presents itself. In all other cases, you can even eliminate half of the work if you want, ignore the dimensions prescribed in the cross-sections, and decrease the number of trench guard positions. Provided that one conforms to a well-ordered design which takes advantage of the elevation of the terrain, you can boldly ignore a textbook approach [*une certain regularité*], as long as you don't make mistakes or expose the men needlessly.[91]

Lest his point be missed, he further cautioned that: "it would not be reasonable to take more precautions against a *bicoque* than against a strong place, nor to treat a weak garrison as if it was strong, or a fortress short of munitions as if it was well-supplied."[92] The 300 pages of his final primer allowed him to develop the theme further, including thirty pages examining how to besiege eight different types of irregular fortresses.[93] Rather than presenting a strict recipe of steps to follow, which would result in inefficiency and wasted effort, he offered his readers a menu of options to choose from. "There is no place where this method cannot be used," he summarized, "but it is more advantageous at some places than others, according to the situation and quality of the terrain."[94] Relying on a stereotyped, rigid view of his siege attack discourages us from exploring this variation.

Deciding which techniques to use required intelligent application, and this in turn came only from experience. Vauban did not expect his engineers to act like automatons, but to intelligently assess each individual situation and apply the guidelines he laid out as consistently as the chaotic arena of combat would allow. His model siege of Ath illustrated clearly this. Since he approached the covered way without a third parallel, he kept a particularly close eye on the situation, "in order to risk nothing and so as to be in a position to

[91] Vauban, *Mémoire pour servir*, pp. 129–130.

[92] Vauban, *Mémoire pour servir*, p. 41. *Bicoque* could refer to either a small building or, in a martial sense, a weakly fortified town.

[93] Vauban, *Traité*, pp. 205–232. He spent, proportionally, a similar number of pages on the topic in his earlier manual. Earlier fortification manuals would also spend much time discussing how to fortify irregular places, while the marquis de Quincy admitted that no manual could encompass the entirety of siegecraft: engineers were to rely on their own experience to cover the topics that he had omitted. Charles Sevin marquis de Quincy, *L'art de la guerre ou Maximes et Instructions sur l'art Militaire*, (Paris, 1740), p. 178. The later French engineer Louis de Cormontaigne would also emphasize the experience requirement. Langins, *Conserving the Enlightenment*, p. 339.

[94] Vauban, *Mémoire pour servir*, p. 127.

return to the rules in case of an accident."[95] Not only did Vauban forego the third parallel, but he also made little use of the trench cavaliers he had built, since ricochet fire made them "needless in the event."[96] More generally, a close engineering associate of Vauban tried to disabuse the director-general of French fortifications Michel Le Peletier de Souzy of the idea that sieges were rigidly mechanical. Explaining a proposed training program for engineers, he explained that: "it is not true as is commonly believed that mathematics are the foundation of our *métier*, they are simply the key, which only good judgment [*le bon esprit*] can use."[97] Given the scarcity of such judgment, engineers needed to sustain a hands-on style of management throughout the siege. Here again Vauban provided the model at Ath, managing the details of the siege by visiting "the Trenches constantly twice every morning to inspect and examine the Work of the night before, in order to see what was further to be done." He even maintained his post after being struck by a musket-ball, giving engineers an example of how to behave after being wounded.[98] Self-sacrifice and relentless attention to detail were the norm.

Despite Vauban's frequent calls for intelligent application, he continues to be portrayed as the genius who turned siege into rigid science, laying down infallible mechanical laws for the attack. Not only did Vauban stress the opposite, but this belief is particularly odd considering how far historians go out of their way to argue the exact opposite for his fortress designs—to dispel the myth that there were 'rules' of fortification, and to emphasize that each place was to be built according to its own unique circumstances.[99] The debate over

[95] *Relation du siège d'Ath*, p. 29.

[96] *A Journal*, pp. 116–117. It was on top of one of these works where Vauban was struck by the ricocheting musketball.

[97] Hué de Caligny quoted in Virol, *Les Oisivetés*, p. 109. See also Vérin, *La gloire des ingénieurs*, pp. 193–194. Such lessons were needed since Le Peletier de Souzy was not an expert on the art of engineering, but rather a bureaucrat from a Parisian robe background whose expertise lay in his financial management skills. His experience with fortifications (and his relationship with Vauban) began when he was made *intendant des finances* of newly-conquered French Flanders at the end of the War of Devolution. In this post he was charged with supervising the renovation of the fortress of Lille and the construction of its citadel.

[98] *A journal of the siege of Ath*, quote on pp. 100–101, also 116; and p. 117 for Vauban's working while wounded.

[99] On the myth of Vauban's three systems of fortifications, see Charles Albert Samuel Lecomte, "Du service des ingénieurs militaires en France pendant le règne de Louis XIV," *Revue du génie militaire*, pp. 25–26 (1877), p. 113; Lazard, *Vauban*,

the existence of Vauban's three 'systems' of fortifications was conceded by the nineteenth century (Fontenelle's 1707 eulogy declares as much, while later 18th century engineers accused their opponents of espousing hidebound systems), but modern scholars still spend twice as much time belaboring this long-conceded point as they do considering the siege attack.[100] Tellingly, Vauban had made it clear that the rules for a siege were exactly the same as those that apply to fortification design: "The principles on which I have founded my method are drawn from those of fortification itself, which assumes a regular system as the most perfect, and all that one can do is strive to approximate this perfection as closely as the different situations allow."[101] In both design and attack, Vauban believed in Method applied to the specific, not a system of rigid theory imposed on reality. That the literature considers theoretical ideals more attainable for frightened and tired humans digging in the dark under enemy fire than for engineers and masons working with inert rock in the security of peacetime is an indication of how neglected the history of the siege attack has been when compared to the motionless walls it targeted.

The ideal manager sought further improvement through evaluation of past sieges. Always with the goal of improving the next siege, Vauban encouraged his readers to conduct post-combat analysis in order to improve future sieges. Most tellingly, Vauban encouraged them to consider how they might have improved even successful sieges, as well as the results of other sieges whose progress they had followed from afar. Whereas people with a less efficient outlook might assume that a successful attack was a well-conducted attack, we find Vauban revising his own successful dispositions when there were no obvious indications of problems. We see this illustrated most clearly in the several pages he spent in his early manual revisiting his successful siege of Lille in 1667—where he exhibited an almost obsessive compulsion for improving efficiency by the smallest of increments.[102]

pp. 371–394; Salamagne, "Vauban et les fortifications du Quesnoy," 50; Blanchard, *Vauban*, pp. 387ff., especially pp. 391–396; and Lynn, *Giant*, pp. 561–563.

[100] On the later eighteenth century debate over fortification 'systems,' see Duffy, *The Fortress in the Age of Vauban*, pp. 149–163; and Langins, *Conserving the Enlightenment*, pp. 51ff.

[101] Vauban, *Traité*, p. 202.

[102] Vauban, *Mémoire pour servir*, pp. 22ff., Chapitre IV: Exemple demonstratif, pour servir de preuve a ce qui a été dit ci-devant. DeJean describes Vauban's pursuit of

In this attack his sovereign was pleased to have the widely-held fears of a lengthy siege dispelled. Vauban acknowledged that it "has received great commendation," only to undercut his success by continuing that its conduct was the "least inept attack [*le moins fait d'inutilités*] we have made in a long time" (p. 22). He argued, for example, that they failed to dispose the cavalry guard properly. But such insight required a reliance on theory—his general principles—since no harm had in fact befallen the horse: "However, *if the enemy had noticed this*, they would have forced us to withdraw or else we would have suffered serious inconveniences."[103] Next we learn that their approach trench was too narrow: "which would have put us in an extreme peril, *had the enemy had the skill and confidence to take advantage of this*."[104] Vauban even conscripted client-patron rhetoric into the effort, attributing the success of the siege more to the King's presence than to the plan of attack. In the process, he further illustrated the attitude a conscientious engineer was to adopt: "hopes of our success were no less rash than for the rest of this attack, which would without doubt have exposed us to great dangers without the King's Fortune" (p. 25). The implication: we cannot always rely on the King to save us from failure, therefore we must prepare for that eventuality. Though he doled out criticism to all involved (including himself), through all the self-flagellation we are never told that his faulty siege had only required nine days of open trenches and cost the French perhaps 600 casualties.[105] Even this low a figure was unsatisfactory for one obsessed with efficiency, for Vauban had discovered that a poorly-sheltered entrenchment had cost the King the lives of eighty to one hundred French soldiers. Rather than rest on his laurels, he labored to uncover more examples of wasted powder, needlessly-sacrificed men and lost time even when others were flush with victory.

---

efficiency as never-ending, an asymptotic quest for perfection, *Literary Fortifications*, pp. 51–52.

[103] Vauban, *Mémoire pour servir*, p. 23, my emphasis.

[104] Vauban, *Mémoire pour servir*, p. 24, my emphasis.

[105] Casualty figures for Lille 1667 are almost completely missing in the secondary literature. The 600 man figure comes from Léon Lecestre (ed.), *Mémoires de Saint-Hilaire*, (Paris, 1903), vol. 1, p. 49, and is supported by Louis' rebuke of Turenne's later storm of Aalst (Alost) that cost 500–600 men: the King complained that he lost more troops at Aalst than he had lost in any of his sieges. Camille Rousset, *Histoire de Louvois et de son administration politique et militaire*, (Paris, 1864–1864), vol. 1, p. 111.

Evaluation also required keeping *au courant* with sieges beyond view in order to not miss a single learning opportunity. Vauban, unlike many around him, was even less impressed with the great costs the Allies expended to capture Namur (Namen) in 1695. Emphasizing the importance of measuring input versus output, he belittled their accomplishment:

> I am persuaded that they have consumed more powder and bombs and destroyed more cannon than we have done in any four of our largest sieges, and that they have lost more men than we have lost in ten. Let them continue on in this manner; as long as the King provisions the places as well as Namur was supplied, we will soon wear them down.[106]

In this Vauban was once again following the lead of past engineers. The importance of such after-action analysis had been emphasized decades before by De Ville in a more general vein:

> Science prepares the mind to take advantage of those things that have been seen and experienced, because it is not enough just to have found yourself in several sieges and combats. You must reflect on everything that has happened, note the mistakes and inconveniences that have occurred, profit from exemplary actions, remember the advantages which resulted and relate them all to the maxims of the science that you have learned, and explore how they relate to one another. From this you can draw conclusions which will yield significant insight into how to act not only in similar situations, but in all those which present themselves, even if they are different from what we have seen.[107]

Here too, Vauban was repeating the wisdom of past generations, though in a more detailed and comprehensive way.

Constant improvement also meant reassessing historical practices in light of present circumstances and future expectations. The young Vauban who encouraged Louvois to occasionally forego a regular attack morphed over the decades into the old Vauban that felt the need to rein in the freedom he had previously allowed his pupils. The account of Ath was careful to emphasize the wisdom of such precautions, noting that even though the trenches were spacious and

---

[106] Rochas d'Aiglun (ed.), *Vauban*, vol. 2, p. 439 to Le Peletier de Souzy, Brest, 25 September 1695.

[107] *De la charge des Gouverneurs des Places*, p. 102. In his *Les fortifications* he echoed Vegetius in his frequent reliance on both personal experience and Classical history to support his points.

admired by all, Vauban assigned additional workmen to perfect them even further, being "desirous to preserve the air of neatness throughout the whole" (p. 108). Vauban's refusal to attack the covered way prematurely was justified by his desire "to run no hazards in the mean time, but to be ready to fall into the proper measures in case of an accident."[108] Echoing the language of Vauban's early treatise, the journalist also noted with approval the cautious decision to extend their lodgment "to favour us in our passing the Ditch; for, although the Enemy made no great firing, it might not be so safe to expose ourselves" (p. 134). The siege also illustrated the validity of such pre-emptive measures by highlighting how carelessness and arrogance had a price. In one case, the Spanish fired on the French batteries, "which were a little carelessly made, as too frequently happens where one has contempt for the Enemy," resulting in the deaths of several gunners and a good artillery officer (pp. 135–136). In general, the account emphasized how Vauban's trenchworks followed his plan whenever possible (e.g. pp. 108, 115, and 118). After praising the "great exactness" with which the trenches were conducted, he wrote that "all this was carried on with so much method on our side, and with so little interruption on the side of the Enemy, that it seemed rather the representation of a siege than a siege itself" (p. 113). Later he went even further, contending that: "it was the great Guns, Shovels and Pick-Ax which took [Ath]; for the Marshal Catinat scarce made use of any other means than the industry of Monsieur de Vauban who so well conducted the Trenches and Artillery, as hardly left the King's Troops any opportunity to signalize themselves" (pp. 120–121). Here was the ultimate engineering compliment. Not only had the engineers proved themselves Louis' most valuable servants, but, more fundamentally, the implementation of Vauban's efforts at Ath essentially achieved the theoretical ideal, allowing intelligence and *industrie* to supplant unthinking force.[109]

[108] *A Journal of the siege of Ath*, p. 116. This is almost verbatim what Blanchard attributes to *Vauban*, p. 351. The *maréchal* highlighted his more general aversion to unnecessary risks in another letter to the King: "do not abandon the certain in order to chase after the uncertain." Rochas d'Aiglun (ed.), *Vauban*, vol. 2, p. 463, 16 June 1697. For an earlier example, see Vauban, *Mémoire pour servir*, p. 107. De Ville had stated the same sentiment (practically verbatim) seventy years earlier in *Les fortifications*, p. 271.

[109] See also p. 129, where the ardor of "unthinking Zealots who knew no better" had to be restrained from attacking a ravelin that surrendered a few hours afterwards.

Vauban supported these sentiments wholeheartedly, as a comparison of his early and late manuals indicates. The increased attention Vauban paid to close management of the siege in his late work was not only a function of his quest for constant improvement, but was also a response to the increasing resistance his recommendations had met. The 1672 treatise was an immature work meant for the Secretary of War Louvois, an administrator who could not directly interfere in the conduct of sieges from Court, whereas Vauban's 1704 magnum opus capped the 70-year old master's illustrious career and was dedicated to the 21-year old *petit dauphin* Burgundy who had disputed his conduct of the previous year's siege of Breisach. Experiencing several such rejections in 1703, Vauban sought to more tightly control the attack by spelling out its nature in greater detail while he still had time. To take one example, his earlier suggestion that a weak fortress allowed one to accelerate the attack and omit half of his works disappears in his 1704 manual. In its place, we find the twenty-third rule: "Never reject or discard the rules under the pretext that a fortress is not strong, for fear of allowing a weak place to defend itself like a strong one."[110] Rather than encouraging his readers to break the rules as long as they made no mistakes (admittedly setting the bar quite high from the beginning), his late work becomes much stricter. It provides thirty rules (*maxims*) to obey, with language far more constricting than previously: never do X (maxims 3, 7, 8, 9, 10, 12, 16, 23), always do Y (maxims 2, 4, 5), break the rules as little as possible (maxims 21, 30), and do not rely on luck to capture the covered way (maxim 7). He also warned that only experienced individuals can apply the rules effectively (maxims 18, 28, 29), and that less-than-strict observation of the rules results in greater difficulties (maxim 24, 30). At the end of his *Traité de l'attaque* he summed up the art of reconciling the need to preempt the almost infinite universe of an enemy's potential moves with the need to minimize unnecessary effort:

> There would be many other things to say about the attack of all types of fortresses, but one would never finish; because as there is not a single one that resembles the design nor the situation of another, each place forces us to deviate somewhat in order to address the specific conditions; and where the observation of the rules becomes impossi-

[110] Vauban, *Traité*, pp. 264–265.

> ble either totally or in part, only good sense [*bon sens*] can help us, but *always with the goal of breaking the rules as little as possible.*[111]

With a lingering sense of unfulfilled potential, decades of additional experience, time for reflection, and with his life drawing to a close, Vauban felt the need to codify his techniques and insist on closer siege management by the engineers. Only through careful preparation, intelligent application and critical self-evaluation could these managers ensure the most efficient siege possible.

5. *The Vaubanian Siege as Science*

Though military engineers were preoccupied with efficiency, we should be careful not to interpret the Vaubanian siege as an overly mathematical or theoretical endeavor, as the rhetoric of siege-as-science often does. Although Vauban's rationalization of the attack coincided with the age of the Scientific Revolution, he was only minimally influenced by the findings most noteworthy to us today. In this he was no different from his predecessors and even his immediate successors; Janis Langins has recently illustrated the phobia military and civilian engineers had in the first half of the eighteenth century for such exotic beasts as algebra and excessive Euclidean geometry.[112] Contemporaries not surprisingly differed over the extent to which war was an art or a science (or craft), while our expanding appreciation of early modern science emphasizes the permeability of the boundary between the theoretical and the practical. An analysis of Vauban's discussion of the siege attack, as well as his other works, indicate that he drew very little from the contemporaneous advances in natural philosophy. His methods show little influence of 17th century mathematical and scientific trends: algebra and Descartes' analytical geometry, Pascalian probability, the astronomical observations of a Galileo, the circulatory system of Harvey, Leibniz and

[111] Vauban, *Traité*, p. 231, my emphasis. He then describes several rules that are "almost always" applicable, such as using parallels.

[112] Langins, *Conserving the Enlightenment*, chapters 1–3 and especially pp. 144 and 229–233, which describes the corps' lukewarm reception to Bernard Forest de Bélidor's attempt to mathematize their *métier* with the 1729 publication of his *La Science des Ingénieurs*.

Newton's calculus.[113] Nor were new scientific instruments, devices such as air pumps, microscopes and the like, shedding much light on siegecraft.[114] Even the telescope, promoted for its utility in sieges by early inventors, was of limited use to Vauban: his manuals did not mention them and instead directed engineers to reconnoiter the fortresses from very close up, risking capture himself on many occasions. Searching his works, we find little reflection of the mathematical formalism so often attributed to him. His treatises mimicked formal geometry in their organization, but in this he only mirrored the many military manuals of earlier in the century, proceeding from definitions and axioms to examples of contemporary bad practices, and then on to how to improve these procedures through application.[115] His geometry was Euclidean in spirit, and would have been familiar to educated people of the Renaissance, of the medieval world if not of ancient Greece. He hid this theoretical foundation, though, behind a façade of rules of thumb that any traditional engineer-craftsman could apply. We find only two examples in his attack treatises where he adopts an overtly geometric approach, when describing how the trenches were to be laid out and when discussing how to determine the powder charge needed for a mine. Even here, he boils the geometric postulates (at their most complicated using notations

[113] Even 17th century advances in physics had an extremely limited impact on ballistics until the middle of the 18th century. For the argument that practice was much simpler than theory, consult: A.R. Hall, "Gunnery, Science, and the Royal Society"; and Steven Walton, "The mathematical and military sciences in Renaissance England," *Endeavour: A Quarterly Magazine for the History and Philosophy of Science*, vol. 24 (2000), 152–156. On the interrelationship between artillery and ballistics more generally, see John Guilmartin, Jr., "Ballistics in the Black Powder Era: a cursory examination of technical factors influencing the design of ordnance and of the emergence of ballistics as an applied science," in R. Smith (ed.), *British Naval Armaments*, (London, 1989), pp. 73–98; Michel Blay, "Le développement de la balistique et la pratique du jet des bombes en France à la mort de Colbert," in L. Godard de Donville (ed.), *De la mort de Colbert à la révocation de l'édit de Nantes, un monde nouveau? XIVe colloque au Centre méridional de rencontre sur le XVIIe siècle*, (Marseilles, 1984), pp. 33–50; and Brett Steele, "Muskets and Pendulums: Benjamin Robins, Leonhard Euler and the Ballistic Revolution," *Technology and Culture* 35 (1994), pp. 348–382.

[114] See the online catalog of 'The Geometry of War,' an exhibition of Oxford University's Museum of the History of Science, currently at http://www.mhs.ox.ac.uk/geometry/content.htm.

[115] Virol, *Les Oisivetés*, pp. 32–33. On pp. 81–90, she illustrates how his manuals' organization, use of history, maxims and definitions, as well as his concern with cost-effectiveness all follow in the footsteps of earlier engineers. Rothrock's translation of the *Mémoire pour servir d'instruction* destroys this effect by moving the section on contemporary faults to the end of his edition.

such as 'line AB') down into memorable aphorisms rather than formal equations. As such, his presentation was accessible to engineers trained in the old-fashioned 'constructive geometrical' school, rather than the more theoretical and academic branch of 'practical geometry.' As a result, some mid-eighteenth century reviewers of Vauban's first published attack treatise (1737), who were abreast of more current attempts to apply modern science to military engineering, complained of the lack of geometric apparatus they had come to expect in more recent treatises on siegecraft.[116] Even rarer in Vauban's work was any kind of algebraic notation. Vauban and other engineers provided tables with redundant information in order to illustrate basic relationships: instead of simply instructing the reader to add ten workers for every additional gun added to a battery, Vauban included a table to show the number of workers needed to construct a battery of four guns, the number needed for a battery of five, of six, and so on.[117] Few engineers of the early eighteenth century thought of their craft in mathematical terms, a conservatism that engineering reformers would struggle against up through the Revolution.

More often, Vauban's calculations relied upon basic arithmetic (again presented in tabular form)—calculating materials needed for battery platforms, the dimensions of trenches, lengths of siege lines, sizes of guard posts, pounds of gunpowder or bread, ratios of trench guards to garrison forces, days of labor, and so on.[118] Most of his numbers were concerned with the less precise matters of troop dispositions (where to place the workmen and in what numbers, how many troops commanded by how many officers of what rank should be used in an attack...) and construction advice, e.g. how to reinforce a trench with gabions and fascines. In other words, he put on paper

[116] Virol, *Les Oisivetés*, pp. 47ff. We can contrast Vauban's sparse presentation of geometry with the much more formal pedagogical treatises seen earlier in the 17th and later in the 18th centuries, primers that usually focused more on reviewing basic geometry than on the siege attack. Virol argues that Vauban refused to provide his manuals with geometric notation for his fortress designs so as to protect such important state secrets (231).

[117] Vauban disposed of these tables in his 1704 treatise, while also eliminating much of the detail regarding the construction of batteries. For a similar Dutch list that spells out the numer of cannon a garrison needed for a given number of bastions rather than describing the simple rule of adding two additional guns for each additional bastion, see Algemeen Rijskarchief (ARA), Collectie van der Hoop (CvdH) #118, Lyste van de nodige Crijgsmaterialen....

[118] On Vauban and arithmetic, see Virol, *Les Oisivetés*, pp. 253–299.

the kinds of indispensable logistical and operational calculations that commanders, engineers, and quartermasters had been performing mentally for centuries.[119] Mathematics served siegecraft only as an ideal.[120]

Not that we should be surprised at the relatively simple level of mathematics Vauban made use of. He had, after all, learned his trade in the artisanal tradition.[121] From his green-thumbed father he had received a knowledge and appreciation of nature, learning the practical knowledge of nature required for a person growing up in the wilderness of the Morvan in Burgundy. Without family money to draw upon, he could afford little formal education and likely attended a nearby *prieuré* where he would have learned from a *curé* some basics of the humanities (including some Latin as well as history), and more importantly for his future path, the rudimentary amount of mathematics, fortifications (i.e. applied mathematics), and draftsmanship expected of a poor, young nobleman. Recalling his sparse education, he admitted only "a fair smattering of mathematics and fortification, and also drawing not at all badly."[122] His real education in the military arts came with his enlistment at the age of eighteen as a *cadet gentilhomme* in the prince de Condé's army fighting against the young Bourbon king (still a minor) in the princely Fronde. With his elementary knowledge of fortifications, he again relied on practical experience rather than book-learning, being assigned to repair threatened and recently captured fortifications as well as assist at Condé's sieges, perhaps a dozen or more over the course of his service with the prince. Thus, Vauban learned much of the specifics and application of siegecraft like most other military engi-

---

[119] Erik Lund writes of this 'operational level' of military practice, the everyday history of how an army in the field marches, camps, forages, etc., as well as its basis in the agricultural economy of the period, in Lund, *War for the Every Day*. For a comparative example of the mathematical skills available to early medieval military planners, see Bernard Bachrach, "Charlemagne and the Carolingian General Staff," *Journal of Military History*, 66 (2) 2002, pp. 338–341.

[120] Indicative of his sparse reliance on modern mathematics prevalent in the modern social sciences is how rudimentary it appeared in a "lettre sur la manière de faire les statistiques," which limited itself to a discussion on how to collection information on a region's basic demography and economic potential. Rochas d'Aiglun (ed.), *Vauban*, vol. 1, p. 590, 1 February 1706.

[121] On Vauban's early life and education, of which little is certain, see Pujo, *Vauban*, pp. 17–22; and Blanchard, *Vauban*, pp. 45–75.

[122] Quoted in Hebbert and Rothrock, *Soldier of France*, p. 14.

neers, in apprenticeship under fire rather than from an academy or erudite theoretical treatises.

Nor did Vauban have much leisure time for such secondary pursuits throughout his career. His family's financial situation had denied him the resources that would have allowed him to be a gentleman scholar, and he was too busy doing the King's business in any case. He collected a wide variety of books, but the modern natural sciences (excluding philosophical and geometrical tomes) were few in number. Although he was appointed to the *Académie royale des Sciences* in 1699, this was an honorary appointment well after his formative years.[123] From what we know of his library (extensive in comparison to most military officers), he had very few books on the subject of natural philosophy, unlike other polymaths such as the Imperial botanist-general Luigi Ferdinando Marsigli.[124] Vauban likely had little time to attend many sessions of the *Académie* and keep up with the latest scientific debates beyond what could be learned in various *précis* and *comptes-rendus*, given his peripatetic life of royal service. In addition to the many campaigns in which he actively participated (1651–1658, 1667–1668, 1672–1678, 1683–1684, 1688–1697, 1702–1703) and the mass of *mémoires* he penned and correspondence he maintained, his post as the King's first engineer required he draft hundreds of fortress plans, as well as travel from one corner of France's nascent hexagon to another inspecting fortifications and preparing them for defense. Unlike generals of horse and foot, in winter quarters he found himself just as busy with royal business as during the campaign season. As a result, he spent an average of only 22 days per year at home in the Morvan.[125] From 1651 to 1667 he averaged only 865 kilometers per year in travel, from 1668–1677 some 4,280 km on average, and from 1678 through 1703 fluctuating between 4,000–5,000 km per year—an average of almost seven miles per day. The vast majority of these journeys were far beyond his home or Paris or Lille. With the onset of the War of the Spanish Succession he found himself in forced semi-retirement, a period of

[123] On Vauban's attendance and his intellectual interests more generally, see Virol, *Vauban*, pp. 289–302.

[124] For the existing inventory of his library, see Virol, *Les Oisivetés*, pp. 85–88 and Annexe 6. On Marsigli, see John Stoye, *Marsigli's Europe, 1680–1730: The life and times of Luigi Ferdinando Marsigli, soldier and virtuoso*, (New Haven, 1994).

[125] On Vauban's many travels, see Blanchard, *Vauban*, chapter 17 "Le vagabond du roi."

which allowed him to put his papers in order—he lamented that he was now a man "with little to do"—and undoubtedly this gave him more freedom to explore his other interests.[126] Overall, however, his many military responsibilities to the Crown consumed most of his adult life, and while his interests extended far beyond the purely military realm, they rarely encompassed 'scientific' debates as we understand the term today.

But Vauban was undoubtedly affected by the broader philosophical underpinnings of his age, combining the Cartesian mechanical philosophy so prevalent in 17th century France with Baconian empiricism. From his brief education Vauban would have learned Descartes' reductionism, whereby all things are reduced to their simplest component parts.[127] Applying this to siegecraft, he reduced a siege to each of its stages, and within each stage, their constituent tactics. To this implicit theoretical framework, however, Vauban attached an appreciation of empiricism in order to bridge the gap between siege theory and reality. Beyond his emphasis on observation, for example his insistence on close personal reconnaissance of works to be attacked, this empirical attitude manifested itself most visibly in his use of historical examples.[128] Illustrating mistakes made in the past, he justified his authoritative judgments by drawing on the experience gained in well over forty sieges, as well as calling upon his study of one hundred others. Experimentation in a laboratory could not approach real-life conditions, but Vauban did make use of hypothetical 'thought experiments' such as revising his attacks on Lille, combining both his general principles and his vast experience. He also experimented with siege tools and techniques, such as *pierriers* (stone-throwing mortars) and cannon-casting methods, as well as his efforts to determine the rate of advance by sap.[129] He was also interested in topics removed from fortifications and the military, the beginnings of the social sciences. Although his analyses were largely limited

---

[126] Fontenelle mentions this in his eulogy. Virol, *Les Oisivetés*, p. 483.

[127] On Vauban and Cartesian reductionism, see Virol, *Les Oisivetés*, pp. 194–204.

[128] Virol, *Les Oisivetés*, pp. 85–88 on the use of examples in his military writings, pp. 223–229 on their use in his non-military works. See also the predominance of history works among an inventory of his book collection (Annexe 6).

[129] For his investigation of cannon founding, see Pujo, *Vauban*, pp. 283–241. On the sap's progress, see Vauban, *Traité*, p. 80 note 1. Undoubtedly there had been earlier experiments with such weapons as *pierriers* and small mortars, e.g. Raimondo conte Montecuccoli, *Mémoires de Montecuculi, generalissime des troupes de l'empereur*, Nouvelle édition, revûe et corrigée (Amsterdam, 1752), vol. 1, pp. 56–58.

to an arithmetic approach, he studied not only architecture and engineering but also agronomy, demography, economics and politics—topics falling under the contemporary rubric of 'political arithmetic.' As with siegecraft, these were all fields that directly affected his efforts to improve the security, strength and prosperity of France.[130] Despite his geometrical theorizing and a marshal's baton, he was at heart a practical man who identified more with the peasants of his beloved Burgundy than with the intellectual elite. For Vauban mathematics were only a means to an end.

Vauban's ideas were not completely original, but he improved their application in order to maximize the efficiency of siegecraft. While he has received the credit for breaking the defensive supremacy, a cursory look at earlier sieges suggests that he was only codifying and expanding practices that had already been developed earlier. To the extent that he 'perfected' the siege attack, he did it not by bringing mathematics to the attack, for many besiegers before him had worked with this apparatus, hence the pre-Vauban appearance of zigzag approaches, trench cavaliers, and parallels in various forms. Nor did perfection mean a strict application of parallels, trench cavaliers and ricochet fire. Nor did the ideal siege require a slavish adherence to the stages of the model siege. What Vauban did do, by the end of his life, was to create a rigorous, coherent presentation of geometrical maxims in codified form, testing and collating the many practical details of 17th century siegecraft according to a guiding philosophy that sought to reconcile low casualties and short durations. Efficiency through preparation and attention to all the details of a siege was his mantra.

Whether or not Vauban was the first engineer to combine a variety of 17th century offensive techniques based on geometrical principles into a cohesive whole, his unique position and prolific quill guaranteed that he would receive the attention of posterity—if nothing else Vauban was the greatest popularizer of a more systematic type of attack. No other similarly-talented engineer survived as many sieges as he did, nor had the fortune to work for a bellicose king who ruled one of Europe's wealthiest kingdoms for as many decades, whose quest for military *gloire* manifested itself in a shared appreciation

[130] See his *Oisivetés* for a sampling of his range, catalogued and excerpted in Rochas d'Aiglun (ed.), *Vauban*, volume one.

of sieges, and whose lands were conveniently located near several of the most heavily-fortified cockpits of Europe. At the very least, it was this convergence of royally-granted authority, a rich resource base, and a king's willingness to wage an offensive war of positions against a declining Spanish foe that gave Vauban his opportunity to shine. The extent to which others followed his example is the focus of the remaining chapters.

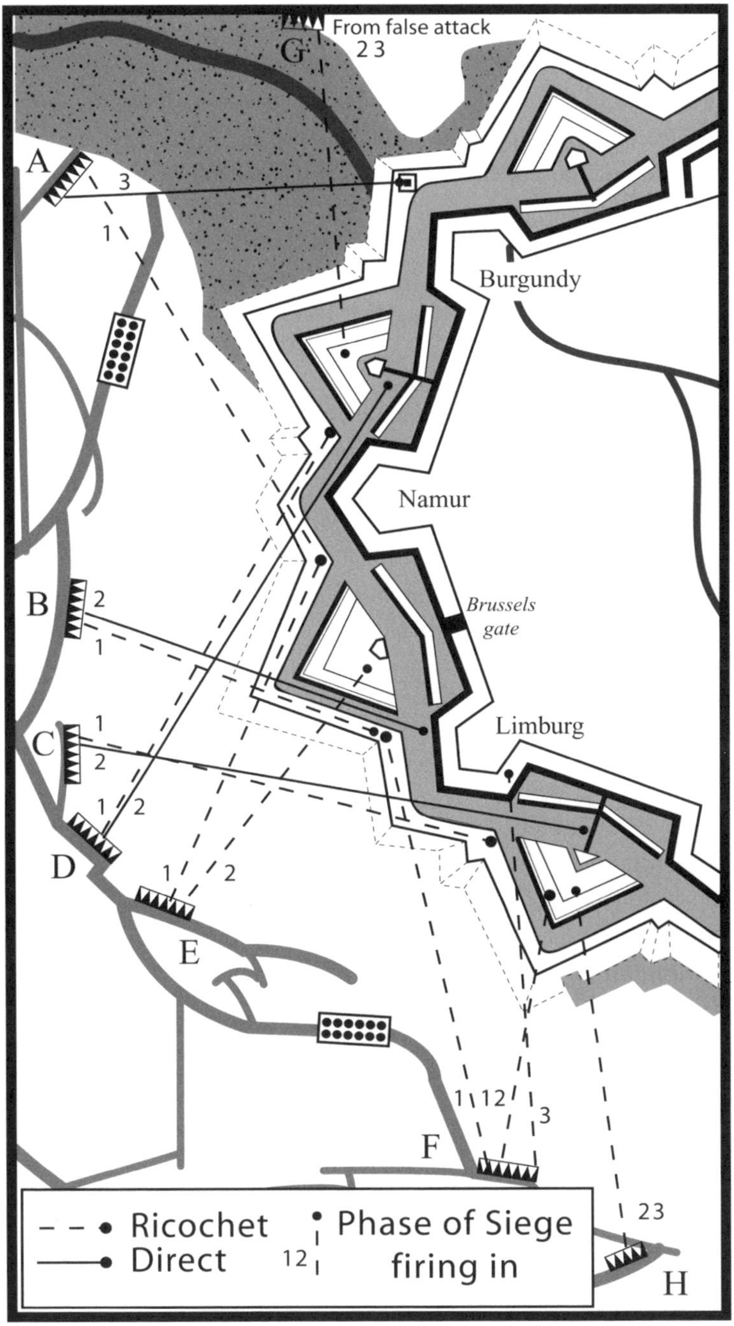

Map 3: The Efficiency of Vauban's Batteries at Ath

CHAPTER FOUR

THE TRANSITIONAL WAR OF THE SPANISH SUCCESSION

The efficient siege of Ath was the pinnacle of Vauban's art. The peace treaty signed after Ath's Fall allowed Vauban to return to his peacetime duties, interspersed with occasional sojourns in Paris, Lille (he was long governor of the citadel of Lille) and at his home in the Morvan. These would be short, for as with so many other peace agreements, towns along the border were constantly changing hands. In the final version of the treaty the French ceded a series of border towns to Spain, from Dixmuide near the coast to the strong fortresses of Namur and Charleroi on the Meuse. The negotiators had, Vauban would later complain, "taken the best [frontier] places from us and given them to the enemy, while also obliging us to raze several others... our frontiers are in very bad shape because we have not a single place that can be said to be totally finished."[1] Joseph Sevin, chevalier de Quincy and younger brother of the marquis Charles agreed wholeheartedly with his first point: "I will never understand the politics of Louis XIV to have fortified so many places so perfectly and then not keep them, or at least not dismantle their works before turning them over."[2] To return the frontier to a semblance of order, Vauban had first to shore up the exposed eastern edge of the *pré carré*—the anchor Dinant had been dismantled so Charlemont-Givet had to take its place on the flank of the *pré carré*'s first line. From 1699 to the outbreak of the next war in 1701, the *grand ingénieur* had little repose as he expanded his tour to encompass the fortifications around all of France, from Flanders to Alsace, to Champagne, to Provence and beyond.

Even this 'respite' was short lived, for in November 1700 the sickly and severely inbred Habsburg King Carlos II of Spain finally died at the end of his thirty-eighth year. The moment European mon-

[1] Rochas d'Aiglun (ed.), *Vauban*, vol. 1, p. 497, Vauban's *Projet de paix assez raisonnable*, written in February 1706.

[2] Lecestre (ed.), *Mémoires du chevalier de Quincy*, vol. 1, p. 43.

archs had planned for over the past three decades had arrived. All their various projects for partitioning the sprawling Spanish empire were nullified, however, by Carlos' testament, which insisted on an intact Spanish Empire going to only one candidate. With the end of the Spanish Habsburg dynasty, all of Spain's dominions were thus offered first to Louis XIV's grandson Philippe de Bourbon, duc d'Anjou and a grand nephew of Carlos' first wife. Louis accepted the proposal on behalf of his grandson, a decision motivated in part by the alternative: another unification of Spanish and Austrian Habsburg branches through the Austrian candidate Archduke Charles. War with Austria over Bourbon-occupied Spain was now all but assured, but it was not yet clear who else Louis would have to fight. The year 1701 saw the first shots fired as France and Austria renewed their 200-year old rivalry for dominance in Italy, while the rest of the western European states prepared for a likely war. Traditional partnerships were renewed, with France finding steady allies in Joseph Clemens Wittelsbach (Archbishop of Cologne and Bishop of Liège) and his brother, a former enemy of the Sun King, the Elector of Bavaria and governor-general of the Spanish Netherlands Maximilien II Emmanuel Wittelsbach. Both Portugal and Savoy would initially ally themselves with Louis as well, yet by 1704 both would abandon him. The critical question of whether the two maritime powers of England and the United Provinces of the Netherlands would enter the fray was still uncertain. Representative assembles in both countries (Parliament and the States-General) were not in a gambling mood, and voted funds for a wartime buildup even before war was declared by the anti-Louis coalition in May 1702. Louis' tendency to undermine his diplomatic efforts with brash dramatic moves surfaced over the course of 1701, and his inflammatory gestures quickly pushed even neutrals in England and the Netherlands into the war camp. He refused to remove Philippe from the French line of succession and thus raised the specter of a united Franco-Spanish hegemon. In February he surprised and imprisoned the Dutch barrier garrisons in their Spanish Netherlands fortresses and then released them so they could serve as the core of an expanded States Army. Upon the death in September of the old King James II of England, an exile at Louis' Court, France's divine right monarch declared the rightful King of England as the Catholic James Francis Edward Stuart—known as the Chevalier Saint-George and later "the Old Pretender" to his enemies, James III to his Jacobite supporters. This brought once again into question England's Protestant succession, an issue which, it was thought, had been resolved both at Rijswijk where Louis recognized William and codified internally with Parliament's Act of Settlement (June 1701) assuring a Protestant Hanoverian successor. Louis further riled up London mercantile interests by giving the Spanish monopoly to carry the American slave trade (the *asiento*) to French merchants, thereby shutting out British traders. Such decisions might have been justified and/or principled, but they were not particularly realistic. It was too easy to interpret such maneuvers in an uncharitable fashion, especially given his past aggressive history. The war parties in both countries were strengthened as William's long-running feud dovetailed with Dutch mercantile fears of a French merchant class given preferential access to Spanish American markets, as well as concern over assertive Belgian merchants under French protection. Louis' old nemesis, William Prince of Orange, *Stadhouder* of the Netherlands and King of England, was only too ready to convince Europe that yet again the French despot threatened to trample European liberties. The result was the formation of yet another Allied coalition against the King of France, to be fortified by 100,000 German mercenaries. Europe was once more engulfed in war.

The dozen years of conflict that Carlos' death inaugurated would see important transitions away from Louis' past wars and from the Vaubanian paradigm in particular. For the first time, the War of the Spanish Succession would find the Sun King on the defensive in Flanders from almost the beginning of the war until practically its end in 1712. Louis' early offensive wars for glory and revenge would be replaced by stalemate and then increasingly defensive conflicts, while the threat to the territory of France posed by William's coalitions would be most pressing after the Prince of Orange's own death. The War of the Spanish Succession would also be fundamentally different from Louis XIV's previous wars because of the disappearance of Vauban himself. Several years before his death in 1707, and just a few years after his engineering masterpiece at Ath, he would be set aside and it would be up to his pupils to continue his legacy of constant improvement. His successors would be busy since at least 119 sieges would be conducted by both French and Allied engineers over the thirteen years of conflict; Appendices B and C present the data and discuss its collection.[3] These sieges will

[3] The siege dataset I have collected in Appendix C only includes 115 sieges: sev-

allow us the opportunity to assess Vauban's impact on positional warfare by gauging what he passed on to others and how well his call for efficiency was received.

To do so more manageably, we will restrict our analysis primarily to the Low Countries sieges of this war (Map 4.1 at the end of the chapter). This theater was, in many ways, the key one, certainly for the French, British and Dutch, the three powers most capable of conducting and paying for state-of-the-art siegecraft. No theater saw more campaigning than the Low Countries (eleven years of full-scale operations), nor did other theaters witness field armies as large as were mustered here, numbering 100,000 men and more by the end of the decade.[4] Additionally, Vauban's legacy, the application of his methodical style as well as his fortress designs would all be tested most forcefully here where his famed *pré carré* blocked the entrance into the heart of France. With the region's flat terrain and numerous waterways flowing northward, besiegers took advantage of extensive transportation networks to marshal all their resources against enemy strongholds. Bordering three of the four main belligerents' territories, the theater's uncomfortably-close proximity to Versailles and Paris dictated that it would be here that the French would exert their utmost efforts to avoid a collapse, and here also where the Maritime Allies would concentrate their labors. As a result of this combination of factors, the theater witnessed thirty-eight sieges, double or even triple the number seen in Louis' previous Flanders campaigns. In fact, this number even approaches Vauban's lifetime total (around fifty) in only eleven years of campaigning (1702–1712). Vauban remained active until his final year of life, but he would conduct only two sieges during this, the last and most defensive of Louis XIV's wars. It was now his students and his enemies who waged positional warfare throughout the region. The convergence of three of early modern Europe's most famous military men—Vauban, Marlborough, and Prince Eugene of Savoy—further facilitated detailed documentation of its large number of sieges, offering us an opportunity to examine in detail the nature of siege warfare

eral (mostly in Iberia and Germany) had to be excluded because there was not enough clear evidence in secondary sources to indicate their duration accurately.

[4] In general, I will refer to the Low Countries theater and the Flanders theater interchangeably, while recognizing that Flanders referred more precisely to the coastal county of Flanders, divided between French, Spanish and Dutch rule.

in the 'Age of Vauban.' It is only with a complete inventory of a large number of sieges at the end of Vauban's personal tenure that we can begin to measure his true legacy and assess how well his techniques were disseminated to the wider world.

1. *The Flanders Campaigns of the War of the Spanish Succession*

Unlike Louis' earlier wars, the operational details of the War of the Spanish Succession have been narrated numerous times in several languages, so only a brief overview will be provided here, supplemented by Maps 4.2 through 4.6.[5] Map 4.3 indicates the rather clear overall narrative: a seemingly inexorable Allied advance into French-held Spanish Netherlands and then into France proper.

Though Gallispan and Dutch troops faced each other across several hundred miles of frontier in 1701, it would take another year before the Low Countries saw war up close. Direct British and Dutch involvement in the War of the Spanish Succession began on a tragic note with William's death after a fall from his horse in March 1702. Although he and Mary had remained childless, his diplomatic legacy at least would live on in the form of yet another Europe-wide alliance against Louis XIV's France, coordinated on the Dutch side by his long-time associate Anthonie Heinsius, Grand Pensionary (*raadpensionaris*) of Holland.[6] William's untimely exit would, fortuitously for

[5] The most important narratives include: Marquis de Quincy, *Histoire militaire du règne de Louis le Grand*; Vault and Pelet (eds.), *Mémoires militaires relatifs à la succession d'Espagne sous Louis XIV, extraits de la correspondance de la cour et des généraux par le lieutenant général de Vault, directeur du Dépot de la Guerre, mort en 1790 . . .*, (Paris, 1836–1842); the Spanish Succession volumes of the Austrian Kriegsarchiv's *Feldzüge des Prinzen Eugen von Savoyen: nach den Feldacten und anderen authentischen Quellen*, (Vienna, 1876–1891); J.W. Wijn, *Het Staatsche Leger*; John Wolf's *Louis XIV*, (New York, 1968); and most recently, the relevant section of John Lynn, *The Wars of Louis XIV*. Biographies of the various kings and commanders (especially those on Marlborough, Eugene, and Villars) also tend to trail a pike behind their subjects on campaign.

[6] The *raadpensionaris* Heinsius became the single most important Dutch political figure after the *Stadhouder* died. A close associate of William, his central position in government encouraged correspondents from throughout the Dutch military and diplomatic communities. Heinsius' central role in wartime is illustrated in his published correspondence (edited by A.J. Veenendaal, Jr.), and, for the earlier Nine Years' War, in John Stapleton, "Grand Pensionary at War: Anthonie Heinsius and the Nine Years' War, 1689–1697," in Jan A.F. de Jongste (ed.), *Anthonie Heinsius and the Dutch Republic 1688–1720. Politics, Finance and War*, (The Hague, 2002).

the Allies as it turns out, open the door for what would become one of the greatest generals in England's history to march through, John Churchill Earl of Marlborough. But before Marlborough could join the Confederate army in July, the Imperials moved against Joseph Clemens' French-garrisoned town of Kaisersweert on the Rhine, where their allied Dutch participants served as 'auxiliaries' until the Maritime Powers officially declared war halfway through the siege. After a less-than-efficient attack, the town surrendered and the campaign shifted back to the west. A number of operations in the western county of Flanders produced few gains. The Allies managed to capture the small Fort Saint Donas after breaking through the enemy fortified lines but little came of this. The person conducting this siege was the Allies' most famous military engineer, Menno van Coehoorn, a Frisian lieutenant-general of infantry as well as the Dutch director-general of fortifications and master-general of their artillery. Several months later, Coehoorn's rival Vauban failed to capture the forts surrounding Hulst west of Antwerp. The significant breakthrough though came in the territory of the Bishopric of Liège. Boufflers' French army had managed to chase a Dutch force under William's old general Godaert van Reede-Ginkel, graaf van Athlone all the way to the gates of Nijmegen, with the Dutch troops saving themselves by jumping into the covered way just in front of the pursuing French cavalry. With Marlborough's arrival, however, the reinforced Allied army was able to push the French back and invest the fortress of Venlo on the Meuse. The town would fall quickly, and three other Liégeois places—Stevensweert, Roermond and Liège—would fall to siege soon after. In a single campaign, the Allies had managed to turn back an imminent French threat and largely eliminated Louis' Cologne ally by conquering the vast majority of Joseph Clemens' territories. The isolated Dutch fortress of Maastricht was now safe as well, and the States-General could breathe easier with their borders more secure—there would be no repeat of Louis' 1672 invasion of the Netherlands. For his significant role in these events, John Churchill was made First Duke of Marlborough by Queen Anne. Over the winter Joseph Clemens would meet with further setbacks, losing Rheinberg to the enemy. The Spanish would also lose Guelders as well.

The victories of the 1702 campaign convinced both Parliament and the States-General to vote an additional augmentation of 30,000 troops for the war effort. Putting these forces to best use would require some tinkering however. Without the unifying presence of William, the growing Allied army would need to develop an understanding of how their multi-national coalition should be run in the field. The English had agreed that since the Earl (now Duke) of Marlborough would be the overall commander of Allied forces in Flanders, he would seek agreement with the generals in the States service as well as with the Dutch field deputies. These field deputies, some military officers and others from the civilian world, were annually appointed by the States-General to serve as a liaison with the army in the field and to coordinate the Dutch army's administration with their paymasters, the provincial States. Their numbers varied between three and six a year, but the most important province of Holland and the *Raad van State* (the Council of State—a committee in charge of daily military administration among other things) always had their own representatives among them. It would take a number of minor crises and personality conflicts, but by the end of 1705 Marlborough had gained the trust of the Dutch. Sniping against obstructionist Dutch allies would continue in a more muted vein throughout the war, and has been a veritable hobby of the Duke's more modern biographers.

The Maritime Allies began their 1703 campaign with a successful siege of Bonn before returning to the Low Countries theater. The assault on Louis' Cologne ally would continue on the Meuse, though various forces would continue to operate along a wide arc from Huy to Dutch Flanders. Marlborough's "Great Design," an ambitious plan to launch a multi-pronged offensive against Oostende, Antwerp and the French army in front of the Lines of Brabant all at the same time, failed as the various English and Dutch forces were unable to coordinate their actions. After a number of maneuvers near Antwerp led only to the two sides trading minor battlefield victories at Stekene and Ekeren, activity shifted southeastwards. Back on the Meuse, an Allied force took less than two weeks to capture the four hill forts flanking the town of Huy. The campaign ended with a contingent of the Allied army traveling across the rough terrain of the Ardennes to Limburg, where another short siege brought this region too into the Allied fold.

With Liège largely under Allied control, the English and Dutch now turned to the lands of Spanish Brabant and their protective lines constructed in 1702 to defend them. The year 1704 would see a number of minor skirmishes in Flanders and a siege of Fort Isabela in maritime Flanders, with the French Lines of Brabant continuing to define the boundary between the two sides. Most of the Anglo-Dutch army marched off to rescue Austria from the growing Franco-Bavarian threat. Their victory at the battle of Blenheim and its aftermath, including the devastation of Bavarian lands and a number of sieges, led to its elimination from the list of Allied enemies. The Elector of Bavaria would continue to personally serve Louis for the rest of the war in Flanders.

In 1705 Marlborough attempted to bypass the Low Countries theater by searching for a less guarded *entrée* into France. His effort to invade via Lorraine resulted in stalemate as his army spent several unproductive months on the Moselle being blocked by *maréchal de France* Louis-Hector duc de Villars' entrenchments. The Anglo-Dutch force returned to the Meuse to reverse a short-lived French offensive, thankful for the excuse. Marshal Villeroi had retaken Huy and invested Liège in an attempt to draw the Allies from Lorraine. Villeroi lifted the siege as the Allies approached and the Allies once again conquered Huy in a matter of days. Next the Duke used a feint to finally force the Lines of Brabant at Elixhem, catching part of Villeroi's retreating army in the process. The Allies were unable to follow up this success with anything nearly as dramatic. Instead, they would besiege and capture two other small places, Zoutleeuw (Léau) and Zandvliet. Once the Lines of Brabant were pierced, they left the Meuse behind for good; the Allies would not conduct major operations there again, not only because they were discouraged from further advances upriver by the strength of French-held Namur and Charleroi, but also because they were enticed with more tempting targets in Spanish Brabant.

The campaign of 1706 would be one of the most famous of the entire war (Map 4.5). It began almost as soon as both sides entered the field with the Allied victory at Ramillies fought north of Namur on 23 May. The wildly-successful post-battle pursuit opened up the rest of the campaign season (five months) for further advances, prompting Marlborough to write: "The consequence of this battaile is likely to be of greater advantage than that of Blenheim, for we have now the whole summer before us."[7] The Allies chased the French army

[7] Henry Snyder (ed.), *Marlborough-Godolphin Correspondence*, (Oxford, 1975) 1:552 #569. Marlborough was so confident of a successful end to the war that he wrote to his wife to prepare the house for his return. *Marlborough-Godolphin Correspondence*, 1:559 #575. See also 553 #570, 556 #572; *Letters and Dispatches*, 2:536.

remnants westward for sixty miles as Villeroi's rapidly disintegrating force failed to hold a number of fall-back positions. Over the next two weeks Allied arms would accept oaths of loyalty on the behalf of Charles III King of Spain from the towns of Leuven (Louvain), Brussels, Mechelen (Malines), Lier, Gent, Aalst (Alost), Oudenaarde, Damme, Brugge (Bruges), and Antwerp. The harried French army was finally forced to disband in the first week of June, its remnants being sent to various fortresses on the border. Without the Gallispan army's protection, all of Spanish Brabant fell to the Allies. Louis recalled the now-disgraced Villeroi and sent for his successful commander in Italy, Louis-Joseph duc de Vendôme, to rescue the increasingly grim situation. Even though Vendôme was able to reestablish the field army by August, his demoralized force could do little but watch as the Allies turned to a series of sieges against the French-held fortresses of Oostende, Menin, Dendermonde and Ath. The Allies' victories in Flanders would be repeated throughout Europe: the conquest of the Spanish Netherlands after the battle of Ramillies, the expulsion of the French from Italy after the relief battle of besieged Turin, and the occupation of Madrid by Charles III after a failed Bourbon siege of Barcelona.

As Map 4.6 suggests, the rapid advances of earlier campaigns would disappear as the Allies plunged into Vauban's *pré carré*. While Britain was gripped by the final one-sided negotiations over a union of the Scottish and English crowns, the Allied powers had agreed to focus their efforts on a southern thrust into France from newly-liberated Italy. This was possible because the previous year saw Vendôme's recall lead to the rout of a French army besieging Turin, setting into motion a chain of surrenders and evacuations. Over the next several months the French had been forced to evacuate the entire peninsula. To follow up this victory, a joint Anglo-Savoyard-Austrian attack on Toulon was agreed for 1707, but the Allied army and its British support fleet never came close to achieving their objective. The Duke of Marlborough was therefore forced to spend the entire campaign in the Low Countries against a numerically superior enemy force that had received reinforcements from the Italian theater. From June till August the two camps watched each other at a distance before finally marching off into winter quarters at the end of October. The Allies would have nothing to show for the year after Ramillies.

The campaign of 1708 would not begin until almost June, and the Allies would make up for it by campaigning into the beginning of January the next year. The season began when Jan van Brouchoven graaf van Bergeyck, chief minister of the Spanish Netherlands masterminded a revolt in Allied-held Ghent and Bruges. Marlborough's plans to abandon Brabant in order to focus on Flanders had itself to be abandoned as these towns opened their gates to the French, and the rest of the campaign would be spent working around the logistical difficulties these reversals engendered. Disagreements between the French commanders led to defeat in a mid-July encounter battle, where Vendôme's half of the French field army was caught crossing the Scheldt River at Oudenaarde while Burgundy's half watched passively nearby. In the wake of Oudenaarde Prince Eugene reinforced Marlborough's army as Vendôme retreated northward to Ghent, seemingly leaving the way open to Paris. The Allies, however, now found themselves confronted with a conundrum. It turns out the *pré carré* was just as effective a barrier to an advance on Paris as an opposing field force, while Vendôme's diminished army was impregnable behind its field fortifications anchored on the Ghent canal. Counsel was divided over their next move, but Eugene and the Dutch convinced Marlborough to advance once more into the teeth of the *pré carré*, this time attacking Vauban's masterpiece of Lille. The town and Vauban's queen of citadels were defended by a garrison of 6,000 (approximately half of the number frequently cited), and their defense would last for four months. This would be an epic siege for the theater, and would see Louis attempt numerous designs to save this jewel of the north, including a poorly-planned diversionary attack on Brussels. Marlborough and the covering force managed to deflect all their efforts. With Ghent in enemy hands, extra supplies had to be brought to Oostende and then transported overland to the site. The English General John Webb won a small field victory at Wijnendael which allowed a major supply convoy to reach Lille, providing the besiegers with ammunition and powder at a critical juncture of the siege. With the help of such desperately-needed munitions, the citadel fell in early December. Yet the Allies were not yet satisfied, for Marlborough then pushed his army northward to conduct a brief siege of Ghent that only ended with the New Year. The remaining holdout, Bruges, surrendered as well, and the Allies barely beat the chill of winter into their seasonal quarters.

It was fortunate for Marlborough's army that Ghent's large garrison surrendered when it did, for just a few days later the winter of 1708–1709 set in, and it would be one of the most brutal in contemporary memory. France suffered particularly hard, but even the Allies were delayed entering the field until mid-June. Louis decided to switch commanders yet again, this time putting his Flanders army under the command of Villars. Villars was the most aggressive French general of his generation, and the one who would give the Duke of Marlborough and Prince Eugene the most difficulty. Surrounded by starvation, his confiscation of peasant grain allowed him to field a fighting force, but he was not able to prevent the Allies from immediately investing the strong fortress of Tournai on the Scheldt. After the 68-day siege of Tournai's city and citadel was complete, the Allies moved to capture the nearby fortress of Mons. They invested the city but then diverted their troops to accept an 11 September battle against Villars' entrenched force near the villages of Blaregnies and Malplaquet. The resulting bloodbath would put 9,000 Frenchmen out of action and perhaps 24,000 Allied soldiers as well—Villars quipped that if God gave the Allies another victory like that, they would be finished. The Allies were not yet finished, however, as they returned to Mons and reduced it after a month-long siege. After only four months of such exhausting operations, both sides went into winter quarters while negotiators continued to search for a peaceful resolution to the grueling war.

The negotiations continued at Geertruidenberg into 1710. They would eventually break down as the Allies, buoyed by their military situation and fearful of alienating each other, escalated their demands beyond what even a war-weary Louis could bear—an insistence that the Sun King contribute his own troops to help depose Philippe. Despite another bitter winter the Allies managed to stockpile enough Prussian grain to allow them to enter the field in late April instead of the usual mid-May. With most of the French army still in winter quarters, they forced the Lines of Cambrin and settled down in front of the town of Douai. They expected a relatively short siege that would give them time to invest another before Villars' army could even enter the field, but the garrison's unexpectedly stiff defense withstood their efforts for two months. The fortress finally capitulated at the end of June, but the extra month of delay forced Marlborough and Eugene to abandon their intended attack on Valenciennes. Hemmed in by Vauban's line of strong fortresses and refused battle by the enemy, they had little choice but to confront another fortress: they turned westward to widen the breach in the *pré carré* by attacking Béthune. After another thirty-seven days of trenchwork, they continued westward, attacking both Aire and its weaker neighbor Saint-Venant at the same time. These two sieges would bring an end to the 1710 campaign in mid-November. The four Allied sieges had managed to carve out a significant breach in France's northern defenses, but at the cost of 20,000 Allied soldiers and an untold toll on the morale of not only the soldiers, but of their publics back home.

The intransigency of the Allied negotiators and their "No peace without Spain" position had assured several more years of conflict. But already one of the main linchpins of the alliance was wavering, as Britain became increasingly wearisome of the Continental effort. The costs of the continuing war of attrition had already been illustrated when Marlborough's political partner Sidney Godolphin was dismissed in 1710, and further highlighted when the pro-peace Tories won a large majority in Parliamentary elections that fall under their leader Robert Harley. England continued to move towards a separate peace as the Allied army managed to force its way through Villars' Lines of Ne Plus Ultra ("No Further"), but then it turned to a month-long siege of Bouchain which dragged on for an additional month as the fortifications were being repaired. Even before this, the diplomatic situation had been turned upside down when the Austrian Emperor Joseph I died in April; in October Archduke Charles/Charles III of Spain was declared successor to the Austrian Habsburg throne. The Anglo-Dutch fear of a united Franco-Spanish Bourbon hegemony was exchanged for a fear of a united Austro-Spanish Habsburg power bloc. The resulting revolution accelerated Tory efforts to find an agreement with the now-humbled French king. As a prerequisite, the Duke of Marlborough was stripped of his command and offices at the end of the 1711 campaign. Accused of embezzling army funds, he would go into exile on the Continent. It was an inglorious end to a glorious military career.

In 1712 the English general James Butler Duke of Ormonde would take Marlborough's place alongside Eugene, but already the Tory government had worked out a secret cease-fire with the French while the various plenipotentiaries were meeting for peace talks at Utrecht. The English would abandon their allies and sign a separate peace, and they would launch a domestic public relations campaign against their "traitorous" Dutch allies as well. On 21 May Ormonde received the 'restraining orders' which ordered no further English participation in any military actions against the French. The slated Allied siege of Le Quesnoy would continue nonetheless—the town fell in early July after almost a month of investment, but a few weeks later English forces (excluding those mercenary regiments which went into Austro-Dutch pay) marched for English-held Ghent and Bruges. The English had left the war, but the Austrians and Dutch insisted on continuing it. The remaining allies turned next to besiege Landrecies, which had they been successful, would have brought them precariously close to breaking through Vauban's northern defenses altogether. The English withdrawal had weakened their hand, however, and a Dutch entrenched camp under Arnold Joost van Keppel Earl of Albemarle at Denain was overrun by Villars on 24 July. Prince Eugene decided to push on with his siege, but the French capture by siege of the Dutch supply depot at Marchiennes finally forced them to abandon it. It was now France's turn to go on the offensive and strengthen its negotiating position at Utrecht. Villars' army turned next to recapture Douai, and after a 39-day siege Le Quesnoy and Bouchain would also be returned to France's possession by siege. The 1712 campaign ended in late October with significant French momentum, forcing the Dutch to reluctantly sign a peace with them in early 1713. The Austrians refused to submit, so Villars would be needed for another campaign in Germany before the two final protagonists came to peace in early 1714. Later that year the final holdout, the Catalans, would be reduced to recognizing King Philip V as their legitimate sovereign.

2. *Fortifications in the Low Countries*

The outcome of the operations in Flanders was due to numerous factors, but one thing that neither the generalship of Marlborough nor Allied numerical superiority could ignore was the theater's fortifications. As a study of Map 4.3 will indicate, there was a strong correlation between the territorial gains in a campaign and the strength of the fortifications in that area. It is worthwhile, therefore, to discuss this often-ignored aspect of Marlborough's campaigns. Only with a better understanding of the nature of these fortifications can we examine how Vauban's siege attack was applied in the war.[8]

[8] The next several pages include material from Ostwald, 'The "Decisive" Battle of Ramillies, 1706," pp. 668–670.

The Southern Netherlands was already seeing martial preparations all around its borders before armed conflict ever erupted on its soil. Louis had finally achieved a long-standing ambition of dominating his Spanish neighbor to the north (albeit in the name of his grandson) yet this turned out to be a double-edged sword. Spain had managed to hold on to the Southern Netherlands throughout the siege-intensive Dutch Revolt, but by the end of the seventeenth century most of its fortifications had fallen to a very low state indeed, in part because of the constant fighting it had experienced over the past century with its southern neighbor. Close examination shows that every Low Countries fortress that would be besieged between 1702 and 1705 qualified as a *bicoque*, (literally a hovel), Vauban's useful term for small, weak places that were never to be confused with serious fortresses like Lille, Turin, Landau or Ath.[9] Some of these towns might well require a formal siege *dans les formes*, but they could not be expected to withstand one for very long, nor to require the full panoply of techniques available to besiegers. Many Spanish towns still retained their medieval fortifications, while others had not seen renovation for decades. French generals quickly discovered all this when they preemptively occupied the Spanish towns in 1701.[10] The logistician Jacques François de Chastenet marquis de Puységur identified the resulting critical disadvantage that the French would labor under for the next several years: "All the places of the Spanish frontier from the sea to the Rhine are only made of earth and the majority repaired very hastily, and therefore cannot make a long resistance; thus we cannot move our armies far from them without giving the enemy the opportunity to profit from this by declaring war and surprising several of them."[11]

---

[9] The scholarship on both Vauban and Marlborough emphasizes how the Low Countries theater bristled with fortresses. For examples, see: Chandler, *Marlborough as Military Commander*, p. 81; Ronald Love's entry on "Ramillies" in L. Frey and M. Frey (eds.), *Treaties of the War of the Spanish Succession*, pp. 372–373, which was probably drawn from Chandler, *The Art of Warfare in the Age of Marlborough*, p. 245; and John Childs, *Warfare in the Seventeenth Century*, p. 137. His slightly longer description in *The Nine Years War* provides a list of fortresses in the theater that conjures up images of a region thoroughly blanketed by strongholds (pp. 32–33).

[10] The desperate financial straits of the Spanish Netherlands is briefly summarized in Vault and Pelet (eds.), *Mémoires militaires*, vol. 1, p. 16.

[11] Vault and Pelet (eds.), *Mémoires militaires*, vol. 1, p. 473, Puységur mémoire 22 August 1701.

Most directly in the line of fire were the places of Liège and Spanish Guelders. The four Meuse towns of Gelders, Venlo, Roermond (Ruremonde) and Stevensweert alone were expected to cost the French and Spanish 100,000 francs to repair, in addition to what had already been spent.[12] Venlo was the strongest of these *bicoques*, but even here, the Dutch general Jacob van Wassenaer heer van Obdam would later note that the town "had never been considered one of the strongest places."[13] Plans to protect the capital of their new Liège ally revolved around building an entrenched camp for a field army, since the town's citadel and small forts were inadequate by themselves to hold off an attacker for any extended period of time.[14] When Allied forces swept through the Meuse valley in 1702, Vauban was hardly surprised at the results. He had personally inspected these works and knew well that Venlo, Roermond, Liège and Bonn were all no more than *bicoques*, "all weak places worse than the others, and not one of which was in a state to hold out for eight days against well-ordered attacks [*attaques réglées*]."[15] Only Dutch-held Maastricht was a strong fortress, and the French had left its garrison isolated in the rear as it occupied the rest of the Spanish Netherlands. Nor did the 1703 campaign offer the Allies a serious fortification challenge. The town of Huy was defenseless other than several forts and a château along a ridgeline overlooking the town. After its short-lived siege in 1703, the Dutch engineer Lucas Du Mée determined that it was untenable in its current state, and that a whole series of additional works would need to be constructed in order to make it defensible.[16] The other town besieged by the Allies

---

[12] Vault and Pelet (eds.), *Mémoires militaires*, vol. 1, p. 74, Boufflers to Louis, Diest 12 May 1701. The small Pays de Guelders was to provide the pioneers and wagons.

[13] Veenendaal, Jr. (ed.), *Briefwisseling Heinsius*, vol. 1, p. 415 #835, Wassenaar-Obdam to Heinsius, Venlo 6 September 1702. See also Puységur's comments in Vault and Pelet (eds.), *Mémoires militaires*, vol. 1, p. 30, to Chamillart, 21 February 1701. For a corroborating Allied account after the siege, see Veenendaal, Jr. (ed.), *Briefwisseling Heinsius*, vol. 1, p. 448 #916, Wassenaar-Obdam to Heinsius, Venlo 27 September 1702.

[14] William III had used this very tactic in the previous war. In the Spanish Succession, see Vault and Pelet (eds.), *Mémoires militaires*, vol. 2, p. 593, Boufflers to Louis, Tongeren 2 October 1702. For Vauban's opinions of Liège (which he estimated would require 400,000 livres for it to hold out just ten days), see Paul Harsin, "Vauban à Liège en 1702," *Bulletin de la Société royale du vieux Liège* 104–105 (1954), pp. 308ff.

[15] Vault and Pelet (eds.), *Mémoires militaires*, vol. 5, p. 653, Vauban to La Feuillade, 13 September 1705.

[16] Veenendaal, Jr. (ed.), *Briefwisseling Heinsius*, vol. 4, p. 102 #277, Dutch Colonel

in 1703, Limburg, was only slightly more prepared for an attack. Although its earthen fortifications were reported to be stronger than expected, the most difficult part of the siege was hauling the heavy siege guns to the site.[17]

Spanish Brabant's great towns had even weaker fortifications than those in Liège or the Spanish lands east of the Meuse—their shortcomings stemmed as much from lack of maintenance as from outdated fortifications.[18] In May 1701 Boufflers had reported to the king that none of the Spanish places they had occupied had palisades or even covered ways: "I found the places of this country in the same disorder as all the others on the Spanish frontier . . . with the exception of Venlo and Guelders, which are in better shape due to the considerable work that the comte de Coigny had already undertaken to repair them."[19] The logistician Puységur explained to Michel Chamillart: "the Spanish Netherlands is full of places which are not able to withstand major sieges, and Brabant is entirely without defense."[20] On the defensive by late 1702, Boufflers repeated Puységur's complaint about the need to cover all of the Spanish Netherlands with his armies.[21] In 1704 Villeroi warned that the Lines of Brabant were the only thing between them and disaster: "all of Brabant remains open to the enemies . . . [it] does not have a single fortress

of the infantry Isaac Cronstrom to Heinsius Huy 20 March 1705. Villeroi's judgment (backed by Valory) was that only Huy's château was defensible, due primarily to its inaccessible perch. AG A[1] 1835, #262, Villeroi to Chamillart, 12 June 1705. Harcourt also agreed with this assessment: AG A[1] 1835, #269, Harcourt to Chamillart 13 June 1705. On Huy more generally, see Rorive, *La guerre de siège sous Louis XIV en Europe et à Huy*.

[17] Allent, *Histoire du Corps Imperial du Génie*, p. 416; Snyder (ed.), *Marlborough-Godolphin Correspondence*, vol. 1, p. 242 #239, Marlborough to Godolphin, Verviers 9 September 1703; Murray (ed.), *Letters and Dispatches*, vol. 1, p. 177, Marlborough to Field Deputies, Robertmont 12 September; and p. 184, Marlborough to Ouwerkerk, Verviers 21 September.

[18] Duffy, *The Fortress in the Age of Vauban*, pp. 34–35, also emphasizes the differences between fortresses in the Spanish Netherlands and those in France proper.

[19] Vault and Pelet (eds.), *Mémoires militaires*, vol. 1, p. 74, Boufflers to Louis, Diest 12 May 1701.

[20] Vault and Pelet (eds.), *Mémoires militaires*, vol. 1, p. 31, Puységur to Chamillart, Brussels 21 February 1701.

[21] Vault and Pelet (eds.), *Mémoires militaires* vol. 2, p. 572 Boufflers to Louis, Beringen 3 September 1702. Boufflers also warned the king that trying to relieve the siege of Liège would open up all of Brabant to the Allies. Vol. 2, p. 595 Boufflers to Louis XIV, Tongeren 1 October 1702. See also vol. 2, pp. 602–603, Puységur mémoire 1 October 1702; pp. 602–603; John Wolf, "Louis XIV, Soldier-King," p. 212 and his *Louis XIV*, (New York, 1968), p. 526.

which could delay an army for twelve hours without the protection of our army." A little later he repeated his warning that without said army, "the enemies [would be able to] enter Brabant without opposition, run to the gates of Brussels and take whatever towns they wanted in that province, all of them being without defense."[22]

The situation became much worse in 1705 as the Allies punctured France's fortified lines. French commanders renewed their concern over the necessity of defending these "open" towns with their overstretched field army.[23] Yet another French *maréchal* echoed his colleagues' complaints: "This country is full of towns, of which hardly a single one is in a state of defense, yet which are nevertheless so important that we cannot lose even one without receiving a mortal blow, which forces us to spread our troops out across several posts."[24] Villeroi explained further how the Spanish alliance continued to stretch thin his manpower:

> our army is separated by indispensable necessity, because we must guarantee completely indefensible places, only covered by a river or retrenchment. . . . If Louvain, Lierre, Malines and Brussels were places that we could save by giving them large garrisons, I would recommend to the king to separate the army among these towns . . . but they can only be defended by a field army.[25]

When Louis accepted responsibility for the Spanish Netherlands, he placed his armies in an onerous position. It is no surprise then that Ramillies was such a sweeping victory for the Allies. Retreating French forces simply had no shelter to seek.

On the other hand, while the loss of their field army and of all this territory was traumatic, there was one bright spot—now the Allies would be encumbered with these indefensible towns. The French commander of Namur did not expect the victorious Allies

[22] Vault and Pelet (eds.), *Mémoires militaires*, vol. 4, p. 12; also p. 13.

[23] See Vault and Pelet (eds.), *Mémoires militaires*, vol. 5, p. 599, Maximilien II Emmanuel to Duisborg, 20 August 1705; as well as pp. 37, 603. The French commander at Leuven (Louvain) reported in 1705 on the shortage of Spanish funds which were needed to put the town in a state of defense. AG A[1] 1838 #220. Louis acknowledged these complaints in Wolf, *Louis XIV*, p. 526; and Vault and Pelet (eds.), *Mémoires militaires*, vol. 5, p. 606, Louis to Maximilien II Emmanuel 24 August 1705.

[24] Vault and Pelet (eds.), *Mémoires militaires*, vol. 5, p. 588, Marsin to Chamillart, Overijssche 16 August 1705.

[25] Vault and Pelet (eds.), *Mémoires militaires*, vol. 5, pp. 91–92, to Chamillart, 30 September 1705.

to backtrack eastward in order to besiege his fortress, as they would thereby abandon to the French the indefensible towns they had captured in the battle's wake.[26] The Spanish minister Bergeyck planned the next year's campaign with the assumption that the Allies would be hampered, as the French had been, by the need to protect all these towns.[27] In fact the Allies quickly came to appreciate the challenges these towns presented. Dutch observers had already commented on the poor state of their own frontier fortresses at the beginning of the war, a concern that Marlborough was able to neutralize early on as he forced the French back on their heels.[28] After Ramillies these Dutch fortresses were shielded by their occupation of Spanish Brabant, but now the Allies found themselves hindered by their new conquests. In the wake of Ramillies Marlborough feared that there were not enough troops to protect the besiegers at Menin and at the same time defend the open towns of Brabant; Sicco van Goslinga was initially opposed to this siege altogether because of the necessity of exposing their newly-conquered lands in the process.[29]

---

[26] AG A[1] 1936, #248, Saillans d'Estaing to Chamillart, Namur 28 May 1706; and AG A[1] 1936 #251, Villeroi to Louis XIV, Saint-Denis-sous-Gand 29 May 1706 published in Vault and Pelet (eds.), *Mémoires militaires*, vol. 6, p. 48.

[27] Vault and Pelet (eds.), *Mémoires militaires*, vol. 6, pp. 586–587. Bergeyck was particularly critical of the towns of Leuven, Mechelen and Brussels. The baron Karg de Bebenbourg, chancellor to the French ally the Elector of Cologne, added Bruges to the list of towns expected to fall without resistance. Louis Jadin (ed.), *Correspondance du Baron Karg de Bebenbourg, chancelier du Prince-Evêque de Liège Joseph-Clément de Bavière, Archevêque Electeur de Cologne, avec le Cardinal Paolucci, secrétaire d'état (1700–1719)*, 2 vols. (Brussels, 1968), vol. 1, p. 43. After the French had regained Bruges in 1708, they once again acknowledged the need for an army to defend the town against an Allied threat. AG A[1] 2083 #86, Charles comte de la Motte-Houdancourt, Ghent 19 September 1708.

[28] For examples, see Veenendaal, Jr. (ed.), *Briefwisseling Heinsius*, vol. 1, p. 46 #63, Noyelles en Fallais to Heinsius, on Bergen op Zoom's status circa 28 March 1702; p. 149 #241, Noyelles en Fallais to Heinsius, Bergen op Zoom 30 April 1702 referring to Dutch Brabant as a "pais ouvert"; and vol. 4, p. 212 #609. See also Duffy, *The Fortress in the Age of Vauban*, p. 34. The Dutch were still complaining about the "indifferent" state of these same fortresses in 1708, as cited in A.J. Veenendaal, Sr., "The Opening Phase of Marlborough's Campaign of 1708 in the Netherlands: A Version from Dutch Sources," *History* 1950, p. 38.

[29] Snyder (ed.), *Marlborough-Godolphin Correspondence*, vol. 2, p. 608 #617, p. 642 #648; Veenendaal, Jr. (ed.), *Briefwisseling Heinsius*, vol. 5, p. 387 #734, Goslinga to Heinsius, Oudenburgh 13 July 1706. Marlborough also wanted to evacuate the exposed garrison at Leuven, since it was too far away to be protected by the main field army, van 't Hoff (ed.), *Marlborough-Heinsius Correspondence*, p. 254 #420. Heinsius, the Dutch general Salisch and field deputy Goslinga also recognized that the Allies could not afford to garrison all of the great towns of Brabant. Veenendaal, Jr. (ed.), *Briefwisseling Heinsius*, vol. 5, p. 346 #652, p. 361 #678, p. 373 #698, p. 475 #915.

In 1707 Marlborough complained of the "plague of covering Bruxelles and the rest of the great towns" with his field army.[30] The Dutch also mentioned the "embarrassment" of covering Brabant's towns, Lieutenant-General Francis Nicolaas Fagel echoing Villeroi's 1704 judgment: "We don't have in all of Brabant a single town or place of which we can be assured . . . when our army is far from these places they are in fear."[31] By 1708 Marlborough's only solution was to abandon Spanish Brabant entirely—a politically dangerous move that deeply disturbed his Dutch and Belgian allies, and one that became impossible to sustain when the French surprised Ghent and Bruges, threatening to cut off Allied communication with Dutch Brabant. The Duke was quickly forced to abandon his plan in order to maintain the remaining link with Dutch territory.[32] The Allies were just as limited by their need to protect the politically-vital but ill-defended centers of Brabant as the French had been before them.

We should keep in mind, nonetheless, that although the French found themselves in a precarious situation defending a province full of barely-defensible Spanish places, some of these poorly-maintained fortresses would still require a formal siege, and this consumed time that Marlborough felt he did not have. Four such fortresses were besieged in 1706 and their defenses consumed the rest of the campaign season and all of the Allies' siege supplies. The strength of the fortifications the Allies faced would only increase as they pushed further into the *pré carré*—comparing the sieges durations of Spanish versus French fortresses in Map 4.1 illustrates as much. And while France's occupation of the Spanish Netherlands was an expensive burden, it did force the Allies to conduct eleven sieges and delayed an advance into France proper until 1708—six years into the war. The delays would have been far greater, were it not for the fact

[30] Snyder (ed.), *Marlborough-Godolphin Correspondence*, vol. 2, p. 794 #802, Marlborough to Godolphin, Meldert 6 June 1707; also vol. 2, p. 780 #791, Marlborough to Godolphin, Soignies 15 May 1707. Writing from Brussels in 1708, Marlborough's trusted subordinate William Cadogan described the town as "being in a manner an open town and subject to the fortune of a battel." British Library, Additional Manuscripts (Add MSS) 61160 f. 56b; see also Add MSS 61312 f. 182b l'Armentière to Marlborough, Brussels 3 December 1708.

[31] Veenendaal, Jr. (ed.), *Briefwisseling Heinsius*, vol. 6, p. 309 #606; see also p.299 #590, p. 312 #612; vol. 7, p. 90 #195, van den Bergh to Heinsius, Brussels 23 February 1708; vol. 12, p. 244 #407, Goslinga to Heinsius, Lillers 30 July 1711.

[32] On this issue, see Veenendaal, Sr., "The Opening Phase of Marlborough's Campaign of 1708 in the Netherlands."

that resistance was too often cut short defending dilapidated Spanish *bicoques* or poorly-maintained French fortresses.

3. *The End of an Age*

Bicoques may have been relatively simple to capture, but the stronger fortresses of Vauban's northern border would challenge both Bourbon and Allied besiegers in the War of the Spanish Succession far more than Ath had challenged Vauban. Born in 1633 and 1634 respectively, Vauban and Coehoorn died within four years of each other and would conduct few of the many sieges discussed above. As the two sides therefore transitioned from the Age of Vauban and Coehoorn to the tenure of their more mundane successors, some lessons would have to be learned anew. But the transition began even before their deaths, for in their last sieges we see their methods increasingly called into question. This deterioration in their authority is evident from the beginning of the War of the Spanish Succession and it would influence how their successors attempted to implement efficient sieges.

Vauban's success as a besieger derived in large part from the almost unfailing support he received from Louis and his ministers.[33] Conquering a number of Spanish fortresses in the brief War of Devolution, he quickly gained the King's trust. His unhindered conduct at Maastricht in 1673 was a significant step marking the elevation of the engineers—the King's personal presence allowed Vauban to direct the attacks largely as he saw fit.[34] The confidence that Louis had in his first engineer, built on Vauban's unprecedented series of successes and reinforced by the King's personal interactions with him at numerous sieges, clearly identified him as a royal representative. Towards the end of his life Vauban noted the significant improvement in the standing of the engineering corps he had overseen over the course of several decades:

> In times past nothing was more rare in France than engineers, and the few that we had lasted so briefly that it was even more rare to find any who had seen five or six sieges, and even more, who had

[33] Lecomte, "Du service des ingénieurs militaires en France pendant le règne de Louis XIV," 122.

[34] Wolf, *Louis XIV*, pp. 232, 531–532. Other qualifications to his freedom of action will be explored below.

> witnessed as many without suffering many wounds which put them out of service at the beginning or the middle of a siege, which prevented them from seeing its conclusion, and consequently few had the opportunity to learn from the experience. This, in addition to other weaknesses, contributed more than a little to the length and considerable losses of our sieges. . . . The truth is we did not understand grand finesse. . . . But since the King started to wage war in person, his presence inspired courage and good conduct in the armies. As His Majesty recognized how necessary it was to have enlightened [*éclairés*] people capable of serving in sieges and in fortresses, he established and maintained a good number of engineers; many people joined this profession, attracted by his patronage and by the distinction which they could find there. As a result, though many more would be killed and maimed, the King always had enough, and for many years we have not conducted a siege without thirty-six brigades of six or seven [engineers] each, so that each attack could have three which would be relieved every twenty-four hours. Thus the trenches are never without engineers who share the travails of their labor, and the trenches are always progressing and not an hour of time is lost.[35]

This could not last however. After 1691 Louis gave up campaigning, losing this intimate contact with a newer generation of engineers.[36] The King and his close advisors could follow the progress of the siege from afar and still invest Vauban with the mantle of *ingénieur du Roy*, but as Louis aged, as he lost his youthful enthusiasm for martial pursuits, as Vauban too aged and spent less and less time in the trenches (often by royal command), and as the amount of information on the many campaigns increased dramatically, the monarch's contact with the engineering corps increasingly began to be filtered through the information and recommendations that his wife Madame de Maintenon and his chief secretaries and advisors passed on to him.[37] This took a toll not only on the authority of the engineers at sieges, but also resulted in a more general loss of the prestige previously associated with the engineering corps when it had been directly under the King's gaze.[38] As a result, a chagrined

---

[35] Vauban, *Traité de l'attaque*, pp. 56–57.

[36] Lecomte, *Les ingénieurs militaires en France pendant la règne de Louis XIV*, pp. 122–123.

[37] In the War of the Spanish Succession, the French Secretary of War's incoming correspondence (series A[1]) alone fills close to one thousand volumes, almost one hundred thousand individual documents to be read, responded to, acted on (or ignored), and then filed away.

[38] Wolf notes that already at the 1688 siege of Philippsbourg only Louis' royal presence ensured that Vauban's recommendations were followed. *Louis XIV*, p. 199.

Vauban noted that his Majesty's personal absence had led to a reduction in the ranks of the engineers' numbers.[39]

The subtle decline in the French engineering corps by the end of Louis' reign was mirrored by a decline in its leader's status. Though Vauban was at the pinnacle of his art in 1697, it would not be long before he found himself on the sidelines. When war came to the Low Countries, Vauban was charged with inspecting the various Spanish fortifications French forces now garrisoned. In late August of 1702 he was diverted to attack the town of Hulst with a Franco-Spanish force of fifteen battalions under the command of the Spanish general Isidoro Juan Hose Domingo de la Cueva y Benavides, Marques de Bedmar. The fortified town itself was far from intimidating, but its coastal location in the middle of a large inundation made it practically inaccessible, and its only point of access was protected by half-a-dozen small forts connected by field entrenchments. Several of the smaller forts were stormed by Bedmar, but the garrison of eight battalions managed to resist a bombardment from Vauban's batteries while the defenders of Fort Kykuyt repulsed all their attacks, killing or wounding hundreds. After Vauban informed Louis of the additional supplies and reinforcements that would be needed for a formal siege, the King ordered him to end it. The next year would be bittersweet for Vauban, for while he belatedly gained recognition for his services with his promotion to Marshal of France, he would also see his last siege. His last military action was to conduct the siege of Old Breisach on the Rhine River under the aegis of the Duke of Burgundy. After its capture, Louis used concerns about his health to refuse him conduct of the upcoming siege of nearby Landau. Heart-broken and confused, Vauban nonetheless refused to abandon his mantra of preparation and penned a long *mémoire* on how the fortress should be attacked before leaving for home.

Vauban would dedicate the few remaining years of his life to his writings. No longer given permission to go on campaign, he nevertheless refused to retire. He spent most of 1704 and 1705 at his various residences where he wrote, among other things, the *Traité de*

[39] Vauban nevertheless assured Burgundy that there would be enough engineers to allow three per approach. Vauban, *Traité*, p. 56. This number seems quite small, particularly when the inevitable casualties the engineers would suffer is taken into account. Virol, *Les Oisivetés*, p. 104.

l'attaque, a summation of his views on the siege attack which we have already encountered. He also put his papers in order, since he was, as he rather pathetically described himself, "a man with little to do." He was recalled for one last hurrah in 1706 as Spanish Brabant began to collapse like a house of cards after Ramillies. Louis turned one final time to his old friend to personally inspect and shore up the defenses of the newly-exposed northern border. The king would not let Vauban make the ultimate sacrifice, however, and refused the engineer's offer to command a garrison defending one of the fortresses likely to be attacked. Having turned in his report on the sad conditions of the fortresses in the *pré carré*, he turned back to his quill and paper. That year he also penned a second summative work, this time in response to the successful Allied sieges that he had been following from afar. His *Traité de la Défense des Places* would be his final opus on siegecraft, one in which he tried to reset the balance between attack and defense. In late March of 1707, a few days after he learned that his reform-minded tax proposal *Projet d'une Dixme royale* had been banned, he died in Paris. The final step in the transition from Vauban to his successors had been taken.

Vauban's rival Coehoorn would also leave most of the sieges in the War of the Spanish Succession to his successors. Coehoorn's final years were slightly less poignant than Vauban's, although the Frisian would also see others increasingly challenge his judgment. His authority as chief engineer was largely due to his patron William III. Personally leading his armies in the field up until his death in early 1702, William kept in much closer contact with the engineers than his arch nemesis Louis. With incontestable authority in the field he supported the tactics of Coehoorn over those of his titular engineers. Coehoorn rose through the infantry ranks, yet William made him director-general of Dutch fortifications in 1695 (as well as master-general of artillery) and backed his proposed tactics at several sieges, including the (in)famous attack on Namur's citadel in 1695.[40] But Coehoorn's authority was eroded with the Stadholder-King's death in early 1702. Almost immediately the argumentative director-general clashed with the Dutch field deputies over whether to carry out

[40] At one point Duffy argued that the largest difference between Vauban's and Coehoorn's method of attack was that the Dutchman did not have the same authority as the French engineer. *The Fortress in the Age of Vauban*, pp. 64–65.

their planned siege of Venlo in 1702. Slated to direct the attack, the Frisian firebrand quickly changed his mind when the promised supplies were slow to arrive and argued that they should abandon the siege altogether.[41] He was eventually cajoled into continuing the attack, a siege that lasted only eleven days of open trenches. Indicative of the engineer's falling stock, Albemarle attributed its capture more to luck than Coehoorn's wisdom. After seeing its Fort Saint Michel stormed and the defenders put to the sword, the town's garrison quickly beat the chamade when the Allies fired a *feu de joie* to celebrate the capture of Landau in Germany—the French had feared the salvoes were a signal for storming the town.[42]

The Allies, after quickly capturing Stevensweert and Roermond, fell upon Liège. Here noncompliant colleagues further bruised the ego of "the great Coehoorn," as he purportedly referred to himself. According to Coehoorn's son, the *Raad van State*'s field deputy Adriaen van Borssele, heer van Geldermalsen intentionally deprived him of the necessary supplies for the siege.[43] Then, after the preparations were finally ready for opening fire, Coehoorn gave orders to the colonel of artillery Willem IJssel to hold his fire in the morning until the *meester-generaal* could arrive and make last-minute corrections. Awakened by the sound of guns firing without him, Coehoorn rushed to the scene and was told that Geldermalsen had ordered the early bombardment. Hunting down the field deputy, he confronted the *Raad van State* representative and belittled him before storming off. The situation was eventually smoothed over, and the weak forts of

[41] A.J. Veenendaal, Jr. (ed.), *De Briefwisseling van Anthonie Heinsius, 1702–1720*, 19 vols. (The Hague, 1980), vol. 1, p. 415 #835, Wassenaar-Obdam to Heinsius, Venlo, 6 September 1702; hereafter Veenendaal, Jr. (ed.), *Briefwisseling Heinsius*. One of Marlborough's engineers (Holcroft Blood) reported back that Coehoorn was indeed justified in complaining about the lack of supplies. BL Add MSS 61306, ff. 35–36 Blood to Marlborough, August.

[42] Veenendaal, Jr. (ed.), *Briefwisseling Heinsius*, vol. 1, p. 439 #895, Albemarle to Heinsius, Aken, 24 September; p. 437 #890, Wassenaar-Obdam to Heinsius, Venlo, 22 September. The garrison did not, as some scholars suggest, surrender solely due to the shock from the bombardment itself.

[43] Wassenaer-Obdam pinned much of the blame on the belligerent hypochondriac Coehoorn, informing Heinsius that he never let anyone know in advance what supplies he required as well as frequently quarreling with the other generals. Veenendaal, Jr. (ed.), *Briefwisseling Heinsius*, vol. 1, p. 440 #898, Wassenaar-Obdam to Heinsius, Venlo, 24 September 1702; p. 456 #935, Wassenaar-Obdam to Heinsius, Roermond, 2 October.

Liège quickly fell to the Allies in a matter of days.[44] The death of the irascible Coehoorn's in early 1704 may have removed a difficult personality, but it also deprived the Allies of their most experienced engineer. How the successors of both Vauban and Coehoorn attempted to follow in the footsteps of their fading founders is the subject of the next chapter.

[44] J.W. van Sypesteyn, *Het leven van Menno Baron van Coehoorn*, (Leeuwarden, 1860), pp. 36–37.

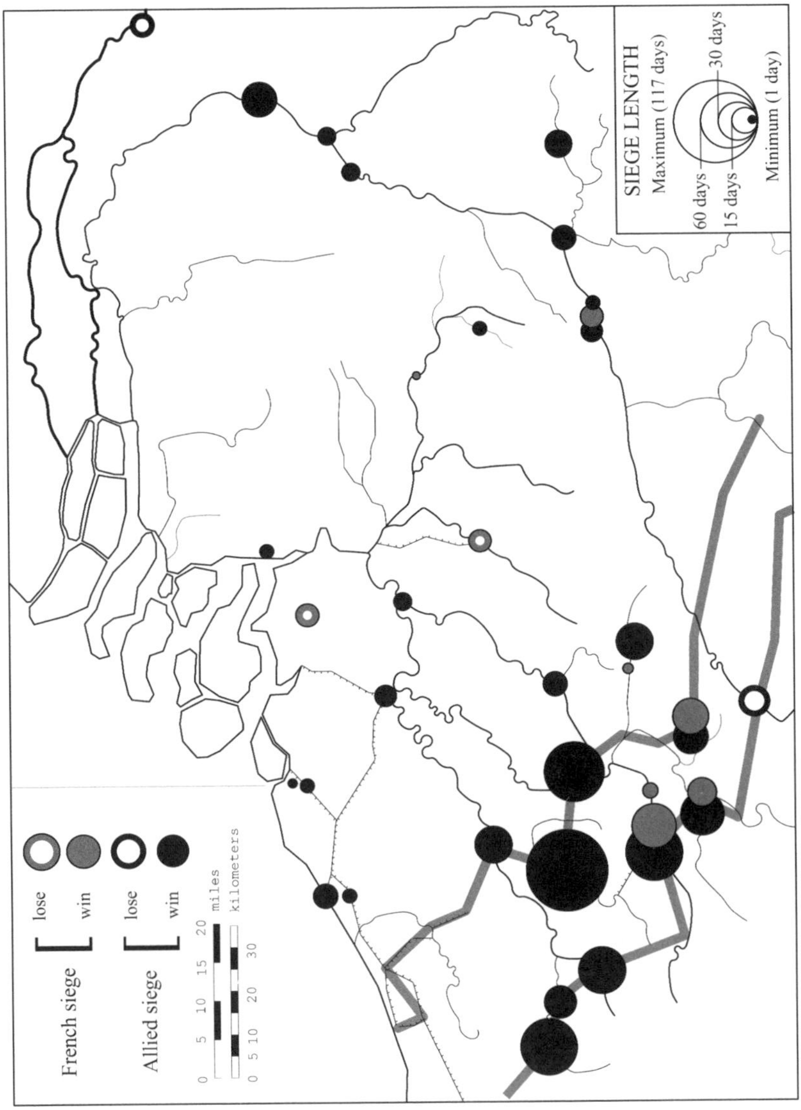

Map 4.1: Low Countries Sieges 1702–1712

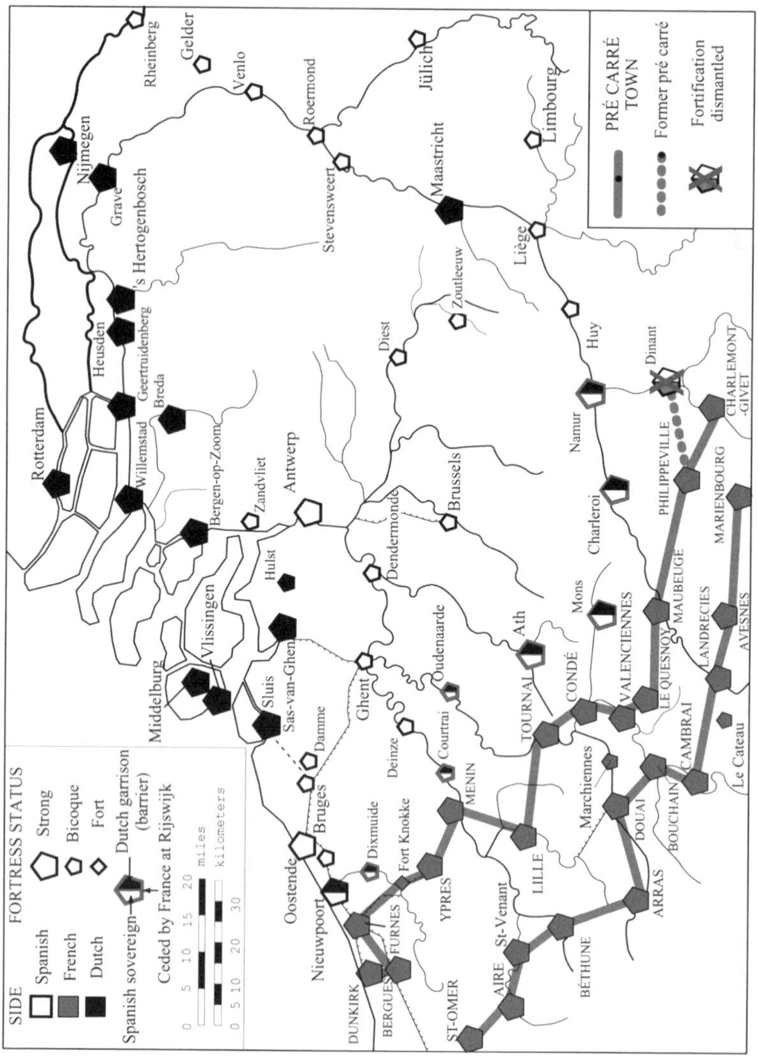

Map 4.2: Fortifications and the Low Countries in 1700

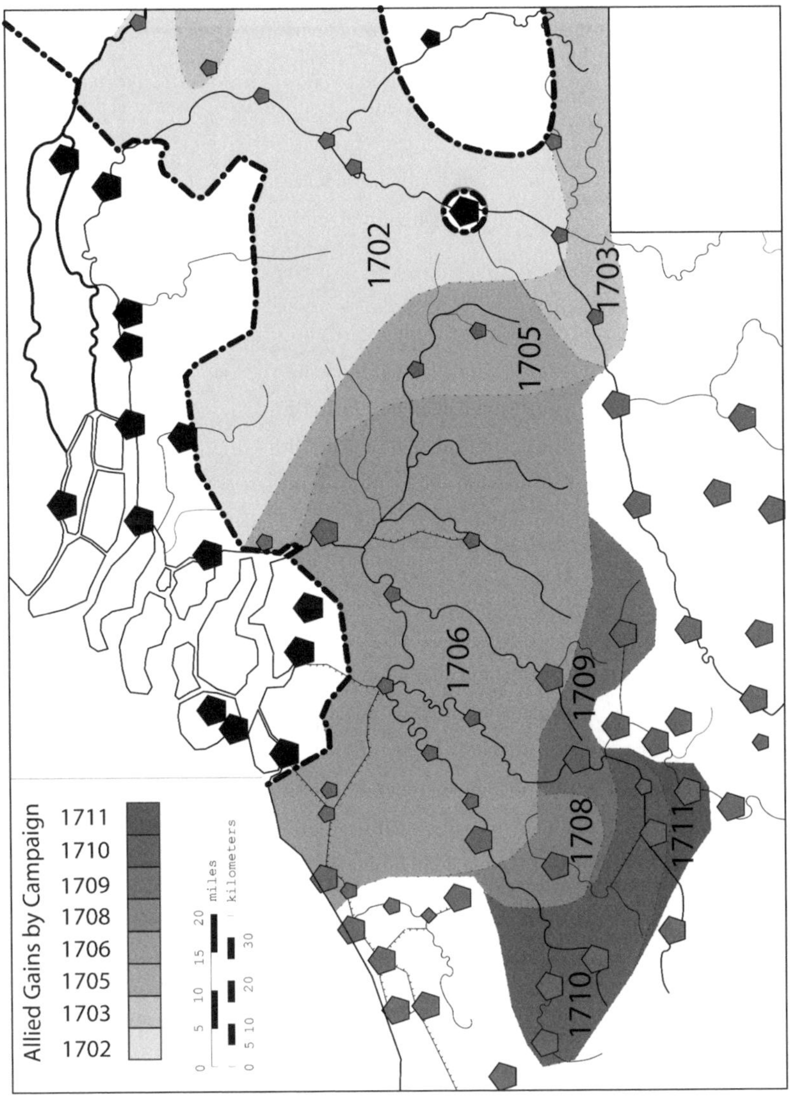

Map 4.3: Allied Conquests in the Low Countries 1702–1711

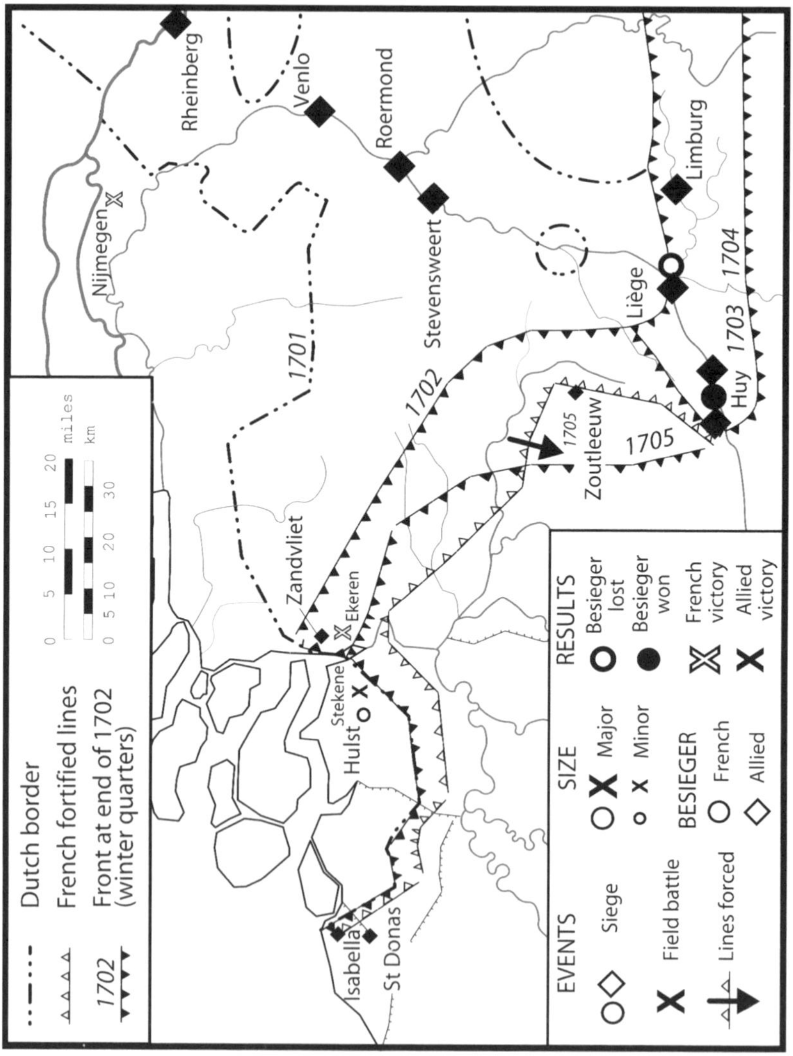

Map 4.4: The Low Countries Campaigns 1702–1705

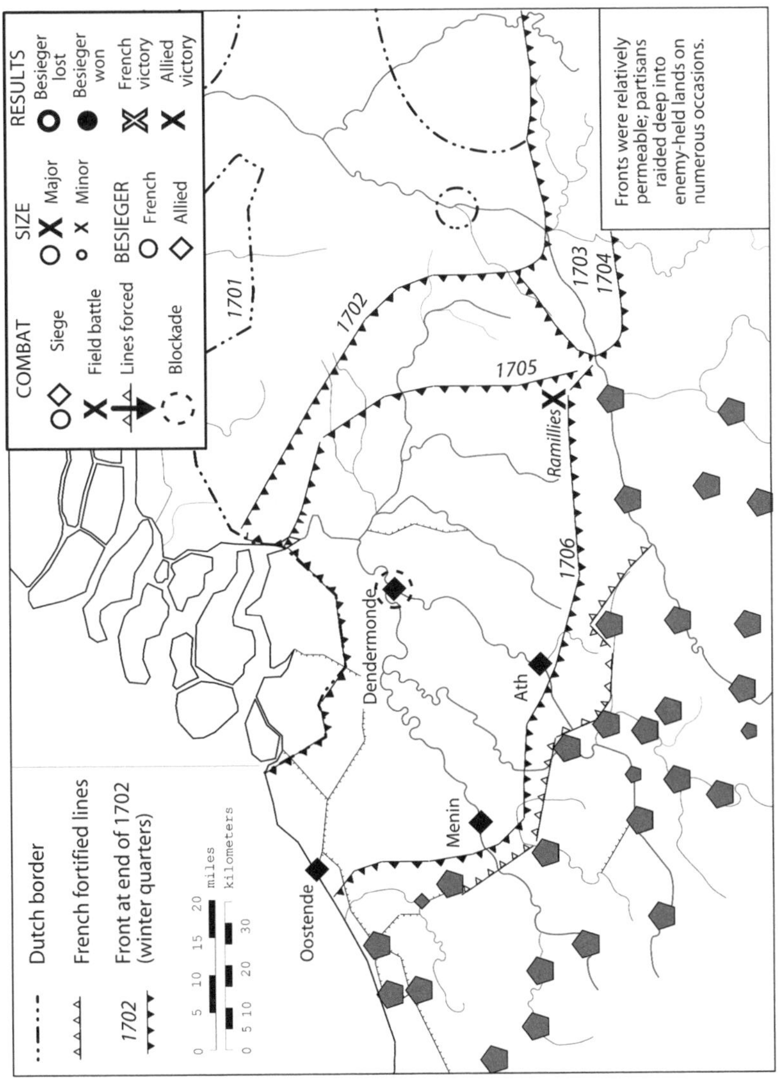

Map 4.5: The Low Countries Campaign of 1706

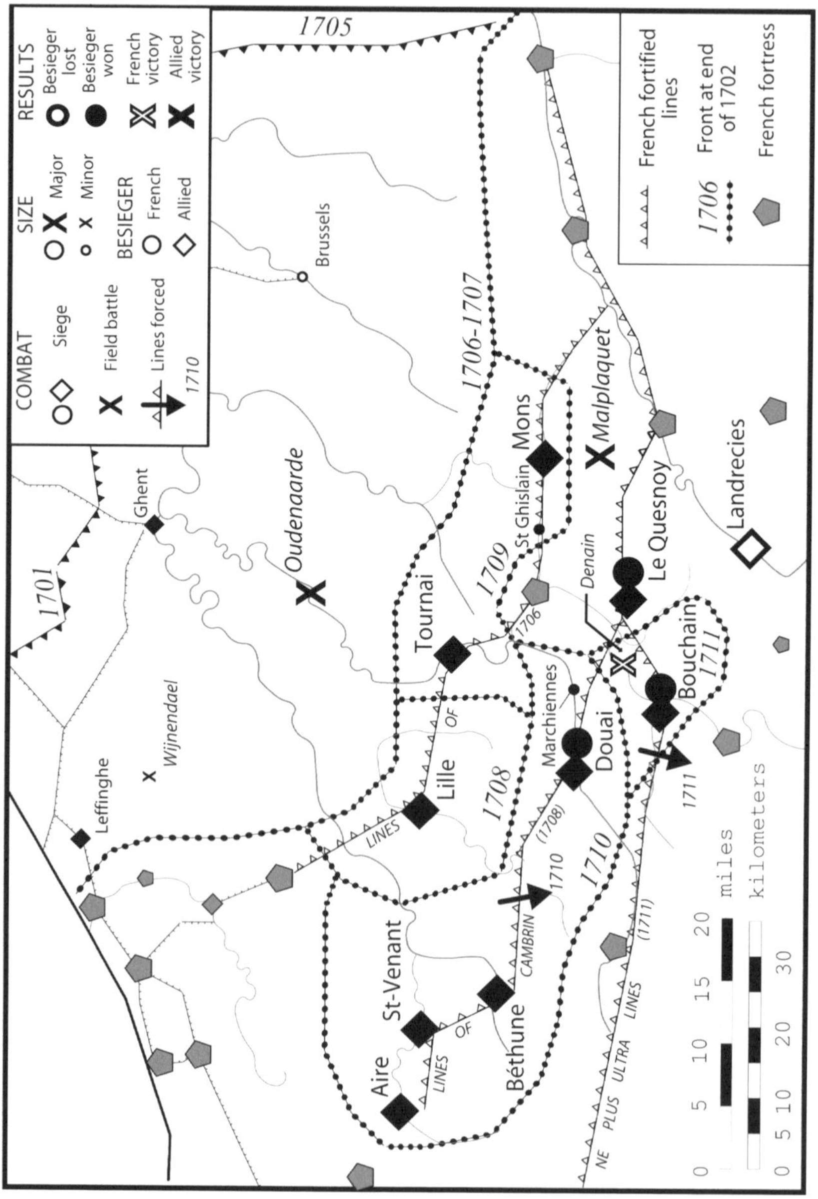

Map 4.6: The Low Countries Campaigns 1707–1712

CHAPTER FIVE

IMPLEMENTING THE PARADIGM SIEGE

As Vauban aged he sought to assure his legacy of managerial efficiency by passing his knowledge and authority on to his fellow engineers, the only ones who could properly conduct a Vaubanian siege. When it became obvious in late 1703 that his long career of active campaigning was at an end, he turned to recording his ideas for posterity in a more systematic fashion, hence his two major treatises on the attack (1704) and defense (1706). After his death, it would be up to others to continue his legacy. Although his treatises and many *mémoires* provided a basis for a curriculum, none of the major European powers established formal educational institutions for their engineers till well after the War of the Spanish Succession, and most waited until the middle of the century. This cost-saving measure undermined their group identity and left the efficiency of sieges in doubt. In spite of all Vauban's efforts, the French were perpetually short of trained engineers, though their enemies were even more so, while individual technicians varied widely in their abilities and knowledge. Even worse, whether following Vauban's strictures or not, the most skilled engineers still faced significant hurdles transplanting their ideas onto the ground in front of a fortress. Every chief engineer's position as siege manager was compromised since he was forced to rely on the goodwill and aptitude of non-engineers to carry out his instructions. When engineers failed to juggle adroitly the many components of the siege apparatus, their attacks deviated significantly from Vauban's ideal. Managing it efficiently required an authority that overworked engineers did not have.

1. *Proto-Professionalism*

The first difficulty in implementing any kind of efficient attack was that it required a large number of highly-skilled engineers who knew how to apply Vauban's maxims to the specific siege at hand—professionally-trained members with an institutional base of support.

Unfortunately for both Bourbon and Allied belligerents, the quality of those available for service varied greatly, for training in a pre-institutionalized age was informal and based on an apprenticeship model. The engineers provided by this informal system failed to supply the numbers needed to fight a major war in several different theaters. The experience and 'good sense' Vauban depended on was a precious commodity, from the start unequally distributed among the engineers and constantly eroded by the casualties of combat.

The uneven quality of engineers made the results of sieges far from certain. Despite Vauban's attempts at rationalizing his craft, the institutional foundations that would assure his offensive legacy would not be established until later in the 18th century.[1] Nor did Louis XIV's peers seek to assure institutional continuity for their engineering corps before the *siècle des lumières* was in full swing. Many countries had long-established schools for pages and young noblemen, but these dedicated only a small part of their curriculum to the technical details of the engineer's craft—especially drafting and geometry.[2] Most engineers of the early 18th century continued to learn their basic geometry from a relative, from Jesuit schools, from individual study, or from tutors in a field camp or garrison.[3] The particulars of their craft were learned by watching and doing rather than by following a standardized, formal course of study in an academic setting intended to inculcate uniform tactical doctrine.[4]

[1] On the early modern French engineering corps, see: Pierre Alexandre Joseph Allent, *Histoire du corps impérial du génie*; Antoine-Marie Augoyat, *Aperçu historique*, vol. 1; Lecomte, "Du service des ingénieurs militaires en France pendant le règne de Louis XIV"; Lazard, *Vauban*, pp. 3ff., 357–359; Vérin, *La gloire des ingénieurs*, pp. 186ff. (pp. 119–128 for the 16th century); Roger Hahn and René Taton (eds.), *Écoles techniques et militaires au XVIIIe siècle*, (Paris, 1986); and Virol, *Les Oisivetés*, pp. 95ff.

[2] H. Wauwermans, "Le marquis de Verboom ingénieur militaire flamand au service d'Espagne au XVIIe siècle," *Annales de l'Académie d'archéologie de Belgique*, 4th series, 10 (1891), p. 286; corrected in P.-E. Claessens, "A propos d'ingénieurs et d'établissements d'instruction dans les Pays-Bas espagnols au XVIIe siècle," *Intermédiaire des Généalogistes*, 88 (1960), pp. 186–191. See also Lund, *War for the Everyday*, pp. 41–43.

[3] For one Frenchman's brief account of his initial mathematical education as a cadet in a garrison town, see Anne-Marie Cocula (ed.), *Mémoires de Monsieur de La Colonie: maréchal de camp des armées de l'Electeur de Bavière*, (Paris, 1992), pp. 51–52.

[4] On the generally rudimentary level of military education in all services, see Lynn, *Giant*, p. 269. Parrott also argues that such early military academies were largely worthless. *Richelieu's Army*, p. 39. See also the brief summary in Christopher Storrs and H.M. Scott, "The Military Revolution and the European Nobility, c. 1600–1800," *War in History*, 3 (1996), 23–24. For an argument that institutions were not absolutely critical to technical expertise, see Lund, *War for the Every Day*, pp. 102–103.

Though without an official academy, the concentration of dozens of engineers in French service gave Louis an advantage over his potential competitors.[5] However, until 1690 this engineer cohort was splintered between the Secretary of War Louvois and his rival the Secretary of the Marine Colbert (officially succeeded by his son the marquis de Seignelay in 1683). The conflicts between these two Court factions exacerbated professional tensions, for each corps had not only its own areas of specialization, but they also shared control of France's fortifications. Engineers in the War ministry were responsible for the fortifications protecting the recent provincial acquisitions of Flanders, Artois, Alsace and Roussillon; this ministry concentrated most of its members on the conduct of sieges. The Marine department under Colbert oversaw the defenses of the older royal domains (the remaining core provinces); its smaller group of engineers was more obviously in charge of designing, constructing and maintaining France's coastal fortifications and ports as well.[6] Serving as *commissaire-général des fortifications*, Vauban answered primarily to Louvois, though Colbert first brought him to the King's attention and granted Vauban's judgment priority over his own Marine engineers on several occasions. In 1690 the two branches of engineers were united; several years later we find Vauban complaining of the many unknown engineers from the Marines. Only in 1697 did personal examination by Vauban become mandatory for all aspiring engineers.[7] Before

[5] Simon Pepper alludes to the benefits of engineering 'dynasties' in "Artisans, Architects and Aristocrats: Professionalism and Renaissance Military Engineering," pp. 140–142.

[6] Jean-Baptiste Colbert, marquis de Seignelay took over his father's post as Secretary of the Marine when Colbert died in 1683. Upon Seignelay's death in 1690, all fortifications and engineers were centralized under Louvois' authority. Louvois died the next year, however, and Louis appointed his son Louis-François-Marie Le Tellier, marquis de Barbesieux to be Secretary of War while giving Louvois' post of *directeur-général des fortifications* to Michel Le Peletier de Souzy. Barbesieux's sudden death in 1701 led Louis to appoint the inexperienced Michel Chamillart (already *controlleur-général des finances*) as Secretary of State for War. On the administration of the French engineers in the first half of Louis XIV's reign, see Trotter, *Marshal Vauban and the Administration of Fortifications under Louis XIV (to 1691)*.

[7] René Taton, "L'école royale du génie de Mézières," and Roger Hahn, "L'enseignement scientifique aux écoles militaires et d'artillerie," in R. Taton and R. Hahn, *Écoles techniques et militaires au XVIIIe siècle*, (Paris, 1986), pp. 559–615 and pp. 513–545; Anne Blanchard, *Les ingénieurs du "Roy" de Louis XIV à Louis XVI: Etude du corps des fortifications*, (Montpellier, 1979), pp. 104–114; Blanchard, *Vauban*, pp. 308–313.

then, even candidates hoping to serve in the War Department's service might be screened by one of his subordinates rather than the *commissaire-général* himself.[8] Depending upon how competent each candidate was in the fundamental mathematics required, he would either be rejected outright or sent on to be tutored by one of the *ingénieurs en chef* in charge of a region's fortresses. If the novitiate showed promise he would be given a brevet as engineer and usually specialize in cartography, hydrology, military architecture, or, most frequently, become a simple *ingénieur de tranchée* dedicated to the siege attack.[9] Those deemed unsuitable for any of these tasks would likely be sent back to the infantry; their knowledge, however, could still be of use during sieges, where they might serve on an ad hoc basis as *ingénieurs extraordinaires*, as could any volunteer showing some aptitude. Thus, much of their training was decentralized and in some respects beyond Vauban's guiding hand, while many of those working in the trenches as extraordinary engineers did not have even this period of apprenticeship. In this environment Vauban instilled his tactical doctrine in the French engineering corps only with much effort.[10]

Compounding the problem of the widely-varying quality of engineers, the numbers of engineers serving during the Spanish Succession were quite small compared to the tens of thousands of soldiers states could muster. As an initial caveat as well as an indication of their low status in early modern society, it is extremely difficult to find information on the members of the late 17th and early 18th century engineering corps, regardless of the country. The names of the engineers employed by most states are often available, yet very little is known about them or their service. While Vauban still held his tenure, engineers were recognized in the more detailed accounts of sieges, but the traces they left behind seem to have diminished in the War of the Spanish Succession.[11] Not surprisingly then, their

---

[8] Lazard, *Vauban*, p. 84. The mathematician and member of the *Académie royale des sciences* Joseph Sauveur began this job in 1702—see Augoyat, *Aperçu historique*, vol. 1, p. 261.

[9] Vauban considered the *ingénieur de tranchée* as the least demanding post, an officer needing only three sieges to learn the basics. Rochas d'Aiglun (ed.), *Vauban*, vol. 2, pp. 379–380.

[10] On engineer training, see Vauban's "Directeur-général des fortifications," in Rochas d'Aiglun (ed.), *Vauban*, vol. 2, pp. 380ff. In the secondary literature, see Blanchard, *Les ingénieurs du Roy*, pp. 118–120.

[11] For example, Estat des ingenieurs commandé pour servir au siege d'Ath (AG

martial exploits are often hidden, for there exist very few lists of which engineers participated at specific sieges, and information on the lower-ranking ordinary and extraordinary (i.e. "volunteer") engineers is almost completely lacking. Contrasted with the many lists of regiments and general officers participating in a siege, it is often exceedingly difficult to even find the names of the directors of the approaches (i.e. those directly under the chief engineer), while most sources never inform us of how many low-ranking engineers served under them (usually numbering in the dozens), much less their individual names.[12] In few siege accounts are specific engineers named even when they are reported killed or wounded, whereas injured officers in the infantry and cavalry are frequently singled out by name in military correspondence and public printed accounts.[13] Otherwise, the individual engineers are largely ignored in the Court's and general officers' correspondence. This is not that surprising considering they were, by and large, a heterogeneous corps composed mostly of bourgeois and recently-ennobled families (four-fifths of the French total in 1691), while only a rare few could trace their noble lineage back more than a century.[14] Without powerful subjects to study and with few traces, few scholars have researched any nation's engineering corps during this period, existing studies being limited largely to recounting the lives of the most prominent of each nation's engineers and their most famous exploits.[15]

A[1] 1400 #155 and #156). For a more general list of the Dutch engineering corps, see Ringoir, *Afstammingen en voortzettingen der génie en trein*, ('s-Gravenhage, 1980). See Blanchard's discussion of the difficulties regarding the French corps in *Les ingénieurs du "Roy" de Louis XIV à Louis XVI*, pp. 13ff. For a rare enumeration of the Allied engineers present at a specific siege (Le Quesnoy 1712), see Wijn, *Het Staatsche Leger*, vol. 8 part 3, p. 416.

[12] A French engineer's manuscript account of the 1712 sieges of Le Quesnoy and Bouchain was particularly conscientious for including the names of not only the regular engineers wounded and killed, but of the infantry officers serving as *ingénieurs volontaires* as well. AG Article 15 Section 2 §1 Douai folder, #19, Siége du Quesnoy en 1712, f. 15; and in the same, Siége de Bouchain, ff. 20–21.

[13] To take an example, in a published account of Lille (*An Exact Journal*) one of the Dutch siege directors named each days' generals of the day but never mentioned his own subordinates, even when they were wounded or killed.

[14] On the French engineers' social composition, see Blanchard, *Les ingénieurs du Roy*, pp. 86ff. Erik Lund and Gregory Hanlon have briefly attacked the stereotype of military engineers as a uniquely bourgeois group in the Austrian and Italian lands. Erik Lund, *War for the Every Day*, pp. 41–47; Gregory Hanlon, *The Twilight of a Military Tradition: Italian Aristocrats and European Conflicts, 1560–1800*, (New York, 1998), p. 347.

[15] The three nineteenth-century authors Allent, Augoyat and Lecomte focused

The leading engineering power of the age, Louis XIV's France, fielded the best organized corps, as well as the only one studied in any detail. France also appears to be the only country that had a large enough cadre of engineers for them to achieve some degree of *esprit de corps* and group identity: the siege accounts that French engineers wrote for their peers appear unique in that they distinguish individual engineers' contributions, as contrasted with the undifferentiated references to "engineers" found in journals intended for non-engineering audiences.[16] From the various documents in this corps' archives we know that late in Louis' reign he could call upon perhaps 300 engineers during wartime; a year after the end of the Nine Years' War some 255 were still on the state's payroll.[17] During the Spanish Succession, the French recruited a total of 329 engineers from 1701 to 1715, these numbers representing a dramatic increase from what they brought in during the Nine Years' War. As in previous wars, most of these newly-minted engineers were released once the Treaty of Utrecht was signed, hampering the lack of professionalism in the corps.[18]

To this total France's allies could add only a few engineers, for Spain's Flemish engineering corps was extremely small.[19] The most

primarily on the campaigns as well as some anecdotal institutional narrative. The standard modern (prosopographical) work on the French engineers themselves is Anne Blanchard, but even here data on these French technicians becomes plentiful only towards the middle of the 18th century, as is the case with the army more generally.

[16] Lecomte notes that the siege accounts from 1644 to 1654 do not mention the engineers involved. "Du service des ingénieurs militaires en France pendant le règne de Louis XIV," *Revue du génie* 25 (1877), part 1, p. 116. For a notable example of such internal accounts from the Spanish Succession era, see AG A[1] 1988, #57 Relation ou journal du siège de Menin. By the middle of the eighteenth century, siege journals put their engineers' contributions front and center by listing their names at the beginning of their accounts, as at the 1746 siege of Mons, suggesting continued attempts by the engineers to legitimize their beleaguered profession. Langins, *Conserving the Enlightenment*, p. 118.

[17] Augoyat, *Aperçu historique*, vol. 1, p. 230; Lecomte, "Du service des ingénieurs militaires," vol. 25, p. 195 mentions 304 engineers in peacetime. Vauban, *Traité*, p. 56. For secondary discussions, see Blomfield, *Sebastien Le Prestre de Vauban*, p. 99; and Chandler, *The Art of Warfare in the Age of Marlborough*, p. 221.

[18] Blanchard, *Les ingénieurs du Roy*, pp. 119–121. Also see Philippe Contamine (ed.), *Histoire militaire de la France*, (Paris, 1992), vol. 1, p. 471.

[19] Josy Muller, "Les ingénieurs militaires dans les Pays-Bas espagnols (1500–1715)," *Revue internationale d'histoire militaire*, 20 (1959), pp. 477–478; P.E. Claessens, "A propos d'ingénieurs et d'établissements d'instruction dans les Pays-Bas espagnols au XVIIe siècle," *Intermédiaire des Généalogistes*, (1960), p. 88; Duffy, *The Fortress in the Age*

notable of these engineers, the Flemish engineer Georges-Prospero Verboom, became suspect after his family and lands were captured by the Allies in the wake of their battlefield victory at Ramillies. Imprisoned by the French, he was eventually released in 1708 and journeyed to Spain to oversee the newly-constituted Spanish engineering corps.[20] France's other main ally, the Elector of Bavaria similarly lacked an experienced engineering corps. The Frenchman Jean de La Colonie left Louis's service in 1702 to serve as chief engineer for the Bavarians since the Elector had few of his own. While fighting with the Maritime powers in the previous war, Max Emmanuel had relied upon his allies for experienced engineers—his own technicians knew little more than how to copy fortress plans.[21] Unable to rely on such allies, France was fortunate to have a well-established corps of its own, even if its composition and numbers fluctuated dramatically from peacetime to war.

The problems faced by the Sun King's engineers were small, however, compared to the weaknesses of the engineering corps of other states. The largest group of Allied engineers came from the United Provinces of the Netherlands, a cosmopolitan republic that had long been a haven for Protestant immigrants seeking to ply their trades.[22]

of Vauban, pp. 98–99. On the Spanish engineering corps more generally, see Martine Galland-Seguela, "Introduction à l'étude du corps des ingénieurs militaires espagnols au XVIIe siècle," *Histoire, économie et société*, 8 (1989), pp. 551–562. Two authors emphasize the critical French contribution: Juan Miguel Muñoz Corbalan, "Les derniers ingénieurs du roi de France en Espagne"; and Manuel-Reyes Garcia Hurtado, "Le siège dans l'art militaire espagnol du XVIIIe siècle," in L. Carle (ed.), *Situazioni d'Assedio. Cities under Siege*, (Florence, 2002), p. 172.

[20] Wauwermans, "Le marquis de Verboom ingénieur militaire flamand au service d'Espagne au XVIIe siècle" *Annales de l'Académie royale d'archéologie de Belgique* 4th series, vol. 10 1891; and Juan Miguel Muñoz Corbalan, "El arresto en 1706 del Ingeniero Mayor Jorge Próspero Verboom," *Aportaciones militares a la cultura, arte y ciencia en el siglo XVIII hispanoamericano*, (Sevilla, 1993). The fortress plans and brief résumé of another Spanish engineer, Jean Boulengier, can be found in Jean-Louis Van Belle, *Plans inédits de Places fortifiées XVIIe–XVIIIe siècle*, (Louvain-la-Neuve, 1989). France also sent brigades of engineers to Portugal (1702) and Spain (from late 1703 on) in order to make up for their shortages. Augoyat, *Aperçu historique*, vol. 1, pp. 259, 284.

[21] Cocula (ed.), *Mémoires de Monsieur de La Colonie*, pp. 167, 213–214.

[22] On the Dutch engineering corps, little has been written, though see Stichting Menno van Coehoorn, *Vesting: Vier eeuwen vestingbouw in Nederland*, ('s Gravenhage, 1982) and the brief J. Sneep, "Het corps ingenieurs en de directie der fortificatiën in het Staatse leger 1579–1795," in F.H. van den Beemt (ed.), *300 Jaar bouwen voor de landsverdediging*, ('s Gravenhage, 1988). Le Normant recommended twenty to thirty engineers for Louis XIII, claiming that Maurits of Orange relied upon fewer than

Their long-standing experience reclaiming polders from the sea and shielding their territorial boundaries with barrier fortresses was buttressed by Louis XIV's repression of the Huguenots, epitomized by the famous revocation of the Edict of Nantes in 1685. Vauban complained of the many engineers lost as a result of His Most Christian Majesty's act of conscience, and many of those targeted joined their fellow expatriates in the Netherlands.[23] Among the engineers of French extraction in Dutch service, two had manned the top engineering post in the Nine Years' War (François Du Puy, heer van Cambon and Charles Du Puy de L'Espinasse), while during most of the War of the Spanish Succession the highest ranking engineers would be two others, Guillaume le Vasseur Des Rocques and Lucas Du Mée. In tandem they would direct most of the Allied sieges in Flanders after the 1704 death of Coehoorn (an infantryman by training) and his ephemeral successor the Liégeois Jean Gérard, Baron de Trognée in 1704.[24] The Dutch also filled out their ranks with German engineers, allowing them to field a good-sized corps over the course of the war, at least 275 engineers of all ranks seeing service during some point of the War of the Spanish Succession. From a high of 111 engineers in their pay in 1696, the number available in any single year of the Spanish Succession conflict ranged between forty and eighty.[25] Their experience with the most recent developments in the

a dozen. *Discovrs povr le restablissement de la milice de France*, pp. 236–237. He also claimed that Maurits of Orange's parsimony forced his experienced engineers to seek greener pastures in foreign service.

[23] Vauban, *Traité*, p. 56. On the military implications of the Edict of Fontainebleau more generally, see Matthew Glozier, *The Huguenot soldiers of William of Orange and the "Glorious Revolution" of 1688: the lions of Judah*, (Brighton, 2002).

[24] On Des Rocques, see Augustus Veenendaal Jr.'s entry in L. Frey and M. Frey (eds.), *The Treaties of the War of the Spanish Succession*, pp. 132–133. The better-known Huguenot author Charles Goulon had also served with the Dutch after leaving post-Nantes France, finishing up his career in Imperial service in Italy. See F.J. Hebbert, "The Memoirs of Monsieur Goulon."

[25] This total figure comes from a count of Ringoir's *Afstammingen en voortzettingen der génie en trein.* In 1696 the *staten van oorlog* (the annual military budgets) listed 111 engineers in pay. F.J.G. Ten Raa, *Het Staatsche Leger: Van de verheffing van Prins Willem III en zijn gemalin tot Koning en Koningin van Groot-Brittanië tot het overlijden van den Koning-Stadhouder (1688–1702)*, ('s Gravenhage, 1950), vol. 7, part 2, p. 393. F.W.J. Scholten's count of Dutch engineers in the *staten van oorlog* fluctuate between forty and eighty per year over the course of the Spanish Succession. *Militaire topografische kaarten en stadsplattegronden van Nederlanden, 1579–1795*, (Alphen aan den Rijn, 1989). Wijn, *Het Staatsche Leger*, vol. 8 part 3, p. 492, for example, mentions 34 engineers first class and 22 engineers second class on their roster.

siege attack, however, was limited at the turn of the century, for the Nine Years' War saw a total of only four Confederate sieges along the United Provinces' borders.

The Dutch shouldered the Allied burden of siegecraft in the Low Countries in much the same way as the French did, as neither British nor Imperial allies could make a significant contribution. The British Isles, protected more by wooden ships than stone walls, had few engineers to send to the Allies' Flanders sieges. To begin with, like the French in the early seventeenth century, the British native school was necessarily small given the limited exposure their troops had to the latest techniques of siege warfare.[26] Unlike the more specialized technicians on the Continent, the English engineering and artillery corps shared a single administrative organization and even shared some of the same personnel.[27] Their expertise in siegecraft (as contrasted with fortress design and construction) was also largely dependent on gaining experience across the Channel: most of the siege warfare conducted in the Irish campaigns of the 17th century, for example, pitted either Irishmen with continental experience against one another or Jacobite-allied Frenchmen against William III's Dutch or Huguenot engineers. Their only sustained native experience with siegecraft had occurred half a century earlier and there was limited opportunity for improvement. The many English strongholds attacked during these English Civil Wars consisted primarily of medieval town walls or hastily-improvised earthworks; most were no more than fortified manors. These sub par fortifications provided English engineers little opportunity to practice their craft against state-of-the-art fortresses.[28] Nor did the English have an institutional foundation that

[26] On England's backwardness in siegecraft circa 1700, see Scouller, *The Armies of Queen Anne*, pp. 173–187; and David Chandler, "Fortifications and Siegecraft," in *Blenheim Preparations: The English Army on the March to the Danube. Collected Essays*, (Staplehurst, 2004), pp. 146–147.

[27] On their names, see Charles Dalton, *English Army Lists and Commission Registers 1661–1714*, 6 vols. (London, 1898), for the years 1702–1707 vol. 5, pp. 115–123 and 5:Part II, The Blenheim Bounty Roll, March 1705, pp. 9–11. For 1707–1714, consult vol. 6, pp. 43–48.

[28] Sheila Mulloy, "French engineers with the Jacobite Army in Ireland, 1688–1691," *The Irish Sword*, 15(16) 1983, pp. 222–232; James Burke, "Siege Warfare in Seventeenth Century Ireland," pp. 258, 276ff.; and Pádraig Lenihan, "Conclusion," pp. 353ff., both in P. Lenihan (ed.), *Conquest and Resistance: War in seventeenth-century Ireland*, (Leiden, 2001). For a Huguenot in English service, see A. Stuart Mason and Peter Barber, "'Captain Thomas, the French engineer': and the teaching of Vauban to the English," *Proceedings of the Huguenot Society of Great Britain & Ireland*, 25(3) 1991,

could match the quantity or quality of Dutch engineers. The Dutch William, as King of England, established the first English company of twenty-eight engineers in 1696 (many of whom were Huguenot), but they had little interaction with one another and their unit was disbanded in 1700, though most continued to serve individually in various capacities. Their numbers remained small in the first decade of the 18th century. The commission lists of English engineers between 1707 and 1714 total only thirty-seven individuals named "engineers" in all theaters, including North America.[29] One indication of how poorly this small corps met its obligations is found in Marlborough's response to a request from the Board of Ordnance (the administrative body responsible for both the British engineers and artillery train) for technicians to see to England's home defenses:

> I . . . am sorry to understand the difficulties the Board is under for want of engineers. They are so scarce with us, that I cannot readily think of one to be sent from hence; when any such occurs I shall not fail to dispatch him away. In the meantime I hope England is not so destitute but that you may find a fit person to be employed in the service required.[30]

England was indeed destitute of suitable candidates, for four years later the Board informed Marlborough that: "We only have four engineers here in Great Britain who are imployed in the several places . . . but wee wish that there coud be found other engineers of abilitys to joyn with them respectively here."[31] The only 'native'

pp. 279–287. More generally, see Whitworth Porter, *History of the Corps of Royal Engineers*, (London, 1889), especially chapter 3. To take another example, many of the coastal fortresses of southern England were renovated by a Dutchman recruited into English service in the 1640s. See A. Saunders, *Fortress Builder: Bernard de Gomme, Charles II's Military Engineer*, (Exeter, 2004).

[29] Dalton, *English Army Lists and Commission Registers 1661–1714*, vol. 6, pp. 43–48. Porter mentions twenty-eight engineers as of March 1697 and another dozen or so in the artillery (*History of the Corps of Royal Engineers*, p. 60). It is possible that William's reliance on his Dutch engineers further discouraged the growth of a native English school from 1688 onward—at the least, there was political controversy among MPs in the 1690s over the number of Dutchmen in the English artillery.

[30] Murray (ed.), *Letters and Dispatches*, vol. 2, p. 105 from Elft, 16 June 1705. The onset of winter quarters freed up at least a few for domestic service: vol. 1, p. 595 Marlborough to the Earl of Leven, St. James, 10 February O.S. Such shortages were systemic, however, as is seen by similar concerns in vol. 4, p. 32, Marlborough to the Board of Ordnance, Bellingen, 28 May 1708.

[31] BL Add MSS 61166, f. 74b, Board of Ordnance to Marlborough, 15 February 1709. An English commander in Spain similarly complained of a severe shortage of engineers in 1704. Chandler, *The Art of Warfare in the Age of Marlborough*, pp. 224–225.

engineers of the period to have merited historical study are the three brothers Michael, John, and Jacob Richards. Indicative of England's backwardness, Jacob followed the standard practice of being sent into foreign service (in this case, Imperial) in order to learn Continental siege techniques earlier in the 17th century—those who remained in English pay also sought experience across the Channel.[32] Forced to travel overseas for employment, all three brothers saw service under a number of different rulers in England, Hungary, Ireland, Flanders, Newfoundland, Venice, Poland, Portugal and Spain.[33] Otherwise, there were few English, or even British, engineers of note serving on the Continent during the Spanish Succession, the most prominent in Flanders being Holcroft Blood (who learned his trade while in French service) and John Armstrong (who served the Duke primarily as a lieutenant quartermaster).[34] If Marlborough's correspondence as Master-General of the Ordnance with the Board of Ordnance is any indication, the Duke interacted very little with the engineers under his command. Given the many hats Marlborough wore, this is hardly surprising. As a result of these weaknesses, English engineers serving with Anglo-Dutch forces usually ended up with the artillery train, a domain where they at least had considerable experience. More often, England sent its engineers to 'subsidiary' theaters such as Iberia to assist allies even more destitute than them. Here their technicians oversaw their new conquests (e.g. Gibraltar and Valencian fortresses such as Alicante) as well as assisting the Earl of Galway on the Portuguese front. This branch of service would continue to be poorly represented in British service throughout the 18th century, for even in the Napoleonic wars their engineers were still found to be in a "pathetic" state.[35]

The Austrians, who contributed substantial numbers of troops and their best commander to the later Flanders campaigns, provided few if any engineers. Though the lack of any systematic records (or organization) of the corps before the mid-eighteenth century is one

[32] Porter, *History of the Corps of Royal Engineers*, pp. 47–48; and Chandler, *The Art of Warfare in the Age of Marlborough*, pp. 218–219.

[33] H.T. Dickinson, "The Richards Brothers: Exponents of the Military Arts of Vauban," *The Journal of the Society for Army Historical Research*, 46(186) (Summer 1986), pp. 76–86.

[34] On Blood's early French service, see Dalton, *English Army Lists and Commission Registers 1661–1714*, 5 part 2, The Blenheim Bounty Roll, March 1705, p. 9 note 1.

[35] Michael Glover, *Wellington as Military Commander*, (London, 1968), p. 171.

indication of their low status, the backwardness of the Austrian corps is perhaps best summarized by one of its own commanders, Jacob de La Vergne. After twelve years of service, he published in 1698 a reform proposal dedicated to Emperor Joseph I. In it he asserted the need for systemic change by consistently denigrating the quantity and especially the quality of his subordinates. He enumerated their many faults: how the service was unable to prevent its rising stars from finding greener pastures in foreign service, how they were therefore left with the least-qualified rejects from home and abroad, how he was unable to punish or reward his own subordinates due to the pernicious effects of patronage, and how as a result of such unchecked insubordination there was a total absence of discipline within the ranks. As La Vergne complained, the Austrian engineers had no single, recognized leader (hence the title of his proposal), and their lack of institutionalization also provided little opportunity to instill *esprit de corps* or uniformity of doctrine among the fractious technicians. As such he itemized far more serious faults than those of which Vauban ever complained. La Vergne's call for reform was apparently ignored, and probably never received a wide hearing in the first place given the extreme rarity of extant copies of this work.[36] He was certainly prudent not to ask for additional money, yet another weakness the Austrian engineers suffered from. The Habsburg Court found it impossible to maintain peacetime levels of funding during war: in 1706 Vienna forced the hiring of such technical personnel onto its local governments due to its own coffers being empty. No surprise then that the Emperor had few skilled technicians to volunteer for Flanders; those few available were busy shoring up Austrian defenses in Germany and their newly-acquired Italian holdings, as well as reconquering rebellious Hungary. In any case, their total number probably only matched those in English service. Lists of

[36] La Vergne's lengthy title is worthing citing in full as it provides a taste of what he had to contend with: *De l'utilité d'avoir un bon ingénieur, ou directeur général des fortifications dans un État: Joint à un bon corps d'ingénieurs, sur lesquels il aye le commandement absolu, et la manière de le composer, sans augmenter les fonds, qui sont déjà fait; Avec un petit Traitté demonstratif de leurs devoirs, chacun en particulier, par lesquels l'on pourroit éviter les grosses confusions et fautes qui sont infallibles dans un corps qui n'a point de commandant tant aux approches qu'aux fortifications ordinaire. Par Jacob de la Vergne Ingenieur en Chef, et Capitaine au Service de Sa Majeste Imperiale,* (Vienne, 1698). For the difficulties in recruiting: pp. 4, 69. On the resulting insubordination, see Avant-propos pp. i [unpaginated], 69–70. On the appeal to start judging engineers according to their abilities: Avant-Propos, pp. i–ii and 3–4, 13–14.

Austrian engineers only become frequent with institutionalization in 1747, but a 1687 list identifies only twenty-seven officers. At Eugene's 1717 siege of Turkish Belgrade we find him assisted by forty-one engineers of *Flemish* origin—apparently the Austrian service was still small enough or inexperienced enough that they immediately drew upon their newly-gained southern Netherlands territories (won at the Peace of Utrecht) to reinforce their native corps.[37]

Nor could the German principalities of the *Reich* offer the Allies much technical assistance. The most powerful German princes provided their subsidy regiments of foot and horse only reluctantly, while even along the Rhine River they could not provide the requirements demanded of major sieges. In both the Nine Years' War and the War of the Spanish Succession, it was up to the Dutch to provide the engineering expertise. "In short," concluded one Englishman, "the Imperialists undertake sieges without cannon, ammunition or engineers with as much assurance as they did a war without money, credit or troops."[38] Marlborough further confirmed the Confederate army's reliance on Holland when discussing potential sieges of Metz and Thionville: "we have seen by long experience that we can by no means rely upon any promises from Germany, so that the whole expense of this enterprise must unavoidably fall upon the Dutch,

---

[37] See John Stoye, *Marsigli's Europe, 1680–1730*, pp. 31ff., which traces the career of another foreigner who quickly rose through the ranks of the Austrian engineers (to which we could also add Goulon), and who possibly became a rival of La Vergne (p. 48). Consult pp. 38–39 for the 1687 list of engineers, including the rapidly-promoted Bolognese Marsigli. La Vergne does not mention the size of the corps, but his recommendation on p. 69 to maintain forty lower-ranking engineers during wartime gives us some idea of the upper limit. See also Prince Eugene's complaint quoted in Duffy, *The Fortress in the Age of Vauban*, p. 25; and more generally Jean-Michel Thiriet, "La fondation de l'Académie des ingénieurs militaires à Vienne (1717) et les Italiens," in J.-C. Allain (ed.), *Des étoiles et des croix. Mélanges offerts à Guy Pedroncini* (Paris, 1995); and Erik Lund, *War for the Everyday*, pp. 44, 62 note 111, and chap. 3, esp. pp. 102, 113–114. Lund contends that institutionalization is not necessarily an indication of quality, though it is of quantity, and dozens of engineers were needed for each siege in addition to generals with engineering experience. Certainly the French corps' success under Louis XIV was largely a function of the personal support Vauban received from his king and Secretary of War over forty years of service.

[38] Royal Commission on Historical Manuscripts, *The Manuscripts of the Earl Cowper*, (London, 1888), vol. 3, p. 48, Richard Pope to Coke, Weissenburg, 8 October 1704. See also *A journal of the Several Sieges of Keiserwaert, Landau and Venlo*, (London, 1702), p. 16. The Prussian failures at Kaisersweert in 1702 are discussed elsewhere. On the heavy Allied reliance on German mercenaries more generally, see Peter H. Wilson, *German Armies: War and German Society, 1648–1806*, (London, 1998), pp. 101ff.

which I can hardly think they will be willing to bear."[39] It fell to the United Provinces to provide the vast majority of engineers for the many sieges in the Low Countries and even for those in Germany.

It is not surprising then that during a period of siege-dominated warfare the short supply of engineers was rarely able to keep up with the demand. France's several hundred engineers were still stretched thin across four theaters of operations: from the Channel to the Ardennes in the Low Countries, on at least two fronts in Iberia, along the Rhine, and in northern Italy and the French Alps. In each region, a dozen or more threatened fortifications had to be put in a state of defense, while dozens more engineers (a major siege could demand fifty or more) would be required for each individual siege that was undertaken.[40] Two factors provided the French with a small amount of consolation as the war turned against them and reduced their overstretched corps. First, the theaters under contention dwindled as their fortunes waned—the Italian and German theaters became secondary to the mounting importance of Flanders. Second, while their defensive posture demanded engineers to cover a large swathe of threatened territory, at least it required fewer engineers than what a single besieging army would demand.

In addition to the need for geographical dispersion, every country's pool of engineers was also kept small by the dangers associated with engineering service. Particularly hazardous were the risks associated with besieging a town. Vauban's dozen sieges under Condé had already taught him the deleterious effect attrition could have on the corps' expertise:

> What we can say on their behalf is that this Science demands much courage and spirit, a very solid intelligence, and above all, perpetual study and a consummate experience in the essential aspects of war. But if Nature rarely combines these first three qualities in a single individual, it is even more extraordinary to see him escape the violence

[39] Murray (ed.), *Letters and Dispatches*, vol. 1, p. 94, Marlborough to the English diplomat George Stepney, Bonn, 9 May 1703. The next year a Dutch general similarly commented on the Imperial lack of preparations for attacking Landau. Veenendaal, Jr. (ed.), *Briefwisseling Heinsius*, vol. 3, p. 349, Reynout [Reinhart] Vincent, baron Hompesch to Heinsius, Croonweissenburg, 22 September. Luh reinforces this point in *Ancien Regime Warfare and the Military Revolution*, pp. 101–103.

[40] Ath 1697 enrolled sixty engineers, and the 1710 siege of Douai required more than fifty engineers. *The Postman*, 25 April O.S. issue #1876, printing an account from Brussels dated 2 May N.S.

> of our sieges in order to live long enough to acquire the other two.... The *métier* is great and very noble, but it requires a special kind of genius and a continual application over many years, something which Nature and the vigor of our sieges rarely allows.[41]

These "martyrs of the infantry" were exposed to constant danger in the trenches, much more so than most other besieging troops, who served in the trenches only one day out of every four or five: at Ath 1697, the guard regiments and generals of the day saw a total of two days worth of trench duty, whereas the engineers saw four to five days each.[42] Besieging engineers, in charge of developing a plan of attack, were to reconnoiter as close to the covered way as possible with only a small detachment (so as to avoid detection), exposing themselves not only to possible capture by enemy parties, but also to the garrison's gunners, who were ordered to fire on those who ventured too near the glacis.[43] In the course of his siege of Ath Vauban reminded the new Secretary of War Barbesieux, of the need to see to the engineers' wages in order to keep them in the service, for "the work is great and the peril extreme," and with the siege complete he reminded Barbesieux, Le Peletier and the King himself of their many sacrifices and the need for just compensation.[44] From the very start of a siege these technicians courted mortal danger, as the chevalier de Quincy illustrated in his account of the French siege of Marchiennes in 1712:

---

[41] Vauban, *Mémoire pour servir*, p. 19.

[42] Siege engineers were also more exposed than their garrison counterparts. The bloody siege of Tournai cost the French garrison four engineers killed and four wounded, while Aire's garrison numbered only eight engineers to begin with. On Tournai, AG A[1] 2152, #145 Estat des officiers, soldats, dragons, invalides, tués et blessés pendant le siège de la ville et citadelle de Tournai; on Aire, AG Article 15 Section 2 §1 Aire Carton 1, #5 Journal du siège d'Aire en 1710 par un père Jésuite qui étoit dans la ville, f. 22. For a general discussion of engineer casualties, see Charles Albert Samuel Lecomte, *Les ingénieurs militaires en France pendant la règne de Louis XIV*, (Paris, 1904), p. 121. This book is a collated version of his earlier journal article series "Du service des ingénieurs militaires."

[43] Siege manuals encouraged garrison commanders to send out parties to capture such reconnoitering parties, e.g. Vauban, *Mémoire pour servir*, p. 68. For attempts at an actual siege, see AG A[1] 1940 #22 Journal du siége d'Ath, 2 October 1706, published in Vault and Pelet (eds.), *Mémoires militaires*, vol. 6, p. 563. For orders to fire on reconnoiterers, see AG A[1] 1939 #204, Journal du siége de Menin, 28 July, published in Vault and Pelet (eds.), *Mémoires militaires*, vol. 6, p. 536.

[44] Rochas d'Aiglun (ed.), *Vauban*, vol. 2, p. 452, 21 May 1697—also to Le Peletier de Souzy.

> The trench was therefore opened the night of the 28th. M. de Villars [i.e. Louis-Joseph de Plaibault de Villars-Lugein], chief engineer and brigadier of the King's armies, was killed with three other engineers as they were observing together the terrain where we were to open the trenches . . . The same cannonball sent all four of them to the afterlife.[45]

Those surviving such freak accidents still had to expose themselves day after day (and night after night) supervising and directing the workers in the trenchworks. Engineering skills were often demanded at the most dangerous point of attack (such as the head of the sap), where only technical knowledge could overcome stiff resistance. As the English veteran Humphrey Bland explained:

> And tho' it is not only the Duty of the Engineers to instruct the Officers, that they may direct their Men in the making of the Works; but likewise to Visit them from time to time, to see that each Part has its true Proportion; yet as those Gentlemen are liable to Accidents, from the Danger they are often exposed to, that may render them uncapable of performing their Duty, by which the Works may be retarded, or ill executed, unless the Officers, from their own Experience, can supply the Defect; I thought the inserting of the above Particulars would be of use to young Officers, by giving them some Notion what Trenches are, and in what manner they are made, that, when they shall be ordered on those Commands, they may not be entirely at a loss how they are to proceed, in case they should fail of the necessary Directions and Assistance of the Engineers, which proves too often the Case at most Sieges; nor is it to be much wondered at, for if the Engineers do their Duty, they are so often disabled, that their Want must be supplied by the Diligence and Skill of the Officers.[46]

The particularly perilous job of establishing a lodgment on the covered way demanded technicians to direct the workers under fire, all the while facing the constant threat of garrison mines and counter-attacks.[47] Building properly-blinded galleries and bridges across ditches required an engineer's oversight as well, placing them once again at the point of most resistance. Expected to keep the big picture in mind while attending to the minutest detail of the trenchworks,

[45] Lecestre (ed.), *Mémoires du chevalier de Quincy*, vol. 3, p. 161.
[46] Humphrey Bland, *A Treatise of Military Discipline*, (London, 1727), pp. 264–265.
[47] Sixteen were killed or injured in the first storm on Lille's counterscarp; see Des Rocques' letter in John Campbell, *The military history of the late Prince Eugene of Savoy, And of the Late John, Duke of Marlborough* . . ., (London, 1737), vol. 2, p. 65. For another case, see *The Postman*, #1899, From the camp before Doway, 17 June 1710.

engineers both high and low necessarily put themselves in harm's way throughout a siege. Those who lived to see a decade or more of sieges undoubtedly had numerous wounds to show for their service. Vauban's forty-plus sieges gave him over a dozen injuries (including a prominent scar on his cheek received at the 1667 siege of Douai), while his younger cousin DupuyVauban claimed sixteen injuries towards the end of his career.[48] Field officers seeking promotions often mentioned a wound suffered on the field of battle as evidence of their faithful service—engineers fortunate to serve the same number of years could count a dozen injuries or more. La Vergne lamented this grim reality and confirmed the risks shared by engineers of all countries when he noted matter-of-factly that "in a large siege many [engineers] are to be found at the beginning, but few usually remain at the end."[49] When the threats posed by powder and lead are combined with constant laboring in the heat or cold in the often squalid conditions of trench warfare, the engineer's life was not an easy one. With such high attrition rates suffered by such a small group, it is not surprising that there were rarely enough skilled overseers in the trenches to manage the attack efficiently.

The many Low Countries sieges steadily bled the engineering corps white. This held true regardless of who conducted the attacks. At the 1688 siege of the Palatinate fortress of Mannheim, Vauban praised a number of his subordinates and reminded Louvois that "these two last sieges [of Philipsbourg and Mannheim] have sidelined more than half of this small group." The capture of Namur in 1692, also conducted by Vauban, still cost the French one-third of their sixty engineers (nine dead, thirteen seriously wounded). At the model siege of Ath only a few hundred Frenchmen in the siege army suffered wounds or death, but Vauban noted that the three engineers that had been killed and the six others wounded were admittedly "not a very considerable number, yet very large compared to the other branches."

[48] Maurice Sautai, "Une lettre inédite du Gouverneur de Béthune du Puy-Vauban," *Bulletin de la commission historique du département du Nord* 28 (1911), p. 518, DupuyVauban to Philippe Bourbon, duc d'Orléans and brother to the King. In addition to his broken leg suffered at Tournai, Des Rocques also broke a bone late in the siege of Douai. ARA, collectie Staten-General (SG), 5185–2, Field Deputies to States-General, Brebières, 8 June 1710. Du Mée was wounded at Menin and killed attacking Tournai's citadel.

[49] La Vergne, *De l'utilité d'avoir un bon ingénieur*, p. 47. Also Landsberg, *Nouvelle manière de fortifier les places*, p. 1.

The bloodier siege of Barcelona in 1697 was not conducted with the same skill and as a result the besiegers saw two-thirds of their forty-eight engineers taken casualty: twelve dead, twenty-two wounded.[50] In the War of the Spanish Succession, the French found themselves again short of skilled men, spread as they were across four theaters. Within Flanders they were particularly hamstrung by the need to shuttle their engineers between the many threatened fortresses.[51] Their losses here were limited in contrast with the Allies, however, since their engineers sheltering inside fortresses were less exposed than those toiling in the trenches. Nonetheless, even after a decade of 'respite' from offensive siegecraft they still lacked enough experts familiar with the Flanders theater when the duc de Villars went back on the offensive in 1712. Villars' hopes of conducting two Flanders sieges at the same time were dashed as a French *maréchal de camp* explained that they simply did not have enough engineers to manage two concurrent attacks.[52] As it was, Villars had to scrounge up engineers and gunners from throughout the theater in order to conduct the siege of Douai. The twenty-five days of open trenches, according to a siege journal, witnessed twelve engineers wounded (one of these would later die of his wounds) and one killed outright. At the end of the siege the chief engineer informed Villars of the need to find a significant number of replacements if he hoped for another enterprise.[53] While half a dozen engineers were busy repairing

[50] On 1688, Rochas d'Aiglun (ed.), *Vauban*, vol. 2, p. 302. For Namur, Augoyat, *Aperçu historique*, vol. 1, p. 183; pp. 223–224 for Barcelona. Ath: Rochas d'Aiglun (ed.), *Vauban*, vol. 2, p. 464.

[51] For example, two engineers managed to sneak their way into Mons right after it was invested (AG A[1] 2153, #188, 25 October 1709). Valory left Ypres too late to slip into Tournai before it was invested (AG A[1] 2151, #166, 8 July 1709) and in the next year he was sent from Douai to inspect Valenciennes a mere week before the former was invested (AG A[1] 2214, #96, 14 April 1710). At Bouchain in 1711 we find similar concerns that harried engineers dash into the town before investment: AG 1 K 299 (Papiers Montmorency-Tingry), folder 7, Le Peletier de Souzy to Christian-Louis, chevalier de Montmorency-Luxembourg, 12 April.

[52] AG A[4] Carton 8 chemise 2 (Flandre août-octobre 1712), Gabriel, chevalier de Hautefort to Louis-Auguste Bourbon, duc du Maine, 29 September. In his discussion of army sizes and siege trends, Lynn highlights the manpower demands concurrent sieges required. To this we should add the difficulty of freeing up nearly one hundred engineers for two simultaneous sieges—this would have required a third of France's entire corps! Lynn, "The trace italienne," pp. 311–312.

[53] Marquis de Vogüé (ed.), *Mémoires du Maréchal de Villars publié d'après le manuscrit original*, 6 vols. (Paris, 1889), vol. 3, p. 161; AG A[1] 2382, #138, the intendant of French Flanders Charles-Etienne Maignart de Bernières to Voysin, 25 August; engi-

Douai's works, the rest of the dwindling corps was sent to the next target, Le Quesnoy. Not surprisingly, the engineer brigades conducting the trench attacks were inexperienced and undersized: "the engineer brigades should have been composed of eight engineers instead of six, and in reality the largest brigades were actually only five [once we take into account the sick and wounded] . . . most of whom were young engineers who have much valor but little experience."[54] Such attrition quickly added up. The butcher's bill for the French corps in all four theaters came to more than fifty-four engineers killed by enemy fire after a decade of siegecraft, or a sixth of its number at the start of the war.[55]

The Allies were even worse off than their opponents, for not only did they have fewer engineers to call upon, but they also attacked many of Europe's strongest fortresses as well. Throughout the war the Dutch field deputies frequently warned the Grand Pensionary Heinsius of how poor the conditions of service were for their engineers. In their very first siege, targeting Cologne's Rhenish town of Kaisersweert, the Dutch quickly discovered that they would have to bear most of the burden themselves. The separate Prussian and Dutch attacks failed to advance in tandem as the German advance lagged far behind their own. The *Raad van State*'s field deputy Geldermalsen advised Heinsius that they had to delay their own plan to storm the counterscarp because their ally lacked engineers and artillery, which the States-General would have to supplement

neer journal in AG Article 15 Section 2 §1 Douai folder, #19, Précis ou journal du Siége de Douay fait en 1712 . . ., ff. 1–10. For the engineer's warning, AG A[1] 2383, #174, Valory to Villars, 9 September.

[54] AG Article 15 Section 2 §1 Douai folder, #19, Siège du Quesnoy en 1712, ff. 12 and 16.

[55] Blanchard, *Les ingénieurs du Roy*, p. 119. Many others would die of natural causes or retire from service during the war. The sieges in the next major war would also claim one-sixth of the force, forty-eight killed between 1744 and 1748 alone, and helped spur the creation of an official École du Génie at Mézières to provide a more regular supply (pp. 189–191). Of 1,490 French engineers between 1691 and 1791 traced by Blanchard, 14% of them were killed in combat, 22% of the lower-ranking engineers and only 8% of the *ingénieurs en chef*. These overall percentages are probably as low as they are due to the decreased frequency of sieges and, more generally, of wars fought by Louis XV and XVI compared to the martial Sun King (half of whose reign is excluded from the count). Blanchard, *Les ingénieurs du Roy*, pp. 292–301.

with their own.[56] The need to coordinate the two attacks, bad weather and garrison reinforcements forced the Dutch to postpone their storm on several other occasions throughout the month of May; in the end, it would be forty-two days before it could finally be implemented, and the town was eventually forced to surrender.[57] The introduction into the theater of British troops later that year and Imperial troops even later in 1708 did little to increase the number of engineers available. The reliance of the Allies on the States' engineers would continue throughout the war, for the Dutch were the only state well-positioned to provide the large number of siege experts, munitions and breaching pieces that the theater's many sieges would require.

The other Spanish Netherlands sieges of 1702 through 1705 provided few challenges to the Allied engineers, but beginning in 1706 they began suffering significant losses.[58] Stronger fortifications and larger garrisons demanded more engineers to oversee more extensive trenchworks; engineer casualties accelerated as a result. The garrison of Menin, despite its poor performance, still managed to kill five Allied engineers during its short defense.[59] By the end of the 1706 campaign, the Frisian field deputy Sicco van Goslinga was already petitioning Heinsius on behalf of the technicians, reporting that of thirty-four engineers participating in the siege of Ath, only fourteen were still available for service after a mere twelve days of open trenches.[60]

Their shortage of engineers became particularly acute as they forced their way through Vauban's *pré carré*. During the two-month

[56] Veenendaal, Jr. (ed.), *Briefwisseling Heinsius*, vol. 1, p. 138 #221 Geldermalsen to Heinsius, Wesel, 28 April 1702; also vol. 1, p. 143 #230, lieutenant-general Walrad vorst von Nassau-Saarbrücken to Heinsius, Kaisersweert, 28 April.

[57] Veenendaal, Jr. (ed.), *Briefwisseling Heinsius*, vol. 1, p. 275 #477, Wassenaar-Obdam to Heinsius, 9 June 1702; Wijn, *Het Staatsche Leger*, vol. 8 part 1, p. 688, Blainville to *maréchal de France* Louis François, duc de Boufflers, Kaisersweert, 10 June.

[58] An English account of Venlo noted, nonetheless, the death of the engineer Peter du Bouchet in the trenches. *A journal of the Several Sieges of Keiserwaert, Landau and Venlo*, p. 45.

[59] ARA CvdH, 142. This list is printed in Wijn, *Het Staatsche Leger*, vol. 8 part 2, p. 759.

[60] Veenendaal, Jr. (ed.), *Briefwisseling Heinsius*, vol. 5, p. 570 #1105, Goslinga to Heinsius, Ath, 2 October. Heinsius shared Goslinga's concerns, and encouraged him to keep an eye out for promising soldiers that might be willing to join the engineers, "the most important people in the army," vol. 5, p. 581 #1127, Heinsius to Goslinga, The Hague, 6 October.

siege of the town of Lille in 1708, sixty-five engineers were wounded or killed.[61] Goslinga reiterated his earlier pleas on their behalf:

> Our infantry is weak and exhausted, half of our engineers are dead or wounded. We lost three just this night, among others the director Longuené. On this subject I must repeat again what I have said many times, that we do not adequately reward them. They are almost all in dire straits, they have few posts, often poorly paid if at all. And with this situation there is no hope of them exerting themselves, the highest ranking must limit themselves to the rank of lieutenant-colonel. They are much better treated in France and everywhere else. What will happen in the future if no one wants to serve in a corps where one is almost certain to meet death on one side and neither profit nor honor on the other?[62]

In his public account of this bloody siege, the director Du Mée justified his conduct in the following terms:

> I am sure the Publick will forgive the errors I may commit upon this Account, or have already committed, in the management of that part of the siege under my care, which I discharged to the best of my capacity, considering the hardships the directors of that siege were forced to undergo by the loss of so many engineers, which were more considerable for their number and experience, than has been known to be lost in any siege these hundred years.[63]

But the campaign was not yet finished, for Marlborough hoped to conduct yet one more siege after Lille's citadel fell, that of Ghent. For this task he informed Heinsius of the need for even more engineers, "for we have very few left."[64]

Goslinga repeated his petition on behalf of the engineers once again as they lost their lives and livelihood the next year before Tournai. He warned Heinsius that the same combination of poor wages, few opportunities for advancement and high casualties would decimate their ranks: "Allow me to advocate to you the interests of

---

[61] Wijn, *Het Staatsche Leger* vol. 8 part 2, p. 789. Another account mentioned thirty engineers put out of service. "La campagne de Lille, contenant un journal de ce qui s'est passé au siége de cette importante place," in Elie Brun, *Les Sept Sièges de Lille*, (Paris and Lille, 1838), p. 366.

[62] Veenendaal, Jr. (ed.), *Briefwisseling Heinsius*, vol. 7, p. 519 #1130, Goslinga to Heinsius, Lille, 13 September.

[63] de May, *An exact journal of the siege of Lille*, The Dedication, p. 2.

[64] Van 't Hoff (ed.), *Marlborough-Heinsius Correspondence*, p. 413 #693, Marlborough to Heinsius, Beerlegem, 5 December.

the engineers, you will lose them completely if you do not compensate them better with both promotions and money; it is a terrible job, there are already a good number dead or wounded; consider this as soon as possible, I beg you."[65] The highest ranking among them had already been put out of service: Des Rocques broke his leg before the trenches at Tournai were even opened, while at the end of August Du Mée suffered a fatal head wound during the attack on the citadel. Marlborough was pressed to write to Heinsius for "a generale of the artillery" to replace the two engineers.[66] The *Raad van State*'s field deputy Philip Frederick Vegelin van Claerbergen echoed Goslinga's concerns: "There are no other directors here, and as for the engineers, of which there are only a few, their insubordination is increasing daily, and I fear that this cannot continue much further if we do not encourage them with promotions or some other improvement in their service."[67] The English chronicler Millner estimated this siege of both town and citadel cost their engineers eight dead and another nine wounded.

The 1710 campaign exerted even larger strains on an engineering corps that somehow managed to (barely) replace its losses. Winter recruitment was disappointing, so they began the campaign with only twenty-nine engineers for the field army and another twelve responsible for various fortified places.[68] These technicians congregated around Douai for an early siege, and were reinforced by a like number of extraordinary engineers. By the end of the frustrating siege Vegelin van Claerbergen repeated the by-now-commonplace complaint, warning that they had twenty to thirty engineers wounded or killed here, and no more volunteers were stepping forward to replace them.[69]

---

[65] Veenendaal, Jr. (ed.), *Briefwisseling Heinsius*, vol. 9, p. 94 #190, Goslinga to Heinsius, Tournai, 25 July 1709. He reiterated his request a week later, counting ten already out of service, vol. 9, p. 115 #236, Goslinga to Heinsius, Tournai, 1 August.

[66] Van 't Hoff (ed.), *Marlborough-Heinsius Correspondence*, p. 461 #784, Marlborough to Heinsius, Tournai, 31 August. Recall the English system, where the engineers and gunners were united under a single command.

[67] Veenendaal, Jr. (ed.), *Briefwisseling Heinsius*, vol. 9, p. 238 #497, Vegelin van Claerbergen to Heinsius, Tournai, 7 September. Marlborough also warned of the dwindling number of engineers at the siege. Murray (ed.), *Letters and Dispatches*, vol. 4, p. 528, Marlborough to head of the *Raad van State* Johan van den Bergh, Tournai, 4 July.

[68] ARA RvS 172, resolutie, f. 787b, 25 March 1710.

[69] Veenendaal, Jr. (ed.), *Briefwisseling Heinsius*, vol. 10, p. 478 #956, Vegelin van Claerbergen to Heinsius, Brebières, 26 June.

Marlborough attributed their unexpected delays before the fortress to both the "obstinacy of the garrison" and "the want of engineers."[70] By the end of the siege, the Dutch counted twelve dead and another thirty-three wounded, with many of the survivors owing their lives to their dented cuirasses.[71] The shortage of subordinate technicians continued at the next siege, Marlborough admitting to one correspondent: "I am almost ashamed to tell you that we are still before Bethune, but we lie under a great misfortune of being obliged to carry on the war here by sieges almost without engineers."[72] Encamped before Aire a month later, Marlborough asked the Dutch Deputies at Brussels to gather together as many Spanish technicians as they could find. Such reinforcements were inadequate, however, as Des Rocques still complained that after only one week of open trenches he had barely enough engineers (six at each attack) to rotate their duty.[73] The number of engineers available for service steadily declined as the Allies smashed headlong into the *pré carré*.

The slower pace of the final two years of conflict provided the Allies little relief. In the 1711 campaign, the capture of Bouchain did not approach the bloodletting of the previous year's marathon of sieges, yet it still cost the Allies nine dead engineers and thirteen injured.[74] By the last year of Dutch participation, their engineering

---

[70] Murray (ed.), *Letters and Dispatches*, vol. 5, p. 44, Marlborough to the English ambassador at Berlin Thomas Wentworth, Baron of Raby, Douai, 7 June.

[71] By the end of the siege, the casualties reached forty-six engineers (30% of these were killed). C. Hipssich, *Spanischer Successions-krieg. Feldzüge 1710*, (Vienna, 1887), p. 607 #16 Verluste der Alliirten wahrend der Belagerung von Douay, vom 4 Mai bis zum 25 Juni, C. See also Wijn, *Het Staatsche Leger* vol. 8, part 2, p. 805. On the many damaged breastplates needing replacement, see ARA Raad van State (RvS), 688–1, Vegelin van Claerbergen to the *Raad van State*, Douai, 9 July.

[72] Murray (ed.), *Letters and Dispatches*, vol. 5, p. 104, Marlborough to General James Stanhope, Villers Brulin, 18 August. The dispositions for the siege called for 10–12 engineers at each approach. ARA RvS 1897, #8 Mémoire pour le siège en question fait le 13 juillet.

[73] Murray (ed.), *Letters and Dispatches*, vol. 5, p. 126, Saint André, 4 September; Veenendaal, Jr. (ed.), *Briefwisseling Heinsius*, vol. 11, p. 159 #291, Des Rocques to Heinsius, Aire, 18 September.

[74] This total comes from Wijn, *Het Staatsche Leger*, vol. 8 part 3, p. 309. The English participant John A. Millner claimed nineteen engineer casualties in *A Compendious Journal of all the Marches, Famous Battles, Sieges And other most noteworthy, heroical and ever memorable Actions . . .*, (London, 1733), p. 313. For specific examples, consult *The Post Boy*, 4 September 1711 O.S. #2546, Journal of the Siege of Bouchain continued; and a further continuation of this journal in *The Post Boy*, 11 September O.S. #2549.

corps had been decimated by a dozen major Flanders sieges. Those available for the 1712 campaign included the three directors (Christiaan Frederik Hertel, Zas van den Bosch and Bruijn), as well as thirty-six ordinary engineers, all headed by Des Rocques. Another thirty-five extraordinary engineers would join them for the season's first siege of Le Quesnoy, where the Dutch had to provide trench directors for not only their own approaches, but for the Imperial attack as well.[75] After this single, short siege, almost one third of those available at the beginning of the campaign had been killed, wounded or deserted.[76] As quickly as they could fill their ranks, the rigors of sieges depleted them.

In short, a decade of sieges resulted in constant attrition, while such dangers, coupled with poor compensation, discouraged volunteers from filling the depleted ranks of an already small force. Over the course of the Spanish Succession, sieges had cost the Dutch engineering corps—the one Allied belligerent able to support a large and experienced engineering establishment—several dozen dead and probably two or three times as many wounded, many suffering injuries on multiple occasions. As both the French and Allies knew, the first requirement of Vauban's 'scientific' tactics, technical experts to conduct them, was a precious commodity. It slowly bled away just when they were being put to good use.

2. *Command and Control Friction*

Even if a besieging force could muster an adequate number of competently-trained engineers, their knowledge and expertise would count for little if their orders were either ignored or badly applied. The battle-centric modern historiography gives little reason to expect such resistance to the inexorable march of the siege machine, while the rhetoric of siege history (in part a by-product of misreading the engineers' prescriptive manuals) also portrays the siege as unwinding according to the dictates of the chief engineer. One of Villars' biographers furnishes us with the conventional wisdom: "A war of siege,

[75] Wijn, *Het Staatsche Leger* vol. 8 part 3, pp. 159, 161.
[76] Wijn, *Het Staatsche Leger* vol. 8 part 3, pp. 416–418.

however adventuresome and honorable, left little initiative for the individual commander."[77] The Duke's chroniclers concur:

> Marlborough could do little to speed up siege operations. The British army still did not have the organization and personnel with the experience and technical expertise to carry out a major siege on its own, but had to rely on Dutch engineers. They were nothing if not methodical, meticulously following conventional methods.[78]

Despite this impression, several sources of friction steadily eroded the engineers' ability to manage the siege according to Vauban's strict standards of efficiency. Organizational factors in particular worked against those trying to implement an efficient attack—Vauban's famous appeal to "Burn more powder and spill less blood" was in fact a reactive attempt to reaffirm the importance of efficiency aimed at those who grew impatient with the length of his siege of Charleroi in 1693.[79] Engineers were poorly integrated within the command structure of the regular army, thus they had to struggle to compel recalcitrant non-engineers to follow their advice. First and foremost, general officers' varying opinions of the engineers and their profession necessarily influenced the willingness to accept their recommendations. Conflicts between the engineers themselves could further complicate the situation by presenting siege commanders with a cacophony of voices on the best tactics to pursue. And although the artillery was integral to the siege attack, its officers also sought to maintain their independence from the engineering corps. As a result, the engineers' 'expert' opinions were sometimes contradictory and always only one consultative voice among many competing for the attention of the commander. With a plan of attack finally agreed upon, engineers then had to rely on non-engineers to translate their sketches onto the terrain. When officers cooperated in this effort, the workmen digging the trenches might have their own reservations, adding yet another hindrance to the smooth implementation of an engineer's designs. Together, these many sources of resistance slowed down the attack and help explain Vauban's constant quest for ever-greater efficiency and ever-closer management of the attack—engineers

---

[77] Claude Sturgill, *Marshal Villars and the War of the Spanish Succession*, (Lexington, 1965), p. 23.

[78] Jones, *Marlborough*, p. 189.

[79] Rochas d'Aiglun (ed.), *Vauban*, vol. 2, p. 396.

had to compensate for inefficiencies introduced by those beyond their control.

The first source of friction was the difficulty engineers of all ranks had in getting their superiors to follow their advice, for technical experts served in a strictly advisory role even in sieges. On the one hand, assuring that an engineer's precise directions were followed was often difficult because engineering posts were not the same as a regular army commission.[80] Thus the ranks of director-general, director of approaches, first, second, and third engineers, down to the extraordinary engineers were all outside of the normal military chain of command, infantry or cavalry. As a result, some held an engineer brevet at the same time as they held a commission in the foot—Des Rocques, for example, received a titular lieutenant-colonelcy in 1707 and became a brigadier of infantry in 1710. The lower ranking pluralists among them might at times be expected to perform the duties required of both branches, even if these were in different theaters![81] La Vergne bemoaned his corps' impotence in Austrian service and theorized that it might be ended if the chief engineer was given his own regiment to be officered by his subordinate engineers, "and by this means he will be able to maintain his rank and also his authority; this will make it more easy for him to hold his ground against another Colonel on campaign, since he too commands his own regiment."[82] In addition to lacking a clear position in the command hierarchy, there was also a long-standing gulf between the engineers and the general officers, those truly in charge. Serving as a noble volunteer in France during the English Revolution, James Stuart Duke of York provides insight into the relationship between generals and their engineers in their mid-century sieges, when Vauban first began his career. In his discussion of the chief engineer's conduct

[80] Lecomte, *Les ingénieurs militaires en France pendant la règne de Louis XIV*, pp. 113–118; Chandler, *The Art of Warfare in the Age of Marlborough*, pp. 219–220; Blanchard, *Les ingénieurs du Roy*, p. 292. As Langins notes in his *Conserving the Enlightenment*, integration with the other services did not substantially improve their status (pp. 189–190).

[81] Blanchard, *Les ingénieurs du Roy*, pp. 106–108. In an English case, Marlborough had ordered one Colonel Bennett to sell his infantry company because of his long absence, but the Board of Ordnance wrote to remind him that Bennett had been serving as an engineer in Iberia since the beginning of the war, and that he should be excused for this reason. BL Add MSS 61166, f. 124, Board of Ordnance to Marlborough, 11 May 1710.

[82] La Vergne, *De l'utilité d'avoir un bon ingénieur*, p. 11.

at the siege of Mousson, he lamented the chevalier de Clerville's unsuccessful attempt to undermine one of the town's towers—only its collapse through cannonfire allowed France's most powerful engineer to "escape a cudgelling." As if the threat of physical violence (a complaint Allied engineers would raise again in the War of the Spanish Succession)[83] wasn't emasculating enough, the future King of England's memoirs went on to praise "the extraordinary care and pains which the Generall Officers usually take on such occasions, to which I cheifly attribute their speedy taking in of Townes." His explanation highlights the extent to which French engineers of mid-century were utterly disregarded by their commanders, Turenne in particular:

> They [i.e. the generals] trust to no body but themselves to view, and make their observations; Mr. de Turenne went in person to view all the ground about Mousson [Mouzon], taking with him Mr. de Castelnau, when, as in another Army, I have seen the Generalls trust a Sergent de bataille or some inferior Officer to do it, so that they were wholy guided, and in a manner govern'd by the eyes and advice of other men: but Monsr. de Turenne made use of his own judgment, where he thought it most proper to break ground, and which way to run the Trenches; when night came, he himself was present at the opening of them, and continued there allmost till break of day: Besides it was his constant method, during this whole Seige to go into the Trenches both morning and evening, in the morning to see if the work was well perform'd, at evening to resolve what would be the work that night, having in his company the Lieut Genl: and some of the cheif Officers who that night were to command in the Trenches, to instruct them himself what he expected to be done. Again after supper he went to see them begin their work, and would continue with them more or less, as he found it necessary for the carrying on of the present design. . . . the Commander in cheif, is not only thus diligent, but all the inferiour Officers are obliged to be as carefull in their severall stations: particularly in all the time of this present Seige, in our side of the attack wee had not so much as one single Ingeneer, nor did I ever observe them to be made use of at any other place, but only as overseers of the work, most of the Officers understanding very well how to carry on a Trench, and to make a lodgement. And not only from

[83] The engineers at Philips Karel von Lottum's attack on Tournai threatened to walk off the job unless an end was put to the *brutalitez* (of an unspecified nature) that they suffered at the hands of the officers. Veenendaal, Jr. (ed.), *Briefwisseling Heinsius*, vol. 9, p. 103, #209, Vegelin van Claerbergen to Heinsius, Tournai, 28 July 1709.

> my own observations, but by what I have learn'd from others who haue had more experience and seen more seruice then myself, I find and am settled in my opinion, That no Generall ought wholly to confide in any Ingeneer for the carrying on of a Trench, it being not reasonable to beleeve, that one who is to be allways there, will hazard or expose himself as far as Officers, who are to take their turns, and who are push'd on by emulation of each other to make dispatch, and carry on the seruice with all diligence.[84]

Here we find the exact opposite of Vauban: the general, not the engineer, should be managing the works; the officers are the true experts of siegecraft, the engineers superfluous at best; the engineers' constant exposure to danger is a rationale for divesting them of authority rather than a reason for investing them with greater influence; while the competition among the officers for glory is to be praised rather than villified. Vauban would spend the rest of his life trying to refute claims such as these. Yet long after Vauban had replaced the less-skilled Clerville and molded the French *Génie* in his own image, this distinction between the branches was still observed at the very highest level of military service. As Vauban himself admitted at the turn of the century, his marshal's baton was different from the others since his expertise and experience involved the design of fortifications and the conduct of sieges, rather than the maneuvering of armies in the field or on the battle plain.[85] Unused to taking orders from engineers in the field, many officers were hesitant to blindly accept their recommendations, even in sieges.

As the future King James II's quote suggests, the overall responsibility for a town's capture was entrusted to the highest-ranking general, usually of the infantry.[86] Ideally for Vauban, the commander would solicit advice from the engineers on the attacks and follow

[84] James occupied himself as a noble volonteer in French military service before he and his elder brother returned to England in 1660. James Stuart, *The Memoires of James II: His Campaigns as Duke of York, 1652–1660*, (London, 1962), pp. 151–153. James went on to report that many of his fellow officers noted that when engineers had control of the conduct of the trenches, as in the Dutch army of the time, the officers quickly lost their facility in making trenches.

[85] Hyrvoix de Landosle, *Vauban: Lettres intimes (inédites) addressées au marquis de Puyzieulx (1699–1705)*, (Paris, 1924), p. 83 to Puyzieulx from Saint-Malo, 2 November 1699. See also Louis's *mémoire* of 29 July 1703 in Vault and Pelet (eds.), *Mémoires militaires*, vol. 3, pp. 892–893.

[86] Lecomte, "Du service des ingénieurs militaires en France pendant le règne de Louis XIV," 27ff.

their recommendations, but the decisions were ultimately his to make.[87] Vauban's early treatise argued against the prevailing view, and sought to make the engineer's consultative role clear: "It is up to the Director to regulate the plan of attack, it is up to him to show the General the layout of the siege camp, and to inform the General of his sentiments and the reasons for his proposals. The General will then order what he thinks appropriate."[88] Several pages earlier he had highlighted the necessity of reconciling the engineer's plan with the means at hand, writing that the camp's layout "must be made in concert with the lieutenant-general, because it is he who knows most intimately the order of battle, the ranks of brigades, the regiments and the size of the entire army."[89] Vauban gained significant headway in this matter over the course of his career, though it was always tenuous without command authority. Reliant on royal support, Vauban could at times be undermined by his own supporters. Louis and his ministers may have generally placed their faith in Vauban's technical judgments, but this did not guarantee the engineer unquestioned authority, even regarding decisions requiring technical expertise.[90] The Secretary of War Michel Chamillart delineated the chain of command quite clearly in a 1703 letter: "in order to convert Vauban [to the need to attack Freiburg], the duke of Burgundy must speak to him as the master; Vauban has the King's orders, he knows the King's wishes and the necessity [of the siege]."[91] The most extreme case of indifference comes from Louis d'Aubusson, duc de La Feuillade, son-in-law to Chamillart and a commander in Italy, who haughtily rejected Vauban's recommendations for a planned attack on Turin: "Trust in me and both you and the King will be better served than by all the engineers in the world. Some men are

---

[87] Earlier in the century the non-engineer Renée Le Normant warned generals to not blindly follow the engineers. *Discours pour le retablissement de la milice de France*, p. 226.

[88] Vauban, *Mémoire pour servir*, p. 53; also p. 49. La Vergne decried Austrian generals who "imagine that they know more than [their engineers] . . . which often makes them commit great mistakes, killing many men unnecessarily." *De l'utilité d'avoir un bon ingénieur*, pp. 7–8.

[89] Vauban, *Mémoire pour servir*, p. 47.

[90] On the relationship between Vauban and Louis, see Lecomte, "Du service des ingénieurs militaires en France pendant le règne de Louis XIV," p. 195; and Wolf, "Louis XIV as Soldier-King," 199.

[91] Vault and Pelet (eds.), *Mémoires militaires*, vol. 3, pp. 441–442, Chamillart from Versailles, 30 August. In the end, the fortress was not besieged.

born to command, and these others [i.e. the engineers] are made only to execute the orders that they are given."[92] Expertise was not the same as authority: from the very beginning of a siege the engineers depended on general officers for military authority.

One of the most important decisions in a siege was determining where to attack a town, and here the engineer's consultative role is quite evident. Fortification manuals might create for pedagogical purposes an imaginary regular fortress with each side sporting the same defenses in a flat plain, but few fortresses ever approached this ideal, forcing besiegers to weigh the pros and cons of each front. An engineer's detailed understanding of the strengths and weaknesses of fortifications was called for, but early in Louis' reign Vauban acknowledged that, ultimately, it was the commanding general who decided where to open the trenches.[93] Nonetheless, by the turn of the century commanders more often than not followed their experts' advice on the approaches to make. The standard consultative procedure is illustrated at the 1706 attack on Menin, where the directors reported their recommendations to a council of the commanding general, lieutenant-generals and field deputies; together this body concurred on the approaches to be made.[94] Commanders and generals often granted such acceptance only begrudgingly, however, for they bitterly complained of their engineers' choices at the same time as they allowed them to determine the attacks. Villars had to quell rumors that they were attacking Le Quesnoy at its strongest side rather than its weakest, and reaffirmed his trust in the engineers on their choice of approach.[95] Later when he was forced to pick between two projected attacks, he settled on his least-favorite choice (proposed by the chief engineer) in order to avoid potential obstructionism: "if I were of a different opinion from [the engineers], I would still go along with their ideas anyway because it is too dangerous to make them do something against their wishes. . . . [Brigadier of infantry Erasme] Con-

[92] Quoted in Augoyat, *Aperçu historique*, vol. 1, p. 303.

[93] Vauban, *Mémoire pour servir*, p. 68.

[94] A.J. Veenendaal, Jr. (ed.), "De Mémoires van Sicco van Goslinga. De ontbrekende gedeelten van 1706 en 1709," *Mededelingen van de Sectie Militaire Geschiedenis Landmachtstaf*, 11 (1988), p. 27.

[95] AG A[1] 2384, #106, to Voysin, 25 September 1712. The language used in his memoirs highlighted the central role he played in deciding on the attacks with the engineers. Vogüé (ed.), *Mémoires du Maréchal de Villars publié d'après le manuscrit original*, vol. 3, pp. 192 and 217.

tades says that the engineers are like wet nurses [*nourrices*] who cannot be denied anything they want."[96] In 1713 at Freiburg near the Rhine, competing projects surfaced once again. Villars personally agreed with one sieur de la Battue, who had previously commanded in the château, but he allowed the chief engineer Charles-Guy Valory to conduct his attack on a different sector nonetheless: "Thus it is that one is often forced to give in to the reasoning of those who are directly charged with the attack, because if you force them to do otherwise [*quand on leur fait violence*], they are more than happy to make sure that nothing succeeds."[97] The willingness to follow the experts was fickle, often based on mistrust that boded ill for future cooperation. In a few extreme cases, commanders might be moved to add their own attacks rather than refuse the engineer: Eugene's restlessness at the progress of the two approaches against Tournai, for example, led him to start a third one on his own initiative.[98]

If commanders and generals were rarely willing to overrule the choice of approaches, they did restrict the engineers' autonomy in later phases of the siege. Vauban's advice for the siege commander was undoubtedly a direct response to the model Turenne had used at sieges such as Mousson:

> it is very important that the general-in-chief visits the trenches, but only occasionally and not every day; because his visits necessarily being long, they will cause too many distractions and delay the siege. . . . He only needs to visit from time to time with only a small retinue, to be personally informed of what is being done.[99]

Vauban pleaded for a hands-off commander who supported the chief engineer's plan of attack, yet commanders long after Turenne felt confident enough to meddle in their affairs. Ever one to take charge and reversing his earlier fears of obstructionism, Villars imposed himself on the engineers in 1712. Ignoring Vauban's advice, he described

[96] AG A[1] 2384, #1, Villars to Voysin, 16 September 1712. See also AG A[1] 2382, #101, Jean-Robert LeFebvre d'Orval to Voysin, 22 August.

[97] Vogüé (ed.), *Mémoires du Maréchal de Villars publié d'après le manuscrit original*, vol. 3, p. 217. Given the apparent acceptability of cudgeling engineers, we should perhaps take Villars' use of the phrase "quand on leur fait violence" more literally.

[98] Veenendaal, Jr. (ed.), *Briefwisseling Heinsius*, vol. 9, p. 115 #236, Goslinga to Heinsius, Tournai, 1 August 1709. Landsberg claims the same at Béthune 1710. *Nouvelle manière de fortifier les places*, p. 48 and this is confirmed by AG Article 15 Section 2, #4bis, Journal du siège de Béthune.

[99] Vauban, *Traité*, p. 234.

his hands-on approach at Douai: "I'm going to lodge myself as close to the head of the trench as I can for the best view, because if there is a task where the commander's eye is necessary, it is there."[100] Generals leery of their technicians sought to grasp the reins of command more firmly whenever possible.

The commanding general was not the only challenge to an engineer's undisputed management of the siege—lieutenant-generals offered even greater sources of friction in the trenches.[101] Once the chief engineer and his associates had drawn up the plan of attack and had it approved by the commander, its implementation would be entrusted to each attack's commander (usually the next-most senior general), each of whom would have a director of the approaches as his engineer assistant. In Vauban's later work he enumerated how the lieutenant-generals and their subordinates were to post the troops, regulate the detachments, supervise their service in the trenches, and furnish the necessary number of workmen.[102] Daily siege dispositions drawn up by the engineers illustrate this dependence, as they provided the infantry generals with lists of the number and types of soldiers to be provided when and where for what service.[103] The details of each day's work in the trenches would be overseen by the general of the day (usually a lieutenant-general), seconded by their majors of the trench (*majors de la tranchée*). The workmen and guards in the trenches rotated daily and were drawn from the siege army regiments, commanded by their regimental officers, with the various engineers marking out the locations of the trenches and batteries and supervising the construction of the ramparts. Vauban's emphasis on the need for generals to follow the engineers' plans, however, shows his recognition that this was not often the case:

> the lieutenant-general of the day commands the cavalry, infantry and artillery, engineers and miners, and generally everything related to the security and progress of the attacks; but he must consult with the director of the trenches, and neither undertake nor decide to do anything

[100] AG A[1] 2382, #124, Villars to Voysin, 24 August.

[101] On the French tradition of lieutenant-generals' independence, see Lynn, *Giant*, p. 296.

[102] Vauban, *Traité*, p. 233. For an English interpretation, see Bland, *A Treatise of Military Discipline*, p. 260. For the duty roster at Tournai, see ARA CvdH 143, Mémoire au sujet de l'attaque de Tournay fait le 3e de juillet 1709.

[103] For examples, see Coehoorn's dispositions for Liège 1702 in BL Add MSS 61187, ff. 41–50b; for Bonn 1703 see ff. 56–80.

> without the engineer's participation; because this last is the soul and the prime mover [*la véritable mobile*] of the attacks.[104]

So much for the theory.

Unfortunately for Vauban's Method, the generals did not consider themselves nearly as dependent on the engineers as the engineers were on the generals. Vauban had first recognized the dangers of the noble officers' independent mindset during his service in the Fronde. Whereas James II saw the competition between officers in the trenches a benefit, Vauban focused on its inefficiencies:

> The competition that exists between these generals often makes them expose their troops to little purpose and demand that their men do more than is possible, so that the officers concern themselves not at all with the loss of thirty men, as long as they advance four steps further than their comrades. By their own authority they order the trenches in whatever direction they please, breaking the design and all the measures that the engineer had taken by their failing to observe the proper conduct [*conduit réglée*] that would lead to a good end. The engineers are forced to serve as an instrument of the generals' different caprices. I say different because one general commands one day in one fashion and the general that relieves him the next day orders differently. And as they are rarely imbued with the greatest capacity for this kind of work, God knows the shortages and outrageous expenses they cause, as well as how much needless blood they spill by prolonging the length of a siege.[105]

Such competition required Vauban to give constant reminders, thus we find his dispositions for Ath 1697 reinforcing yet again that the brigadiers in the trenches were to coordinate every day with their peers in the other attack, so that they would not deviate from the carefully-designed plan. Despite such admonitions, the chevalier de Quincy, himself with a noble pedigree, illustrated how difficult it was to eliminate all such examples by the end of Louis' fifty-year reign:

> It will not suffice to leave to the desires and caprice of the general officers of the day in the trenches to attack the covered way or other works that we besiege; because we can say, to the credit [*à la louange*] of the majority of these Messieurs (we have already remarked on it far too often), that they are extremely ignorant on the day they find themselves in the trenches.... To remedy this abuse, it will be necessary that the chief engineer be called by the generals of the trenches

[104] Vauban, *Traité*, p. 232.
[105] Vauban, *Mémoire pour servir*, pp. 16–17.

> to determine if the work that is to be attacked is indeed ready to be assaulted. Certainly, this precaution would save many men and we would no longer make such stupid mistakes [*et on ne feroit point de cacade*].[106]

If the ranking generals did not wish to provide the engineers with the requested number of troops, there was little that the engineers (and even the commanders of the approaches) could do other than complain.[107] In addition to informing the generals of the number of men necessary for the trenches, the engineers also drew up lists of all the supplies that would be needed (fascines, gabions, tools, wooden planks . . .). But it was up to the army administrators—whether French provincial and army intendants or Dutch field deputies—and their civilian subcontractors to transport the supplies on site, where the lieutenant-generals and their subordinates were charged with bringing them into the trenches. Besieging the Cologne city of Bonn in 1703, Coehoorn had to draft a *Memoire preparatif pour l'ouverture de la trenchée* to remind the generals to provide the tools, fascines, wagons and munitions that he would need for the trenchworks—his authority did not allow him to requisition the supplies directly, nor did he have the engineers to transport these materials to the trenches himself.[108] In every aspect of implementation, engineers had to rely on men who cared little for Vauban's efficiency.

The engineers' corporate authority was weakened even further when they engaged in internecine arguments over seniority and even tactics. Coehoorn had difficulties with his superior officers from William's death onward, and when Coehoorn himself died in early 1704 the Allied engineers were left without an obvious successor. A replacement would not be named until five years later. Vauban's

[106] Lecestre (ed.), *Mémoires du chevalier de Quincy*, vol. 3, p. 256.

[107] For several examples: at Menin 1706, BL Add MSS 61310, f. 3, Blood to Marlborough, 2 August; at Ghent in December 1708, BL Add MSS 61312, f. 226, Bengt Oxenstierna (a Swedish lieutenant-general in Dutch service) to Marlborough; at Béthune, BL Add MSS 61245, f. 120, Schulenburg to Marlborough, 14 August 1710.

[108] BL Add MSS 61187, ff. 54–54b, May 1703. Unlike the orders sent by commanders, Coehoorn's dispositions were often phrased as a request for the generals to take care to provide the necessities, for example: "Les generaux commandants les attaques sont prié de faire toute leur possible de fournir les travailleurs necessaire . . ." BL Add MSS 61187, f. 65, Disposition pour le 6 de Mey. For an example of French engineering complaints about the shortage of trench materiel, see AG A[1] 2381, #114, LeFebvre d'Orval to Voysin, Cambrai, 7 August 1712.

long tenure likely allowed him to shape the French corps in his own image once the Marine and Army corps were united in 1691, although this conventional wisdom deserves further analysis. The Allies' engineers, composed of many nationalities, came from varying traditions and did not always mesh well together. The example of the Austrian service is perhaps an extreme example, but instructive nonetheless. La Vergne lamented that as a result of widespread insubordination "each [engineer] thinks himself in command . . . no matter who the commander or what the orders, one will reverse what the other has just finished, so that instead of advancing we go backwards, to the extent that if we need a trench they'll make a redoubt, if we need a redoubt they'll make a trench." The frustrated technician also spent several pages contrasting his motley crew with the "well-paid, well-respected and well-honored" French corps, led by the properly-vested Vauban, a leader who held his subordinates "in fear, respect, good order and friendship, and rewarded them according to their merit."[109] We will soon enough discredit La Vergne's overly optimistic view of what French generals thought of their own technicians, but this underlying tension between engineers was real enough and is most visible in Dutch service during the War of the Spanish Succession. Upon Coehoorn's death in early 1704, candidates began jostling to win the posts he left behind. With no single individual firmly in charge of the engineers, it was not until June 1709 that the States-General appointed director of approaches Des Rocques as director-general of Dutch fortifications.[110] Disputes over status festered in the interim. At the siege of Menin in 1706, Marlborough informed Heinsius: "The reason of our siege going on so slowly proceeds from all our Inginiers being on a leavel, so that no one has the directions,

[109] On the resulting insubordination, see La Vergne, *De l'utilité*, Avant-propos, pp. i [unpaginated], 4–5 (quote on moving backwards), 69–70. The contrast with the French corps is on pp. 8–11. Michael Hochedlinger's brief discussion of Austria's abysmal technical corps confirms the tenor of La Vergne's criticisms. *Austria's Wars of Emergence: War, State and Society in the Habsburg Monarchy 1683–1797*, (New York, 2003), pp. 123–125.

[110] Oddly, the top engineering position in English service (First Engineer of England) went similarly vacant from the death of Sir Martin Beckman in 1702 until the promotion of Michael Richards to the post in 1711. It is surprising that during a period as dominated by positional warfare as this, the institutions of those two branches most crucial to it were so ignored in both English and Dutch service—is this an indication of the hostility aimed at the technical branches, or just neglect?

which occasions many unreasonable disputes."[111] These disagreements notwithstanding, the town capitulated the next day, one of the strongest fortresses in Spanish Flanders lasting only eighteen days of open trenches. At the siege of Lille in 1708, the Allied dispositions were designed to minimize these continuing quarrels: the junior Du Mée was charged with conducting the attack on the town, while Des Rocques was slated to lead the subsequent effort against the citadel.

This conflict between Des Rocques and Du Mée came to a head during the siege of Tournai in late July the next year. Des Rocques had still not received his new rank, which only fueled the continuing feud with the younger director. The resulting disputes precipitated a field deputy's report stating that Du Mée flatly refused to serve under the newly-appointed director-general, complaining to anyone who would listen that his older rival was being far too cautious in his attacks. With the experts squabbling among themselves and the siege dragging on, it was with evident relief that Vegelin van Claerbergen could report at the end of his 28 July missive that the town had finally beaten the chamade.[112] Goslinga's suggestions to craft a compromise that would appease both Des Rocques and his junior became moot when the younger engineer suffered a fatal head wound during the attack on the citadel. Despite the elimination of this particular conflict, complaints continued to filter in from the Dutch administrators the next year. At Béthune, the *Raad van State*'s deputy complained of the animosity that now surfaced between Des Rocques and another director of the approaches, Hertel.[113] Infighting and accusations of incompetence weakened the corps from within and gave generals yet another reason to hesitate before following their advice.

The artillery proved just as difficult to manage as the generals.[114] With a separate chain-of-command (except in England), and headed

[111] Van 't Hoff (ed.), *Marlborough-Heinsius Correspondence*, p. 259 #429 Marlborough to Heinsius, Helchin, 21 August 1706.

[112] On these disputes at Tournai, see Veenendaal, Jr. (ed.), *Briefwisseling Heinsius*, vol. 9, p. 103 #209, Vegelin van Claerbergen to Heinsius; and p. 115 #236, Goslinga to Heinsius, 1 August.

[113] Veenendaal, Jr. (ed.), *Briefwisseling Heinsius*, vol. 10, p. 559 #1117, Vegelin van Claerbergen to Heinsius, Bevri, 21 July 1710.

[114] On Louis XIV's artillery, see Susane, *Histoire de l'artillerie française* (Paris, 1874), chapter 4; Ernest Picard, *L'Artillerie française au XVIII*[e] siècle (Paris, Berger-Levrault, 1906); Michel Decker, "Louvois, l'artillerie et les sièges," *Histoire, économie et société* 15(1) 1996, pp. 75–94; and most recently and comprehensively Frédéric Naulet,

in France by the *grand-maître d'artillerie* and in the Netherlands by the *meester-general der artillerie*, this autonomous arm guarded its privileges as closely as did the other branches.[115] Its mission also differed from the engineers and thus they held divergent interests. Whereas the engineers' wider interests encompassed military architecture, cartography and hydrology, the artillerists' purview included not only the siege trains and garrison pieces, but also extended to the use of lighter cannon in field battles, their movement and safeguarding. They were further charged with the design, construction, provisioning and maintenance of artillery and all the infrastructure such responsibilities required—cannon foundries, arsenals, artillery schools and gunpowder mills. As with the military engineers, the artillery of France has received the most attention from European historians and its structure illustrates the many potential areas of dispute they shared with their sister service. In addition to those who commanded the corps, the cannoniers and the administrative personnel, the corps of artillery also included 'fire-workers' (i.e. pyrotechnicians), bombardiers to man the mortars and *pierriers*, fusiliers to guard the gun trains, various metal-workers and carpenters, and even miners, those expert in the underground use of gunpowder.[116] Unlike the military engineers, the French artillery established their own schools under Louvois' supervision as early as 1679, and these included an extensive curriculum that focused on casting, firing, transporting and repairing a wide variety of artillery pieces. In this formal setting gunners developed a professional identity that stressed their autonomy from their sister service. Pierre Surirey de Saint-Rémy's two-volume textbook for aspiring gunners was full of technical details on guns, their

L'artillerie française (1665–1765): Naissance d'une arme, (Paris, 2002). On the Dutch artillery, see F.H.W. Kuypers, *Geschiedenis der Nederlandsche artillerie van af de vroegste tijden tot op heden*, (Nijmegen, 1869). For the English, consult F. Duncan, *History of the Royal Regiment of Artillery*, (London, 1872); and David Chandler, "The Guns," in *Blenheim Preparations*, pp. 109–137.

115 Blanchard, *Les ingénieurs du Roy*, p. 137. The Marine department had their own gunners, which on occasion might see service at a terrestrial siege and further complicate the chain of command. For an overview of the office grand-master of artillery, see Pinard, *Chronologie historique militaire*, vol. 3, pp. 468–469; and more recently Lynn, *Giant*, pp. 99–100.

116 In the 1670s Louis created one company of miners for the Royal-Artillerie regiment (augmented to three by 1695) and one company of sappers in the 1680s. These units would only be transferred to the engineering corps in 1758. M. de Briquet, *Code militaire, ou, Compilation des ordonnances des rois de France concernant les gens de guerre*, (Paris, 1741), vol. 1, pp. 232ff.

carriages and artillery parks, but mentioned the critical topic of artillery-engineer coordination in sieges only long enough to discourage gunners from following engineers' orders to place their batteries in the trenches rather than construct separate batteries.[117] Even Vauban's good friend Vallière sought to stress the independence of his artillery in his post-war reflections. Though he highlighted the need for both chief engineer and artillery general to work in concert, he underscored that the gunners should rely on their own judgment rather than give in to the engineers' demands.[118] Not surprisingly then, relations between these two branches were problematic, as is illustrated by the abject failure of a 1755 French experiment to combine the two sister branches under a single administration.[119] For all these reasons, gunners were hesitant to surrender their initiative and follow the orders of engineers.

Vauban, despite his close relationship with Louis, would experience this conflict first hand. Before rising to prominence in His Majesty's armies, he had already discovered examples of the artillery's ignorance as they consistently failed to "site the batteries according to a well-designed plan, [instead] battering whatever first comes into view without paying attention as to whether the fire is direct or oblique. However, many of the shots intended to breach the walls are not direct and thus have little effect as they glance along the revetment."[120] At a number of subsequent sieges he lamented his continued inability to convince the gunners of the error of their ways; his lack of authority over them prevented him from fully systematizing the attack in spite of his friendship with several of their highest-ranking officers and with Louvois, who served as the *de facto* head

[117] For the duties of artillery officers during sieges, see Surirey de Saint-Remy, *Mémoires d'artillerie*..., (The Hague, 1741), vol. 2, pp. 255–269. In the Preface this *commissaire provinçiale de l'artillerie* described his *opus* (duly approved by some of the most important French gunners of the day) as an attempt to standardize the widely-varying views taught by numerous artillerists within the French service.

[118] Jean-Florent Vallière, *Mémoires d'artillerie à l'usage du sieur St. Vallier[e], officier dans le Regiment de Royal artillerie du battaillon de [Corigny?]*, (1717), f. 12. This manuscript from the collection of the Society of the Cincinnati Library (MSS L2000F374 Bound) focuses primarily on the use of artillery in the siege attack, unlike Saint-Rémy's much wider-ranging work.

[119] For the results of this experiment, see Briquet, *Code militaire*, vol. 1, pp. 120ff. For an example of disputes between the artillery and engineers in a later war, see Langins, *Conserving the Enlightenment*, p. 447 note 42.

[120] Vauban, *Mémoire pour servir*, p. 15. Note again the emphasis on planning.

of the corps until his death.[121] Besieging the hexagonal fortress of Charleroi in 1693 Vauban identified several weaknesses that eroded the siege's efficiency:

> I have moreover suffered greatly from the ignorance of the cannoniers and bombardiers, who, with the exception of a small number, I cannot be pleased with. The proper use of artillery and bombs requires an exact art with precise rules that not a single one of these people know. They do things as they have always done them, and have primarily their own interests in mind, which means that all the batteries are defective and imperfect. Furthermore, as none of them understand the principles of fortifications, they know neither where nor how to fire against the works. They are not familiar with their own guns, which does not surprise me, because they are only infantrymen that serve the cannon and who obey the artillery officers only when it pleases them; and since there is usually only one artillery officer for several pieces and because the officers cannot keep an eye on everything when they are aiming a piece, it is often the case that more than half or two-thirds of the rounds are squandered or miss the target completely.[122]

Once Louis-Auguste Bourbon, duc du Maine and a legitimized bastard son of the King acquired the post of *grand-maître d'artillerie* in 1694, he moved to further assert its independence from the engineers.[123] The journalist of Ath 1697, echoing this divide, described

---

[121] Vauban, for example, co-wrote a proposal for artillery reform (bemoaning the corps' low standing and its lack of training) with the artillerists François Frézeau marquis de La Frézelière, lieutenant-general of French artillery during much of Louis XIV's reign, and Armand de Mormes, sieur de Saint-Hilaire. Printed in Vauban, *Traité*, "De l'artillerie."

[122] Rochas d'Aiglun (ed.), *Vauban*, vol. 2, p. 399 Vauban to Le Peletier de Souzy, Charleroi, 13 October 1693. Saint-Rémy confirmed the reliance on conscripted soldiers to serve the cannon, much as engineers were forced to oversee regular soldiers as sappers (see below). Saint-Remy, *Mémoires d'artillerie*, vol. 2, p. 265.

[123] Lecomte, "Du service des ingénieurs militaires en France pendant le règne de Louis XIV," pp. 123, 127ff. He exercised his patronage, for example, by choosing one sieur de Magni over the universally-lauded Vallière to command the artillery at the 1712 siege of Douai. AG A[1] 2381, #114, LeFebvre d'Orval to Voysin, Cambrai, 7 August. Vallière would rise through the ranks and later spearhead an attempt to standardize France's artillery. The only biography of Maine, W.H. Lewis' *The Sunset of a Splendid Century: The Life and Times of Louis Auguste de Bourbon Duc du Maine, 1670–1736* (New York, 1954) provides little useful information on his tenure. See a brief *précis* of his career in Pinard, *Chronologie historique militaire*, vol. 3, pp. 500–503 and his perogatives as spelled out in the 1694 ordinance in Briquet, *Code militaire*, vol. 1, pp. 22–29. Rowlands, *The Dynastic State*, pp 344–345 also discusses his role in this post.

the gunners' traditional ineptitude and initial conservatism in very similar terms to what Vauban had described thirty years earlier:

> Every body knows that the common practice of the officers of the train is to fire at all before them, with all the fury that is possible, ruining without distinction all objects that present themselves to view, whether works, batteries, towers, or cavaliers; they even forget themselves so far as to fire upon any building which affords a fair mark, as gates, bridges, corps de gardes, sentry-boxes on the walls, sometimes at houses and steeples, for the mattrosses [gunners] only want to make havock appear, which was not Monsieur de Vauban's taste, who had so frequently seen the fruitlessness of it at so many sieges. . . . Bounce and clatter and readiness for action had hitherto composed the whole merit of the train at sieges.[124]

This success was exceptional, however, as the Ath journal identified: "Monsieur de Vauban had appointed commissaries, men of understanding, to each of the batteries, to whom he had given directions for the conducting and serving 'em, without any alteration to be made, or being reliev'd themselves, during the whole time of the siege; by this means every one was well informed as to the intent of his own battery, and with the range of his pieces" (pp. 105–106). As engineers were rarely afforded the opportunity to choose the artillerymen, little had changed by the end of Vauban's life, witness his admission that ricochet fire remained poorly known almost a decade after its introduction. The same old complaints of poorly-constructed battery breastworks and poorly-served guns can also be found in his final siege, (Alt-)Breisach 1703, where he was driven to write:

> if the artillery had done its job, we would very soon be inside the town; but it is infinitely difficult to direct them. They are all men who have hardly ever seen a siege and who know only how to fire straight ahead; they do not even know how to construct proper ramparts. I must exclude from such ignorance the person who commands them [La Frézelière], who is a man of *qualité* and very well cultured [*honnête*], who has more ability than his age suggests.[125]

[124] *A Journal of the Siege of Ath*, p. 103—note the vigorous tone of the last sentence. Goulon also recognized this shortcoming when advising garrisons to frequently move their pieces, for "the besiegers instantly cease firing on those parts which do not immediately incommode them, without troubling themselves about what may come from thence afterwards." Goulon, *Memoirs of Monsieur Goulon*, p. 10.

[125] Rochas d'Aiglun (ed.), *Vauban*, vol. 2, p. 525 to Chamillart, Biesheim, 3 September. Vauban is referring here to the 31-year old Jean-Angélique Frézeau,

In his final treatise on the attack he concluded that "Nothing is more important in a siege than the proper use of cannons; but it is very rare to see them well served, much less used how they should be. One is quite rightly surprised at how much variation there is from one volley to the next as well as their small effect, but few people see the problems."[126]

At the end of his career Vauban was forced to acknowledge he had been unable to change the employment of the siege guns or the training of its crews. Nor were such complaints limited to the *commissaire-général*. Bombarding Huy's hill-top forts in 1705, the French commander noted that the gunners' profligate consumption of powder was having very little effect; he concluded that further improvements were needed in the artillery schools.[127] Evidence of continued friction between engineers and artillerists comes from the Secretary of War's reiteration to an artillery officer of the rather fine division of labor between the two branches: "All your capacity must be confined to promptly constructing the batteries and serving them with as much dexterity as diligence; but it is up to the engineers to indicate where they want them to be placed."[128] Forced into intimate dependence on each other, it was not often that commanders could write, as the *maréchal de France* James Fitz-James Stewart duke of Berwick did in 1707, of the *assez rare* coordination between the two arms. After Vauban's death, the French author Guignard noted in 1725, the lieutenant-generals of artillery quickly reclaimed their control over the construction and siting of siege batteries.[129] Pierre d'Artaignan, the comte de Montesquiou and newly-minted *maréchal de France* personally attested to the resurgence of bickering at Douai in 1712:

> This siege is going too slowly for my taste, there is too much division between the engineers and the artillery and since no one person is

marquis de La Frézelière, brigadier of artillery and son of the late François, also a brigadier, who died in 1702. On the amorphous connotations of *honnêté*, see Louise Godard de Donville's entry on "Honnête Homme" in Bluche (ed.), *Dictionnaire du grand siècle*, pp. 728–730; and more broadly, Jolanta Pekacz, *The Conservative Tradition in Pre-Revolutionary France: Parisian Salon Women*, (New York, 1999), chapter 1.

[126] Vauban, *Traité de l'attaque*, p. 238.

[127] AG A[1] 1835, #269, Henri duc d'Harcourt to Chamillart, 13 June.

[128] Quoted in Lecomte, "Du service des ingénieurs militaires en France pendant le règne de Louis XIV," 131, Chamillart to d'Houville, 3 August 1705.

[129] Guignard, *L'École de mars*, vol. 2, p. 182. See also Lecomte, "Du service des ingénieurs militaires en France pendant le règne de Louis XIV," p. 26.

> charged with overall command of the siege, the service is often delayed. To negotiate such dialogues we need a man who follows each step and who has complete command, but as no one has overall authority each general officer in the trenches follows his own ideas.[130]

From an engineering perspective, the efficiency of Vauban's vaunted attack was eroded by the intransigence of the artillery; the gunners naturally contended the reverse. Without expertise in both areas, impartial observers were forced to agree with one contemporary's sensible opinion that "the lack of cooperation between these two corps" was as much to blame as incompetence in either service.[131]

The Allied artillery provided a frequent source of friction in their sieges, and this was further exacerbated by the mix of nationalities in the confederate force. At Kaisersweert in 1702, field deputy Geldermalsen complained of the gunners, "It is a pity to see how slowly the gunners are working, they are my principal complaint, but in general I do not know if it is these four years of peace or the large number of new recruits, or new regiments, or because they are without a commander who would be able to discipline and reward them."[132] At the next siege, General Obdam continued to bemoan the poor quality of their gunners.[133] Rapid turnover of personnel only exacerbated the confusion in the service. After Coehoorn's death in March 1704, Johan Wijnand van Goor briefly held the post of *meester-generaal der artillerie* before being killed later that year in front of the entrenchments of Donauwörth on the Danube. The position of head of the artillery remained unfilled for the rest of the war, with the predictable result of further disputes among the artillery officers. When the colonel of artillery Willem IJssel was wounded at Oostende in 1706, field deputy Goslinga complained of the resulting insubordination among the remaining gunners, while the siege commander Field Marshal Hendrik van Nassau, heer van Ouwerkerk

[130] AG A[1] 2382, #173, Montesquiou to Voysin, 27 August. An engineer's complaint of the marine gunners, who only "push their batteries forward instead of battering the works," can be found in Article 15 Section 2 §1 Douai folder, Précis du journal du siege de Douay fait en 1712 . . ., f. 23.

[131] Referring to the artillery-engineering dispute at the 1697 siege of Barcelona. Naulet, *L'artillerie française*, pp. 234–237, quote on 237.

[132] Veenendaal, Jr. (ed.), *Briefwisseling Heinsius*, vol. 1, p. 111 #169 Geldermalsen to Heinsius, Kaisersweert, 21 April.

[133] Veenendaal, Jr. (ed.), *Briefwisseling Heinsius*, vol. 1, p. 420 #840, Wassenaar-Obdam to Heinsius, Venlo, 8 September 1702.

lamented the resulting delays to battery construction.[134] During the bombardment of the town of Dendermonde, located at the confluence of the Dender and Scheldt rivers, the English Lieutenant-General Thomas Meredith expressed exasperation with the Spanish bombardiers and some satisfaction with the Dutch, though these last abruptly decided to "work [no] longer without being refreshed" with bread, beer and brandy—he was forced to search for others with experience firing batteries.[135] Later, at the more demanding siege of Tournai, we find the leaderless artillery officers clashing with the generals in charge of the approaches. The resulting squabbles began to create serious problems for the attackers as described by field deputy Goslinga:

> The good of the service demands that we appoint a general of artillery; subordination is not very great in this corps, but it is above all necessary in order to prevent further squandering of our munitions; we have preached this need over and over to the [infantry] generals, and they follow it as long as we are there, but once we leave, things return to how they were before. Each general, low-ranking or high, competent or ignorant, acts as if he was a general of artillery. If the colonel [of artillery] or his subalterns don't obey them, the generals quarrel and accuse them of sparing ammunition at the expense of the lives of soldiers; they even say such things in the troops' presence, which can only have a very bad effect on their morale. These reasons will convince you, as they have me, that we must fill the vacancy.[136]

A few days later Vegelin van Claerbergen confirmed the ill-effects of inter-service rivalries: "Our artillery are firing furiously here, and perhaps to little effect. The colonel [of artillery] blames the generals, and I do not know if either of them have gone too quickly, faster than the condition of the attack allows."[137] Little had changed by the siege of Mons several months later, where the engineers in their dispositions for the siege had to remind the infantry officers

---

[134] Veenendaal, Jr. (ed.), *Briefwisseling Heinsius*, vol. 5, p. 357 #674, Goslinga to Heinsius, Oostende, 1 July 1706; BL Add MSS 61179, f. 161, Ouwerkerk to Marlborough, Oostende, 29 June. See also BL Add MSS 61309, ff. 163–164, Oxenstierna to Marlborough, Oostende, 29 June.

[135] BL Add MSS 61163, f. 223, Meredith to Marlborough, 24 June 1706.

[136] Veenendaal, Jr. (ed.), *Briefwisseling Heinsius*, vol. 9, p. 94 #190, Goslinga to Heinsius, Tournai, 25 July.

[137] Veenendaal, Jr. (ed.), *Briefwisseling Heinsius*, vol. 9, p. 103 #209, Vegelin van Claerbergen to Heinsius, Tournai, 28 July 1709.

yet again to shell only those places that they had identified as targets.[138] The Allied artillery, reluctant to accept engineering leadership and lacking the protection of a general of their own, proved as unmanageable for the engineers as the generals.[139]

Engineers had one additional management challenge that complicated implementation of Vauban's efficient attack even further—the variable quality of the workmen. Getting the officers and gunners (and one's own engineers) to accept the chief engineers' authority was not enough to assure an efficient siege, for the rank and file did the digging for the engineers, just as they manhandled the gunners' artillery pieces into position. Properly-constructed ramparts were only one cause for concern. The most serious problem was flight. Soldiers on guard duty had their weapons to reassure them, whereas the workmen—motivated primarily by money and drink—were armed only with digging tools. With mere pick and shovel to counter matchlocks and bayonets, laboring soldiers often chaotically abandoned their works when sallying defenders approached, and sometimes even when they did not: "nothing is more common than for the workmen to take to their heels."[140] Focusing the troops' attention on constructing the trenches to the exclusion of thoughts about their own personal safety was a challenge for both officers and engineers alike. His parallels were of limited utility against sorties if the workers ignored them in their flight.

Vauban had recognized how this critical weakness impacted the efficiency of a siege. As a result he petitioned the King for independent sapper and miner brigades (rather than the individual companies incorporated in the Royal-Artillerie regiment), soldiers who would become experts in the dangerous task of advancing the trenches under enemy fire and undermining the garrison's works. Just as importantly, they would be trained to abandon the advanced trenches for the safety of the rear parallel in an orderly fashion.[141] The best

---

[138] BL Add MSS 61339, Memoire des qu'il convient d'avoire alataque de la port de Hanere, 9 October 1709, f. 166.

[139] Later engineers also noted the continuing conflicts between the engineers and gunners in the War of the Austrian Succession. Simon L. Lefebvre, *Oeuvres completes de Mr. Le Febvre, major au Corps des ingenieurs de Prusse, membre ordinaire de l'Academie Royale des Sciences & Belles-Lettres de Berlin*, (Maestricht, 1778), Préface, p. iv.

[140] Goulon, *Memoirs of Monsieur Goulon*, pp. 13, 16.

[141] The issue of unreliable workmen is discussed in Lecomte, "Du service des ingénieurs militaires en France pendant le règne de Louis XIV," pp. 124–126.

plans accounted for little without skilled workmen, and in 1672 he warned that his method for attacking a fortress

> demands intelligence both on the part of directors and workmen, I should add that we cannot hope to make the best use of it with those whom we usually employ in our sieges. They are naturally maladroit and surly; besides they are neither trained nor drilled to execute with precision the tasks to which they are assigned. It is, therefore, absolutely essential to form and train a special body of well-versed men, either drawn from several regiments or raised separately, as a corps of engineers.[142]

More than a decade later Vauban reminded Louvois of the impossibility of conducting orderly siege attacks with men "the majority of whom I do not know, who do not know me, and who absolutely refuse to listen to me."[143] His demands rebuffed yet again, Vauban warned of the dire results if his twenty-year old petition continued to be rejected:

> as the King will not make the company of sappers that I have proposed to him many times, it must be accepted that we will always lose a lot of engineers and many more soldiers and officers, and that it will always cost him more time and money to reduce the places; beyond which from the very first day of the siege it will end up costing me my life, because I am forced to be almost constantly in the trenches due to a lack of skilled people, which is killing me with fatigue and exposes me to death one hundred times every day.[144]

After complaining of a shortage of skilled miners at Breisach in 1703, he repeated at the end of his career the same request for three regiments of artillery and a separate company of sappers.[145] Turning to the younger generation, his treatise dedicated to Burgundy reminded the royal prince of how his many sieges had suffered from a lack of trained men and then explained that sapper companies officered

[142] Vauban, *Mémoire pour servir*, p. 142. Earlier he noted that even with small groups of subalterns "the least accident makes them give way, and scare off their comrades, who are easily inclined to fright by the appearance of an immediate threat" (p. 106).

[143] Rochas d'Aiglun (ed.), *Vauban*, vol. 2, p. 249, from the siege of Luxembourg, 6 June 1684.

[144] Rochas d'Aiglun (ed.), *Vauban*, vol. 2, p. 400, to Le Peletier de Souzy, 13 October 1693, regarding the siege of Charleroi.

[145] On Breisach, see Rochas d'Aiglun (ed.), *Vauban*, vol. 2, p. 522 at Biesheim, 22 August 1703. For his later call, see Vauban, *Traité*, pp. 269–295 for the artillery and pp. 296–309 for the sappers.

by engineers could quickly learn the trench skills that would speed up the siege attack and at the same time decrease the number of casualties among the engineers, the artillerists and the common soldiers alike—yet another example of efficient improvement.[146] But his requests were continually rejected for being too costly. Here as well Vauban was unable to eliminate the engineers' reliance on others, a fundamental source of inefficiency. Surprisingly, his complaints sound quite similar to La Vergne's lamentation of the Austrian service just a few years earlier:

> the majority of officers, no less than the soldiers, know nothing of what they must do; this always creates a great confusion in the approaches, it being impossible for the engineer to provide everyone with assistance when there are 500 or 600 men to command, being unable to tell each one what they need to do . . . this has happened to me many times despite all the precautions I have taken.[147]

The concern for efficiency once again lost out to the short-term desire to save money.

Although participants in Spanish Succession sieges rarely commented upon the skills of the workmen, when they did it was in the form of complaints about their tendency to run or slink away in the dark, particularly in long, difficult sieges. The dispositions for the trenches at Mons had to remind the workers yet again to only leave their trenches with the permission of the supervising engineer.[148] The English Lieutenant-Colonel John Blackadder reported his difficulties at Douai in 1710:

> I find the command far less troublesome when the regiment is in [the trenches on guard duty] than with the workers; there is always a great deal of confusion at any business of that nature in the night; and so it was yesternight. We were to make up the lodgment on the other side of the outer fossé [i.e. the advanced ditch at the bottom of the glacis], which we had been chased from the night before; and indeed

[146] Vauban, *Traité*, pp. 296ff.

[147] This warning from another of his works, *Nouveau exercice du gabion et de la fascine*, (Vienne, 1698), pp. 1–2. The journal of Ath suggests that proper trench construction was not self-evident even to the French in 1697, for it went into specific detail about who exactly was to do what and when in the trenches, with what number of siege materials of what dimension provided by whom. *Relation du siege d'Ath*, pp. 12–18.

[148] BL Add MSS 61339, Memoire des qu'il convient d'avoire alataque de la port de Hanere, 9 October 1709, f. 166.

> our workmen did their business very ill, for the French came out several times with great noise on purpose to frighten the workmen; and it had the effect, for they ran away so that it was impossible to get the third part of them kept together. However there was a lodgment made. These commands are exceedingly troublesome, because of the vexation it gives an officer when his men do not do their duty.[149]

The German engineer second-class Johan Landsberg echoed Blackadder's frustration at keeping his own workmen together:

> The enemies chased my workers away four times. Confusion reigned supreme! I say again . . . that a soldier is a coward when he is working in the trenches, no matter how brave he may be anywhere else; I've seen this many times. The enemies' first sortie consisted of only ten or twelve troops and immediately my men ran away over the bridge of the advanced ditch. I sent a few sous-lieutenants to retrieve them, which they did, and I put them back to work. Other times the enemies had only to scream "Kill! Kill!" and my men would flee yet again. I tried to encourage them, telling them that it was nothing, I even posted grenadiers as guards and I stayed with the workers in the trenches myself, but I was the only one to stay put, and when they returned to the trench and I berated them, they told me that they had neither enough cover nor enough support, that a dozen grenadiers on the other side of the bridge was not enough to protect them. The sappers also take flight, and don't return for the rest of the night.[150]

An engineer defending Béthune reported several occasions where their small parties would charge at enemy lodgments shouting *Tue! Tue!*, only to see their opponents flee for safety.[151] Proper supervision was critically important to a constant advance; once officers were killed or incapacitated, their troops often halted their assault under heavy enemy fire, or if working on lodgments or trenchworks fled for the safety of the rear trenches, or might simply put down

---

[149] Andrew Crichton (ed.), *The Life and Diary of Lieut.-Col. J. Blackadder, of the Cameronian Regiment* . . ., (Edinburgh, 1824), p. 386, letter to his wife, Douai, 23 May; and p. 376, 23 May.

[150] Landsberg, *Nouvelle manière de fortifier les places*, p. 43. One report from the Dutch field deputies specifically mentioned Johan Willem Friso's efforts to repair the blinds protecting the trench so that their workers would not desert. ARA SG 9193 #48 Field Deputies to States-General, Brebières, 9 June 1710. See other examples of the difficulties at Douai 1710 in: Crichton (ed.), *The Life and Diary of Lieut.-Col. J. Blackadder*, pp. 385; 387, to his wife; and ARA SG 9193, #76, Field Deputies to States-General, Brebières, 23 June.

[151] AG Article 15 Section 2, Béthune, #4bis, Journal du siège de Béthune, ff. 26, 38v–39.

their spades and refuse to work.[152] Similarly, casualties among their compatriots might also frighten the workers, prompting one journalist at Aire to recount: "after the first volunteers were killed or wounded, we could not find any more no matter how much money was offered."[153] The Allies encountered similar experiences at Béthune and Aire, whose garrisons conducted frequent sorties.[154] So traumatic were some sieges that some workers even deserted *into* the doomed towns in order to avoid the slaughter in the trenches.[155] The unreliability of those digging the trenches was yet one more item in a long list of uncontrollable factors that engineers had to manage in order to succeed.

The Vaubanian siege, as it was implemented, was not a model of mechanical efficiency. The scope of the engineer's managerial challenge is best encapsulated in the initial goals Vauban set for himself in his 1672 work. His second chapter discussed a general list of the more prosaic faults in siegecraft that needed to be corrected, everything from failures of secrecy, camp security, proper investment, locating the artillery park, and proper line construction to assuring communication between the approaches. Such matters were critical to a siege's success, yet Vauban's vaunted three tactics had no relevance and Vauban could only hope to convince generals of the importance of proper conduct. In the next chapter Vauban then focused his attention on the faults committed in the trenches, the

[152] For example, at Tournai 1709 see Johann Mathias reichsgrafen von der Schulenburg, *Leben und Denkwürdigkeiten Johann Mathias Reichsgrafen von der Schulenburg,* (Leipzig, 1834), p. 398, 30 August. Des Forges claimed the same for garrison troops in his account of Béthune.

[153] AG Article 15 Section 2 §1 Aire carton 1, #5, Relation du siége d'Aire en 1710 par un officier de l'armée des Alliés, 23 September. Miners at Douai refused to return to their work after an explosion killed one of their own. *Relation de la campagne de l'année 1710,* (La Haye, 1711), p. 49.

[154] On Béthune, see Veenendaal, Jr. (ed.), *Briefwisseling Heinsius,* vol. 11, p. 17 #24, Hertel to Heinsius, 4 August 1710. For Aire, AG Article 15 Section 2 §1 Aire carton 1, #5, Relation du siége d'Aire en 1710 par un officier de l'armée des Alliés, 20 September.

[155] At Venlo 1702, see Veenendaal, Jr. (ed.), *Briefwisseling Heinsius,* vol. 1, p. 448 #916, Wassenaar-Obdam to Heinsius, 27 September. The most striking case was Tournai: AG 1 K 299 1, Tournai, 20 July 1709 and especially AG A[1] 2151, #263, Villars to Voysin, 26 July. Although this practice also occurred long before our period, it is particularly surprising during the Spanish Succession given the near-certainty that the town would fall to the besiegers in only a matter of weeks. In the capitulations, besieging generals demanded that all deserters be returned.

arena of warfare where the engineers exercised the most control. Here he identified eight mistakes commonly committed by French besiegers on the eve of the Dutch War: poorly-chosen attacks, a lack of quality siege materials, shortages of knowledgeable workmen, poorly-planned trenchworks, poorly-sited artillery, poorly-coordinated responses to sorties, a lack of knowledgeable and patient generals, and finally the weaknesses of the engineers themselves (pp. 12–18). Of these eight, *none* was under the complete control of either the director-general or the *commissaire-général*, much less the chief engineer at a siege. The choice of attacks saw the most progress towards engineering control, as the choice was usually determined by the engineer in the War of the Spanish Succession, though even here it was ultimately the responsibility of the commanding general. The siege materials were constructed and provided by the soldiers or conscripted peasants, the workmen were drawn from the rank and file of the infantry, the regimental officers were the only people with command authority over their men in the trenches and were hesitant to follow an engineer's instructions, the artillery was its own independent corps, and the generals were beyond even Vauban's influence. The engineering corps itself was in many ways at the mercy of funding levels established by the State—Vauban could make few structural improvements to eliminate or educate 'ignorant' and 'incompetent' engineers without additional money for recruitment, training and retention from the Crown. As a result, the variety of complicated tasks required in a siege demanded a wide range of skills that could only be met by experts in several different fields—engineers, gunners, miners, and sappers to name a few. These independent branches naturally struggled with each other for prominence, while the same tension could occur within each service among its own officers, particularly where the hierarchy was dimly elaborated or constantly changing. From his earliest experiences Vauban recognized the many inefficiencies that resulted, and was able to eliminate or neutralize a few of them in French service over the length of his career by receiving overall authority from the King himself, by carefully planning his siegeworks in advance and carefully managing their implementation, as well as by demanding technical competence in his subordinate engineers. His aptitude in all of these fields is well-evidenced, ranging from his recognition and perfection of ricochet fire and experimentation with stone-firing mortars (*pierriers*) to his concern with proper trench construction to his attempts to improve the morale of the infantry workmen with better pay and improved safety.

Vauban could not, however, enlarge the scope of the engineer's authority to include other branches, nor could he convince Louis to fund a more permanent corps, nor could he manage the attack when he was no longer on site. The Allied engineers, for their part, lacked any kind of leader with the authority to lobby for their corps and protect their interests, and were rarely united long enough to present a united front in any case. French and Allied sieges in the Spanish Succession, therefore, suffered to different degrees from fragmented command. The engineers, the 'experts' of siegecraft, had the daunting task of coordinating the efforts of these many branches without royal authority. At the top of the chain of command, the chief engineer had to implement his geometrical trenches and lines of fire against a specific fortress using the artillery's gun crews, relying on military administrators to provide the necessary arms and munitions, counting on foot soldiers lured or coerced into digging the laid-out trenches, while maneuvering against fellow engineers and other officers in order to convince the siege commander to follow *his* projected attack. When disputes arose among the engineers, or between the engineers and artillerists, the generals took control. The full impact this had on Vauban's legacy of efficient sieges is the focus of the next chapter.

CHAPTER SIX

CONTESTING THE PARADIGM SIEGE

As we have seen, realizing Vauban's efficient attack required surmounting a number of hurdles. Military engineers in the War of the Spanish Succession were stretched thin across sprawling theaters, their numbers masking their widely-varying quality. Even the most skilled engineer had difficulty managing independent-minded cogs of the clockwork siege. These sources of friction could be minimized if the chief engineer could win the trust of or receive authority over the commanding general and his approach commanders, but such wide-ranging authority was rarely granted after Vauban and Coehoorn left the scene. Contrary to the assumption held by the scholarship on the late-seventeenth and eighteenth-centuries—that Vauban's method of attack served as the paradigm for most besiegers—there was a fundamental rift within the military community over the nature of the attack. Military engineers criticized the many sources of siege inefficiency, and general officers in turn savagely criticized their engineers for incompetence. Inquiring more deeply to resolve these competing claims, we find that the widespread complaints about the incompetence of the engineers were based on a poor understanding of Vauban's precepts, while the many specific complaints do not stand up to detailed scrutiny.

1. *Ignoring and Criticizing the Engineers*

Vauban's reputation, as impressive as it appears today, did not overawe many generals of his day. Though the Vauban-centric literature does not pay much attention to the details of his offensive method during its formation, it does recognize that at the end of his career his ideas were challenged by several French commanders who either criticized or rejected out of hand his proposals. The King may have valued his judgment, but this did not convince his generals in the field to blindly implement the projects they were sent.[1]

[1] Lecomte provides the best account of this in his *Les ingénieurs militaires en France*

In Germany in 1703, Villars ignored Vauban's proposal when attacking Strasbourg's Fort Kehl; he succeeded against the advice of the majority of the engineers present.[2] Encamped before Landau later that year, Camille d'Hostun, *maréchal* Tallard chose an artillery officer's proposal over that of Vauban.[3] At the Mediterranean port of Nice in 1705, Berwick also refused Vauban's siege plan and went on to capture the town by a different approach.[4] Pressured to accept Vauban's recommendations, head-strong La Feuillade made a point of reminding his father-in-law Chamillart that Vauban considered Nice unassailable where Berwick attacked it, yet the town only held out for twenty-seven days.[5] The blue-blood continued to insist on his independence when besieging Turin: he flatly rejected Vauban's proposals and refused to relent even after the great engineer publicly criticized its conduct.[6] Most other commanders relied not on their elevated lineage, but on the plausible argument against long-distance meddling: "those who see from close up are to be believed over those who see from far away."[7]

Nor did Vauban's reputation protect those who were entrusted to carry on his legacy. A few criticized the engineers for not living up to Vauban's standards. Antoine de Pas, marquis de Feuquières and Lieutenant-General of Louis' armies, for example, denounced the

pendant la règne de Louis XIV, pp. 137–140. Modern scholarship has discussed the matter more superficially: Hebbert and Rothrock, *Soldier of France*, pp. 207–208; Blanchard, *Vauban*, pp. 508–511; Lynn, *Giant*, pp. 575–578.

[2] Vogüé (ed.), *Mémoires du Maréchal de Villars publié d'après le manuscrit original*, vol. 2, pp. 49–61. Villars reproduces his justificatory letter to Chamillart in vol. 2, p. 67. In the secondary literature, see François Ziegler, *Villars: Le centurion de Louis XIV*, (Paris, 1996), pp. 107–108.

[3] Allent, *Histoire du corps impérial du génie*, pp. 418–431; Augoyat, *Aperçu historique*, vol. 1, p. 267.

[4] Rochas d'Aiglun (ed.), *Vauban*, vol. 2, p. 566, Berwick to Le Peletier de Souzy, 6 January 1706; also Berwick, *Mémoires*, vol. 1, p. 190 and 194. In fact, both the chief engineer and head gunner at the siege agreed with Berwick's attack, this bastard son of James II praising their efforts. G. Esnault (ed.), *Correspondance et papiers inédits recueillis et publiés par l'abbé G. Esnault*, 2 vols. (Geneva, 1970), vol. 2, p. 55, the engineer Paratte to Chamillart, Villefranche, 10 October 1705. Hereafter: Esnault (ed.), *Chamillart.*

[5] Esnault (ed.), *Chamillart*, vol. 2, p. 91 #189, to Chamillart, Casale, 8 February 1706.

[6] Rochas d'Aiglun (ed.), *Vauban*, vol. 2, p. 582, to Chamillart, Dunkirk, 23 July 1706 (AG A[1] 1938 #302).

[7] Rochas d'Aiglun (ed.), *Vauban*, vol. 2, p. 566, Berwick to Le Peletier de Souzy, 6 January 1706. Given this, it is no surprise that all of these sieges were conducted in theaters distant from Vauban, who gravitated toward the Flanders theater.

engineer Louis Lapara des Fieux for his conduct of several sieges in the Nine Years' War (Coni 1691 and Barcelona 1697), concluding that: "he was much inferior to M. de Vauban, and his errors can be only imputed to his incapacity." Yet Lapara was widely acknowledged as one of France's most gifted engineers, a good opinion that Vauban himself shared.[8] More often, the corps as a whole was indiscriminately condemned. The young La Feuillade complained of the "crass ignorance of all the engineers that I have here with me." A little later he pledged his unfaltering position to his father-in-law: "M. Pelletier will not compel me [to follow the engineers' advice] with the praise that he gives to those that are here with me; between you and me, they are all asses except for M. Filley. Excepting him, they all lack courage."[9] At Douai in 1712, a failed attack on an outwork prompted the Lieutenant-General Yves marquis de Tourzel et d'Alègre to note that the engineers' broken promises were "nothing new . . . it always happens at every siege." The duc du Maine, commenting in the margin of this letter, echoed both the tension between the engineers and his own artillery officers as well as the declining status French engineers had suffered since Vauban's departure: "We have only too many examples that the engineers are no more infallible than other people."[10] Jean-Robert LeFebvre d'Orval, a member of the French Flanders *parlement* and frequent correspondent on military affairs reported the disgust of French officers at the engineers' poor conduct attacking the same town that they had defended two years earlier: "when I left the siege [of Douai] everyone was murmuring against the engineers, who appear to have run out of ideas, and who can only respond 'we will see when we are able to examine the works,' as if it has been a century since they last saw the fortress. This highlights how little attention these engineers have paid to studying a place when they are in it."[11] The

[8] Antoine de Pas, marquis de Feuquières, *Memoirs Historical and Military: Containing a Distinct View of all the considerable states of Europe* . . ., 2 vols. (London, 1736), vol. 2, pp. 248–250, quote on pp. 248–249; p. 268 for Coni. For Vauban's high opinion of Lapara: Lazard, *Vauban*, pp. 235, 274. See also François Bluche's entry on "Laparra de Fieux" in F. Bluche (ed.), *Dictionnaire du Grand Siècle*, (Paris, 1990), pp. 828–829.

[9] Esnault (ed.), *Chamillart*, vol. 2, p. 10 #147, La Feuillade to Chamillart, Nice, 11 April 1705; and p. 16 #152, from Toulon, 10 May. His favorite engineer Filley would die at the siege.

[10] AG A[4] Carton 8 chemise 2, Alègre to duc du Maine, 4 September 1712.

[11] AG A[1] 2382, #101, LeFebvre d'Orval, 22 August 1712. This probably referred

chevalier de Quincy expected his audience to believe that he single-handedly resolved a dispute among Vauban's heirs (featuring an indecisive Lieutenant-General Claude-François chevalier d'Asfeld, future director-general of fortifications) over how to trace a line to a troublesome retrenchment in 1713. An erstwhile supporter of the engineers, the younger Quincy reported that d'Asfeld immediately sided with his suggestion and declared him *un grand ingénieur*, but he modestly admitted that "only good sense inspired me at the moment."[12] So low had the French engineering corps apparently fallen, that such a story was to be taken at face value.

Allied commanders were even more livid with the conduct of their own engineers. Criticism of their behavior increased dramatically as the sieges became longer, with the attacks following Ramillies providing the first occasion for sustained critiques. The beginning to the siege of Menin was not encouraging:

> Yesterday [Dutch General of Foot Ernst Willem van Salisch] complained to the States of the ingineer [Des Rocques] that he minded nothing nor did nothing, which in truth they doe not for the lines of circumvallation that they have been a doeing ever since we came here and might have been done in half the time are not yet so good as the ditches that was there before they begun.[13]

At the 1706 assault on Ath several Allied participants griped in a similar vein that their engineers underestimated the strength of the defenses at their chosen attack.[14] The sieges of Oostende, Menin, Dendermonde and Ath nevertheless all progressed much faster than at first feared, so Lille in 1708 was the first town to plunge the engineers' aptitude into serious, sustained doubt.[15] Allied generals felt

---

most directly to Valory, who had defended the town in 1710 and was now in charge of the attacks.

[12] Léon Lecestre (ed.), *Mémoires du chevalier de Quincy*, 3 vols. (Paris, 1898), vol. 2, pp. 250–251.

[13] BL, Add MSS 61310, f. 3, Blood to Marlborough, Menin, 2 August 1706. Later in the siege, Marlborough repeated much the same in Murray (ed.), *Letters and Dispatches*, vol. 3, p. 79, to Hedges, Helchin, 12 August. Discussing the 1706 campaign, one of Marlborough's recent biographers repeated that the Dutch engineers relied upon their "often lethargic and overcautious" approach. Jones, *Marlborough*, p. 127.

[14] BL Add MSS 61180, f. 38, Ouwerkerk to Marlborough, Ath, 27 September; Veenendaal, Jr. (ed.), *Briefwisseling Heinsius*, vol. 5, p. 556 #1079, Salisch to Heinsius, Leuze, 27 September.

[15] At Ath 1706, for example, several generals complained of the delays due to

particularly conflicted by their inability to refuse the attacks projected by the engineers. Commanding one of the approaches at Lille, Saxon general Johann Matthias Count von der Schulenburg's initial faith in his engineers decreased steadily as he waited for them to decide where to open the trenches.[16] Several months into the siege, Marlborough was driven to write to his confidante back in London:

> It is impossible for me to expresse the uneasyness I suffer for the ill-conduct of our ingeniers att the siege, where I think everything goes very wrong. It would be a crewell thing if after we have obliged the enemy to quit all thoughts of releiving the place by force, which thay have done by repassing the Schell, we should faile of taking it by the ignorance of our ingeniers, and the want of stores; for we have already fiered very near as much as was demanded for the taking the town and cittadell, and as yett we are not intier masters of the counterscarp; so that to you I may own my dispair of ending this campagne, so as in reason we might have expected.[17]

A few days later he vented his frustration to Heinsius as well, proclaiming the engineers "very ignorant if there be nothing worse. . . . I have the spleen and dare say no more."[18] Summarizing his own experience at the siege, Landsberg proclaimed that: "There are people who believe that when a man is called an engineer, he acts according to the best science, but one would be mistaken and this is obvious from the engineers' conduct at Lille. Those who reflect on the matter will note that when sieges drag on, it is not always due to the good defense of the enemy."[19] Extending the campaign season to squeeze in one final conquest, Marlborough was still frustrated with the "negligence of the engineers, who only stay in the

the engineers' initial underestimates, yet their earliest length estimates were almost exactly on target. Torrential rains also slowed the pace of trenchwork.

[16] BL Add MSS 61245, f. 68, Schulenburg to Marlborough, Lille, 23 August 1708.

[17] Snyder (ed.), *Marlborough-Godolphin Correspondence*, vol. 2, p. 1099 #1100, Marlborough to Godolphin, Lannoy, 20 September 1708. See also Murray (ed.), *Letters and Dispatches*, vol. 4, p. 237, Marlborough to the Earl of Sunderland, Lannoy, 24 September; Christopher Atkinson, "Gleanings from the Cathcart MSS," *The Journal of the Society for Army Historical Research*, 29 (1951), p. 66; and Royal Commission on Historical Manuscripts, *Report on the Manuscripts of the Earl of Mar and Kellie*, p. 464, Nodding to the Earl of Mar, Lille, 20 September.

[18] Van 't Hoff (ed.), *Marlborough-Heinsius Correspondence*, p. 402 #670, Marlborough to Heinsius, Lannoy, 24 September 1708.

[19] Landsberg, *Nouvelle manière de fortifier les places*, p. 43. See also p. 35.

[Duke of Württemberg's] trenches [before Ghent] one hour a night, having in their heads that this can only be a false attack."[20]

At the bloody siege of Tournai's citadel the next year, a fort endowed with an extensive system of countermines, Schulenburg bemoaned the lack of competent people to oversee this new experience of subterranean combat: "This is a siege altogether different from the others that we have undertaken thus far; the most embarrassing is that there are few officers and even engineers that know how to execute them, at least as far as how they are to be attacked."[21] He was also quite severe concerning the chief engineer's subordinates:

> Mr. Du Mee has few engineers and what is worse not one of them is good, the miners cause me more trouble than I can possibly say, the sappers make so many mistakes that I must personally go check everything they do, which requires me to be in the trenches all day long. Such a siege demands far more than this—competent people who take things seriously and who cooperate with each other.[22]

With the siege winding down, Dutch general Albemarle conveyed Eugene's fear of another siege to the Grand Pensionary, stemming primarily from their lack of "capable" engineers.[23]

The chorus of Allied complaints reached a crescendo in 1710. Two months into the campaign the Allied generals had lost all patience with the engineers. Marlborough's secretary Adam Cardonnel echoed his superior's own feelings when he complained that "our siege [of Douai] goes on but slowly for want of good engineers."[24]

[20] Murray (ed.), *Letters and Dispatches*, vol. 4, p. 381, Marlborough to Goslinga and Geldermalsen, Merelbeke, 29 December 1708.

[21] Schulenburg, *Leben und Denkwürdigkeiten*, p. 397, 18 August 1709. At the very beginning of the siege Goslinga had already noted that they had an insufficient number of miners. Veenendaal, Jr. (ed.), *Briefwisseling Heinsius*, vol. 9, p. 9 #21, to Heinsius, Tournai, 3 July.

[22] BL Add MSS 61245, f. 106–106b, Schulenburg to Marlborough, Tournai, August 1709. Like the French, the Allies too lacked trained companies of sappers.

[23] Veenendaal, Jr. (ed.), *Briefwisseling Heinsius*, vol. 9, p. 194 #406, Albemarle to Heinsius, Tournai, 24 August 1709.

[24] BL Add MSS 61401, f. 32b, Cardonnel to Colonel Hales, Douai, 22 May 1710. At Namur in 1695 Cardonnel also criticized the slow pace of the siege. Childs, *The Nine Years War*, p. 282. See Marlborough's own views on Douai 1710 in van 't Hoff (ed.), *Marlborough-Heinsius Correspondence*, p. 495 #850, Marlborough to Heinsius, Douai, 9 June; and Murray (ed.), *Letters and Dispatches*, vol. 5, p. 23, Marlborough to the Austrian major-general Ferdinand Alexander Franz Maffei, Douai, 12 May. Rumors circulated at Versailles that Marlborough had publicly cursed the engineers and their progress on a 6 June visit to the trenches. Edouard Pontal (ed.), *Mémoires du marquis de Sourches sur le règne de Louis XIV*, (Paris, 1882), vol. 12, p. 239, entry of 9 June.

A week later the scribe fulminated to another correspondent: "Our siege goes on still very lamely through the ignorance or timorousness of our engineers, which costs us daily great numbers of poor men's lives."[25] A few days later still, Albemarle averred: "I assure you that we must tremble every time we besiege a place, because we do not have any subjects competent enough to conduct such tasks."[26] Prince Eugene, overseeing the siege, was livid with "the mistaken ambition of the engineers, who try to invest themselves with the same authority that was enjoyed by the late General Coehoorn. I never met Coehoorn, but I know that there can be no comparison between his ability and that of the horrible little men we have with us now."[27] Albemarle penned his frustration yet again to Heinsius: "I have become increasingly aware that sieges do not suit us, because each presents us with difficulties; before we arrived here we were told that the place was very poor, but now we are told that it is very strong."[28] Schulenburg summed up the Allied generals' shared wisdom of Douai, bitterly complaining that had the town been attacked properly, it would have lasted less than two weeks instead of seven, and cost them only 1,000 men instead of 8,000.[29] Douai's conduct did not augur well for future efforts, particularly when officers were unwilling or unable to correct their technical advisers' mistakes. The Saxon general repeated his earlier formula again at the next siege of Béthune:

> These gentlemen . . . want to continue as usual, that is, to attack places at their strongest point and spend two or three months at a siege and

[25] BL Add MSS 61401, f. 41, Cardonnel to the diplomat Frances Palmes, 4 June. To the paymaster James Brydges he wrote of the "want, unskillfulness and timourousness of our engineers." BL Add MSS 61401, f. 41, 5 June.

[26] Veenendaal, Jr. (ed.), *Briefwisseling Heinsius*, vol. 10, p. 408 #801, Albemarle to Heinsius, Douai, 6 June. See also the Utrecht field deputy Nicolaes Pesters' comments in Veenendaal, Jr. (ed.), *Briefwisseling Heinsius*, vol. 10, p. 406 #798, to Heinsius, Brebières, 5 June.

[27] Hipssich, *Spanischer Successions-krieg. Feldzüge 1710*, p. 140, Eugene to Austrian minister Ferdinand Sinzendorff, Douai, 20 June, quoted in Duffy, *The Fortress in the Age of Vauban*, p. 37.

[28] Veenendaal, Jr. (ed.), *Briefwisseling Heinsius*, vol. 10, p. 558 #1114, Albemarle to Heinsius, Viller-Brulin, 21 July. See also p. 564 #1129, Des Rocques to Heinsius, Béthune, 24 July. Landsberg complained of his superiors that "we tell the generals everything will be simple at the beginning and when were are at work and we encounter difficulties, we don't know where to turn." Landsberg, *Nouvelle manière de fortifier les places*, p. 45.

[29] Schulenburg, *Leben und Denkwürdigkeiten*, p. 459, 5 July 1710. Landsberg's post mortem concluded the same, claiming that he had warned the other engineers of their folly. *Nouvelle manière de fortifier les places*, p. 42.

> suffer 8,000 or 9,000 casualties as they did at Douai, so that if you trusted them and let them proceed, they would attack this front for three or four more weeks, another two months at least. . . . Messieurs the Directors, as ignorant as they are malicious, cross me and obstruct me wherever they can, and Mr. Des Rocques having been shown quite clearly the mistakes made at Douai, would be happy if this siege went about the same.[30]

In many sieges Spanish Succession generals vehemently castigated the engineers for their perceived faults.

Though an engineer himself, Landsberg neatly summarized the widely-held judgment of the engineers and their conduct, at the same time ingratiating himself with the commanders: "we have very talented generals which give us the means to besiege; it is only a question of the engineers directing the attacks better and capturing places more quickly than they do at present, that they save lives, and that they do not waste the Sovereign's treasure uselessly."[31] The volume and hostility of all this criticism is even more remarkable considering how completely the engineers were otherwise ignored, and how we find practically no criticism of other officers for brashness or recklessness. Among the thousands of siege-related letters written by generals, there are only a handful of positive comments about the engineers, while even neutral discussion of the individual engineers is almost impossible to find. Engineers were either invisible or culpable to most officers.

As the previous pages attest, generals did not think much of their engineers. The vehement and widespread criticism from generals in both French and Allied service raises a fundamental challenge to the notion of Vauban's paradigmatic siege as the model throughout the eighteenth century. The generals did more than complain, however, as they ignored their advisers and followed their own counsel whenever possible. Vauban historians have advanced a perfunctory explanation to explain the abuse he personally suffered, tied to the specific

---

[30] Schulenburg, *Leben und Denkwürdigkeiten*, pp. 471–472, letter of 13 August. Schulenburg perhaps meant in his last sentence that Des Rocques wanted to see his replacement Hertel fail as he himself had at Douai.

[31] Landsberg, *Nouvelle manière de fortifier les places*, p. 84. Note that while Landsberg attacked his peers, he also acceped the values that engineers encouraged—an efficient siege that saved lives and money. This work was motivated in part, he admitted, as a way to find employment, which may explain the vigor with which he attacked his technical peers.

context of Louisquatorzian France. Paralleling the twilight of the Sun King, some scholars describe the reaction against him as a generational response of relative newcomers seeking to displace his status as one of Louis' most trusted military advisors.[32] Although this explanation has not advanced beyond single-sentence suppositions, it does little to explain how Allied generals had exactly the same complaints about their own engineers. The question of why Vauban's method was rejected is an important one for a larger understanding of the period's warfare.

Vauban's end-of-life troubles make more sense when put in a broader military perspective, one that examines the receptions French generals gave not only to Vauban, but also how Allied generals reacted to Coehoorn, and how both sides treated the successors of these two. Only with this wider view do we see the striking similarities between both French and Allied criticisms; their common language and the context of their criticisms shows that they were engaged in a single discourse over the nature of the siege attack. In the historical literature the Marlburists come closest to recognizing this shared mentality. While they accept the reality of Vauban's 'scientific' attack, they also accept the contemporary complaints against Allied engineers at face value and assume that the technicians were inept because they had not been trained in the French system. Taking a wider perspective than most, Duffy compared the Spanish Succession protagonists in the Low Countries:

> Since [the age of Condé, Turenne and Montecuccoli], the perfection of such mysterious-sounding devices as parallels, trench cavaliers and ricochet batteries had elevated siegework into something of a black art, which became less and less accessible to busy field commanders. France owned a whole generation of engineers who had been bred up by Vauban to put the new techniques at the service of Louis' marshals. The Allies, on the other hand, had to rely on a dwindling, overworked and demoralized band of cosmopolitan experts. Marlborough and Eugene had precious little confidence in their ageing Huguenot engineers, but they could not trust themselves to take over in their place.[33]

---

[32] Both Blanchard (*Vauban*, pp. 509–511) and Lynn (*Giant*, p. 578 and "Vauban" entry, p. 459) briefly mention the role of a rebellious, younger generation without analyzing the matter.

[33] Duffy, *The Fortress in the Age of Vauban*, p. 40; see also p. 37. This is also the view of earlier scholars, e.g. Maurice Sautai, *Le siège de la ville et de la citadelle de Lille en 1708*, (Lille, 1899), p. 275.

While the tenor of his passage is accurate, several important details require clarification and elaboration. First, as we have already seen, Vauban and his students did not enjoy universal acceptance—ignorance of these "black arts" prevented neither widespread complaints nor interference. Even engineers trained under Vauban rarely received *carte blanche* to conduct sieges however they saw fit. More importantly, however, it is difficult to corroborate these complaints since we have only the word of the critics.

With these many complaints fresh in our minds, we should dig a little deeper to assess their legitimacy. In doing so, however, a number of difficulties arise, making it problematic to conclusively determine whether the engineers were as incompetent as their superiors claimed. The existing historiography ignores this fundamental issue of bias in the sources as well as the uncertainty of contemporary judgments. As far as the accusers are concerned, testing their propositions is challenging because they rarely make charges specific enough to be tested; instead we are given only vague, blanket statements of engineer incompetence. When complaints are precise enough to examine, specific individuals are rarely mentioned: not worthy of much individual attention, technicians high and low are lumped together into that derided category of 'engineer,' making it extremely difficult to corroborate specific flaws. Furthermore, the precise conditions at these 300-year old sieges are unknowable to us today, for extant sources do not provide the necessary detail for completely accurate reconstruction, while almost all of the fortifications and their environs were destroyed and built-over by the turn of the twentieth century. Reconstructing military campaigns and judging competing claims is an exceedingly difficult task with the best of documentation, as Clausewitz recognized from his personal experience: "Only now and then, in the memoirs of generals or of their confidants, or as the result of a close historical study, are some of the countless threads of the tapestry revealed."[34]

Just as importantly, we cannot weigh the accounts of generals and engineers against each other and draw solid conclusions, as the accused are largely silent. There are extremely few sources in which

[34] *On War*, p. 112. Later he noted that in war particularly "facts are seldom fully known and the underlying motives even less so. They may be intentionally concealed by those in command, or, if they happen to be transitory and accidental, history may not have recorded them at all" (p. 156, also p. 164).

the engineers bothered to defend their actions, for they were far too busy in the trenches to provide either lengthy justifications of their conduct or even to devote to correspondence during a siege.[35] Vauban is not the only engineer to apologize for being too busy to write to Court, whereas commanding generals usually had plenty of time cooling their heels as the siege moved on.[36] As a representative example from the War of the Spanish Succession, Des Rocques wrote a total of six brief letters to Heinsius during the 52–day siege of Douai in 1710. He relied instead on the approach commanders and the field deputies to keep the *raadpensionaris*, the *Raad van State* and the States-General informed of what was happening at the siege on a daily basis. The harried French chief engineer Valory was only a little more productive in his 1712 campaign, and most of his letters are known to us via third-party copies. After conducting the attacks, those engineers who were lucky enough to emerge unscathed were not yet finished, for they had to spend time inspecting the captured fortifications and supervise their repair, as well as prepare for new attacks. A head engineer was almost as busy in peacetime as in wartime, allowing them little time to write manuals, memoirs or diaries in order to shape posterity's impression of them.[37] What they did record, largely for their peers, provides less information than one would like. The many siege journals—our main source from the engineer's perspective—are consistently 'objective' descriptions, clinical accounts of events rather than analysis, highlighting the daily advance of the attack trenches, the batteries, the sorties, and other

[35] For a rare example of an engineer's self-justification, a letter that blames the slow pace of the siege on the mistakes of the approach commander, see Veenendaal, Jr. (ed.), *Briefwisseling Heinsius*, vol. 11, p. 212 #387, Des Rocques to Heinsius, Aire, 2 October 1710. The specificity of his critique, which contrasts sharply with the vague accusations of most general officers, hints at how thoroughly our view of siegecraft might have to be revised if we had more analytical accounts from the engineering perspective.

[36] See one example in Père Henri Griffet, *Recueil de lettres pour servir d'éclairissement à l'histoire militaire du règne de Louis XIV*, (Paris, 1740–1741), vol. 8, p. 104, Vauban to Louvois, Philippsburg, 15 October 1688. At the earlier siege of Luxembourg Louvois pestered his engineer to provide a plan of the attacks everyday. Rochas d'Aiglun (ed.), *Vauban*, vol. 2, p. 232. By 1697 his son Barbesieux attempted to lighten Vauban's work load by exempting Vauban from the need to send him reports separate from those addressed to the King.

[37] The exceptions from Louis XIV's period are, of course, Vauban and the Allied engineer J.H.D. Landsberg. Contrasted with Vauban, Coehoorn has left almost no documentary evidence of his sieges.

siege actions with almost no mention of internal disputes, alternative courses of action discussed. Nor, for that matter, do they provide much in the way of explanation or judgments about the conduct of the siege in general, hence our interest in the anonymous journalist of Ath.[38] To cite the most striking example, despite widespread criticism of his conduct besieging Lille, Du Mée's public, explicitly self-justificatory account of the siege (*An exact journal*) refused to discuss the options available to them and why he and his rival Des Rocques made the decisions they did, beyond a cursory nod to the high attrition suffered by the engineers. For many reasons then, the detailed information necessary to conclusively evaluate the conduct of a siege was available only to those skilled practitioners who witnessed it in person.

Given the highly skewed nature of our sources, therefore, we are largely left with the prosecution's case. Even with eyewitnesses, however, we need to be careful in accepting extraordinary claims without corroboration. One of Heinsius' Huguenot contacts reported his own difficulties finding out what was really happening in the trenches:

> As I am sure that you see reports from Tournay more recent and substantial than what I could give you, I went searching to discover the state of the siege. Each person is absorbed with their own particular task, and without orders to meet with their colleagues and familiarize themselves with what is happening at the other attacks, those who are at one approach know almost nothing about what is happening at another. It is no great surprise, then, if the news that we receive is often wrong.[39]

Lacking adequate knowledge of conditions at the siege and of the engineers' decision-making process, we almost never know what motivated their specific actions on a particular date, nor what important factors they may have been forced to take into account that non-engineers either remained ignorant of or simply dismissed out of

---

[38] For a discussion of other siege accounts which were intentionally written to create or reinforce a particular identity among its civilian readership, see Michael Wolfe's "Walled towns during the French wars of religion (1560–1630)," in J. Tracy (ed.), *City Walls*, pp. 337–347; and his "Writing the City under Attack during the French Wars of Religion," in L. Carle (ed.), *Situazioni d'Assedio. Cities under Siege*, pp. 197–203.

[39] Veenendaal, Jr. (ed.), *Briefwisseling Heinsius*, vol. 9, p. 35 #73, Caesar Caze d'Harmonville to Heinsius, Lille, 10 July 1709. Vauban had ordered just this at Ath in 1697.

hand. Added to the difficulty of trying to prove counterfactuals (e.g. the claim that an attack against sector X would have been better than the one they did conduct against sector Y), extant sources allow few opportunities for engineers to prove their worth to future generations. It is only reasonable therefore to keep the engineers' silence in mind when judging the critiques of their detractors.

In spite of such methodological difficulties, the numerous complaints from many different officers cannot be perfunctorily dismissed. The question remains: were generals skilled enough in siegecraft to be accurate judges of the engineers?[40] Do these complaints reveal some significant shortcoming on the engineers' part? In answering these questions, we are interested only in judging contemporary claims that the head engineers were systematically incompetent. The sources do not allow us to focus below the rank of chief engineer, while it is hardly damning to admit that individual technicians (including the head engineers) made mistakes, but in this they were no different than their peers. Studying the matter in detail, a number of general observations and specific tests suggest that most of these complaints either had little foundation or were based upon unreasonable expectations. At the least, their merit can only be judged by taking into consideration the underlying mentality that motivated them: a predilection for martial vigor rather than Vauban's *industrie*.

Before examining the specific critiques, we should note that the overall pattern of complaints suggests that a particular dynamic was at work. All these complaints against the engineers were limited to their behavior when besieging a town; unlike their trench counterparts, garrison commanders never blamed their engineers' technical decisions for the failings of their defense, though they were just as obvious a scapegoat.[41] Furthermore, while besieging generals assailed the engineers again and again, they almost never criticized those

---

[40] Erik Lund, for example, argues that Imperial generals were well educated in those technical arts relevant to siegecraft, although he provides little evidence for their dexterity with the technical details of the siege trenches beyond the fact that many who rose through the ranks had some engineering experience. *War for the Everyday*, pp. 42–47, 51–52. La Vergne would strongly disagree.

[41] Jean de Mesgrigny was unique for being criticized for his role in the defense of the citadel of Tournai, which surrendered early due to a lack of supplies. Though few questioned his skills (a close associate of Vauban's, he had served in the army since 1651 and as engineer and miner since at least 1668), some rumored that he had requisitioned the garrison's foodstuffs for his own profit.

lieutenant-generals commanding the engineers, though these approach commanders were in charge of the attacks and, as we have already seen, were quite willing to ignore their technical advisers. It is also surprising to note that even though the harshest critics admitted the shortage of engineers with some regularity, this rarely prevented these same officers from blaming shortcomings on the competency of the individual engineers, rather than address the shortage of engineers, or the ad hoc system used to raise, train and compensate them, or deal with the inter-service rivalries and command friction that hindered their command authority. One did not have to command the engineers as a corps in order to welcome their advice and defend them against their critics, yet we find extremely few examples of such support in the correspondence. In short, when a siege went wrong, the engineers invariably received the blame regardless of their real influence over its progress.

Beyond such generic claims of engineering incompetence, three interrelated charges are specific enough to test: the extent to which engineers were judged out of practice at many of their sieges, their tendency to attack a fortress at its strongest point, and their inability to accurately estimate the difficulty of their sieges. Upon closer examination, the specific charges leveled against these engineering corps fall apart in almost every single occasion; the engineers' conduct does not by itself appear to justify the extreme hostility they endured.

First is the complaint that the engineers' siege skills were rusty. While a plausible critique, it is hardly evidence of widespread incompetence and the language used to express such complaints hints at an alternative motive. Refused the conduct of the upcoming attack on Landau, Vauban acknowledged this problem in general terms. He sent a long project to those in charge: "I am going to make a general disposition of the attacks concerning the works which are usually used, the essentials of which few people understand, due to a lack of practice and inexperience with long sieges."[42] Years later the French army, after spending seven years on its heels in the Low Countries, started slowly once back on the offensive, prompting complaints about the lack of preparations at the second siege of the

---

[42] Rochas d'Aiglun, *Vauban*, vol. 2, p. 532 to Chamillart, Strasbourg, 6 October 1703.

season, Douai. "Apparently," wrote the chevalier de Quincy, "we forgot that [trench tools] are needed in order to open the trenches. It was a long time since we had conducted any sieges, so it was even possible that we forgot how to begin one."[43] Both Villars and the cavalry brigadier François-Marie comte de Broglie surmised that the slow pace of the siege was due to the engineers being out of practice. As Broglie lamented: "it has been such a long time since we have conducted a siege that it can hardly be otherwise in the beginning."[44] With the engineers still deemed out of practice at the third siege of the season, Le Quesnoy, Villars interjected himself yet again: "As we have not conducted sieges in a long time, my involvement was necessary to speed its pace."[45] The Allies also noted a similar lethargy among their technicians (including the gunners) at their earliest sieges.

On the surface this criticism seems reasonable given the transitions both sides went through during the War of the Spanish Succession. The vast store of knowledge and experience that Vauban accumulated over his career was impossible to pass on to succeeding engineers. Coehoorn's experience, though far more limited, was still significant, including directing six sieges over the course of a dozen years spanning two wars. Once both Vauban and Coehoorn had left the scene, it was up to their successors to oversee the attacks, and these subordinates would necessarily have less experienced stepping out from under the long shadow of these two masters. On the French side, their engineers gained the most attack experience conducting several dozen sieges in Germany, Italy and Iberia. Ironically, the resulting losses deprived them of many of their most experienced siege engineers. Vauban's assistant Louis Filley, participant at Ath,

[43] Lecestre (ed.), *Mémoires du chevalier de Quincy*, vol. 3, pp. 160–161. The chevalier does not identify whether this oversight was due to the engineers failing to order the appropriate number of tools, whether the army administrators failed to supply the necessary implements, or whether the infantry generals failed to provide the ordered supplies to the workers in the trenches.

[44] AG A[1] 2382, #99, Broglie to Voysin, 22 August 1712. This is in the context of the batteries being sited too far away from their target, so it is unclear who exactly is to blame here, though we recall that after Vauban's death the artillery moved to assert their authority over the siting of batteries. See also AG A[1] 2383, #102, Villars to Voysin, 6 September. More generally, see AG A[4] Carton 8 chemise 1, Lieutenant-general Antoine de Gramont, duc de Guiche, 28 July.

[45] Anquetil, *Suite des mémoires rédigée par Anquetil*, vol. 2, p. 389.

defender of Kaisersweert and director of the siege of Landau in 1703, was killed two years later at Nice. After successfully capturing the Savoyard fortresses of Vercelli (Verceil) and Ivrea (Ivrée) in 1704, the veteran Guy de Richerand (a *maréchal de camp* who had served under Vauban since 1672, including at Ath) would meet his demise before the walls of Verrua later that year. Once Lapara des Fieux (another veteran of Ath and director of the less successful siege of Barcelona later that year) arrived there in February of 1705, he managed Vendôme's long siege of the mountain-top fortress to a successful conclusion and went on to oversee the capture of Chivasso and Mirandola later in 1705. He had less luck attacking Barcelona in 1706, being killed while directing the trenches against its Fort Monjuich. The French corps gained a measure of reprieve after the debacles at Turin and Barcelona until Villars' 1712 counter-offensive; over these five years the French would not conduct another major siege in the theaters known for their fortresses (Italy, Germany and Flanders).[46] Their return to siegecraft in Flanders was directed initially by Villars-Lugein, one of the few remaining directors with considerable siege experience in the current war. His career included, however, more than the usual number of setbacks. He commanded the engineers in the capable three-month defense of Landau in 1702, but had the bad luck of directing the failed attempt on Gibraltar in 1704, succeeding Lapara at Barcelona only to be ordered to lift the siege, as well as assisting Tardif at Turin before La Feuillade's mistakes forced a headlong retreat. His final misfortune was to be killed by a cannonball in front of Marchiennes. Villars turned then to Valory to finish its capture as well as prosecute the three remaining sieges in Flanders and the attacks on Landau and Freiburg on the Rhine in 1713 as well.

By 1712, then, the French engineering corps had lost many of its most experienced *ingénieurs en chef* in combat and Villars relied therefore upon Valory, who had served as director of Menin's fortifications and whose most recent attack experience was directing the successful attack on Huy's forts in 1705. But Valory was hardly a neophyte

[46] There were a dozen French Iberian sieges over this period, but the theater's difficult terrain gave siege warfare a very different tenor compared with the cockpits elsewhere in Europe—weak fortifications were made immensely stronger by the difficulty of transporting heavy siege artillery to the many sites. Nor did the French engineers conducting these Iberian sieges direct any in Flanders in 1712.

despite his seven-year respite from leading a siege effort. He had over thirty years of service and had risen to the rank of lieutenant-general by 1710. Just as importantly, he and his peers serving under him had considerable familiarity with Flanders siegecraft despite their limited experience as directors. Valory had participated in practically all of the Low Countries sieges during the Dutch War and the Nine Years War, where he was able to watch the master in action. In addition to their training and pre-Spanish Succession service, Valory and his subordinates had also played a large role in slowing down the Allied advance on Paris from 1706 onward. Valory had personally directed the defense of Menin, served under DupuyVauban defending Lille two years later, and also directed Douai's surprisingly vigorous defense in 1710. In other words, Valory had plenty of recent opportunities to practice siegecraft while on the defensive, a useful substitute when we recall how carefully the besieged attended the attackers' motions (as evidenced by their siege journals).[47] The Allies could also call upon men of experience, though their numbers were fewer and their practice less frequent. Neither Des Rocques nor his junior Du Mée had direction over a siege until 1705, but Des Rocques had at least served in a dozen sieges in William's previous war as well as serving under Coehoorn.[48] These engineers, elevated late in the war, were undoubtedly less experienced than their illustrious predecessors, but their limited experience directing attacks did not mean they were unfamiliar with the details of siegecraft.

Nor, for that matter, are such criticisms of inexperience very plausible, particularly those suggesting that Valory and his subordinates were still out of practice in their third and fourth consecutive sieges of the year (first Marchiennes with Villars-Lugein, then Bouchain, Douai, and Le Quesnoy all under Valory). Furthermore, such complaints are difficult to separate from the general lack of preparation time available. The speed of the French shift from being on the defensive to taking the fight to the enemy occurred within a few days—the unexpectedly-successful assault on Denain took place on

[47] The 19th century engineer-historian Augoyat praised Valory's attacks of Landau and Freibourg in 1713 and declared him one of the age's best siege directors. Augoyat, *Aperçu historique*, vol. 1, p. 434.

[48] At the end of the war Des Rocques' résumé included twenty-eight sieges. Veenendaal, Jr. (ed.), *Briefwisseling Heinsius*, vol. 13, p. 462 #706, to Heinsius, Ruesnes, 20 June 1712.

24 July, Marchiennes was invested the next day with the trenches opened three days later, while the second target of Douai was invested a little over a week after the battle. As French provincial intendants scrambled to fill the sudden demands for siege materiel and peasant laborers, both Louis and Villars pressed even harder, sending troops to invest Bouchain before Le Quesnoy had even been captured. It would hardly be surprising then, given this accelerated pace, that the engineers were caught off-guard by the sudden surge of activity and therefore lacked the time to shift from readying the defenses of their fortifications to planning for offensive operations as thoroughly as Vauban counseled.[49] More fundamentally, however, these out-of-practice criticisms skirt around the main issue, for the critics did not share their technicians' ideal of efficiency. Instead, they assumed that if only the engineers had more practice, they would conduct the attack 'properly,' i.e. they would accelerate the pace of the attack by choosing different tactics altogether. It is no coincidence that the engineers' purported lack of practice manifested itself solely in terms of a slow advance, rather than, say, higher casualties. Broglie's complaint at Douai quoted above made this underlying assumption explicit: "nothing here is organized as one might hope, so that this siege might be conducted as quickly as it should be." Being out of practice might indicate a lack of skill among the engineers, but we have only the vague comments of time-conscious generals as evidence.

Out of practice complaints often manifested themselves in specific critiques of the choice of attacks. This rebuke is based on the perceived failure of the engineers to accurately judge the strength of a fortress's works; in their confusion they would attack the front with the strongest fortifications and thereby unnecessarily complicate its capture. We have already quoted a number of such complaints, and as these criticisms imply, choosing an approach should have been an easy thing to do; the historical literature reinforces the idea that almost any contemporary could have precise knowledge of a fortress's

[49] Bernières noted Valory's concern for siege supplies as well as the shortages resulting from their rushed preparations, e.g. AG A[1] 2382, #18, from Hénin-Lietard, 17 August 1712. Lefebvre d'Orval confirmed the shortage of gunpowder and cannonballs in AG A[1] 2382, #102, Valenciennes, 23 August. See Villars' promise to invest Bouchain four or five days before Le Quesnoy would fall in AG A[1] 2383, #223.

works from published plans.[50] The reality is far murkier, however, for the clarity imagined by after-the-fact critics and historians fades when we examine Vauban's own experience with the issue. Even the great Vauban's opinion on choosing attacks seems to have drifted somewhat over his life. Early on, he lectured on the dangers of attacking a fortress at its strongest point. In his expanded manual he wrote that the debate as to whether to attack at the weaker or stronger point remained unresolved, after he noted that it was not at all easy to distinguish the weak from strong parts of a fortress without intensive personal reconnaissance both during the day and at night.[51] He then spent nine pages describing various features that made a side weak or strong. He concluded somewhat unhelpfully: "In conclusion, you must always seek the weak sector of the fortress and attack it rather than any other part; at least as long as there is no other significant feature that would force you to do otherwise." His general maxim is just as cryptic: "Always attack places by the weakest front and never by the strongest, at least when you are not constrained by better reasons which trump the usually-reliable reasons: this depends on the place, the time and season when the places are being attacked, and on the different situations in which they are found."[52] Unfortunately, he never defined what exactly constitutes a weak sector in any concise, objective way, nor how to balance the many, often-contradictory factors involved: the strength of a place's fortifications, the garrison's size and its supply situation (as well as the besieger's), the terrain and inundations, ease of access to the site, the location of the artillery park relative to the trenches, the direction from which possible relief efforts might come, and so on. Selecting the best approach could only be finalized after close reconnaissance on site, and pre-existing siege projects might have to be modified accordingly once the reconnoitering engineers discovered alterations to the fortifications and surrounding landscape. Only an experienced eye camped before a fortress could adequately judge the interaction of these many variables.

---

[50] For an example of this claim, see Childs, *The Nine Years War*, p. 93. In 1704 Vauban advised his readers to take advantage of published maps while warning them of their faults. *Traité*, p. 38.

[51] Vauban, *Mémoire pour servir*, p. 41. De Ville displayed a similar ambivalence in *Les fortifications*, p. 297.

[52] Vauban, *Traité*, pp. 41–50; and 259.

These theoretical complications manifested themselves in the real world, adding further confusion to the seemingly-straightforward task of choosing a front of attack. Elaborating upon his general maxim Vauban reaffirmed his decision to attack Valenciennes' most heavily-fortified front in 1677. The decision was based not on the strength of the fortifications, the most obvious criteria and the one privileged by non-technicians, but on ease of access to the site, the close proximity of wood for the construction of fascines, and the terrain, which provided a secure entrenchment that would allow the King to send many more of his siege troops to the observation army in case of a relief attempt.[53] Given such intangibles, it is not surprising that Vauban's decisions were questioned, as Feuquières did in retrospect with the engineer's successful attack on Charleroi in 1693:

> He was charged with the conduct of the works, under M. de Luxembourg, and yet, tho' he was a man of great ability and prudence, and very cautious in hazarding the mens lives, he attacked the place, in a part strong enough to hold out longer than any other quarter, and in a season far advanced. . . . This good fortune should be no inducement to pursue the same conduct in any future conjuncture, and is only to be ascribed to the incapacity of the officer who defended the place, and to the weakness of the defence, which are accidental circumstances, on which the besiegers should never depend. And the enemies of M. de Vauban have declared, that he had an inclination to discover his capacity, in the conduct of the works for this particular attack.[54]

We should probably forgive Feuquières for not referring to Vauban's early admission to ignore the rules if the place is weak, for Vauban later argued just the opposite, leaving posterity with a puzzle. As Feuquières' mention of "enemies" suggests, complaints such as his surfaced during the siege itself, prompting Vauban to rather testily affirm his counter-intuitive preference as the best choice:

---

[53] Vauban, *Traité*, pp. 259–260. La Vergne more generically affirmed the need to consider the terrain beyond the outworks when judging a front's strength. La Vergne, *Nouveau exercice du gabion*, p. 37.

[54] Feuquières, *Memoirs Historical and Military*, vol. 2, pp. 247–248. Feuquières contradicted himself by noting the occasional need to attack a fortress by a stronger side in *Mémoires sur la guerre, où l'on a rassemblé des maximes les plus nécessaires dans les opérations de l'art militaire*, (Amsterdam, 1731), pp. 340–341. His criticism of Charleroi is repeated almost verbatim in P.P.A. Bardet de Villeneuve, *Traité de l'attaque des places*, (La Haye, 1742), pp. 52–53. On Feuquières more generally, see Jean-Pierre Bois, "Le marquis de Feuquière, stratège au temps de Louis XIV," in *Combattre, gouverner, écrire. Etudes réunies en l'honneur de Jean Chagniot*, (Paris, 2003), 147–160.

> I know well that the speed of this siege is not at all acceptable to either the courtiers nor many cavalry officers who like nothing more than to see their horses suffer and who, moreover, only participate in sieges by transporting fascines to the trenches; but I can say they are incompetent judges and very ignorant of the strengths and weaknesses of fortresses, and that if they followed their misguided opinions and attacked places where they preferred, these men would quickly change their minds with the cold and the rain on their backs. In a word, the courage of a man who has warm feet and who reasons in the comfort of his chamber; it is quite different where there is no peril versus when this same person is in the heat of action. It is one thing to be brave far from danger, and quite another to be brave when under fire.[55]

Too busy to refute the accusations in detail, his self-justification side-stepped the issue rather than addressing the critics' concern with delays, doing little to clarify matters. Even the model siege of Ath 1697 saw intangibles play a key role in the decision of where to attack, such as the soft earth that would be easy to dig in at the Brussels gate. Too many variables had to be considered to justify glib judgments.

The great engineer's self-assured statements leave us with yet another conundrum. When Vauban criticized sieges conducted by his contemporaries, such as the Allied attacks on Menin and Ath, we are hesitant to reject his judgment given his vast experience and his successful prediction of the Turin debacle from afar. Yet we must also recall the cases in which his warnings of catastrophe never came to pass, particularly Kehl 1703.[56] His infallibility is further undermined by his own logic when justifying the Charleroi attacks, for if only those present at the siege are in a position to criticize, then his own judgment away from the scene is of less value than those on site, no matter how familiar he may have been with the fortifications years earlier and no matter what intelligence he was receiving from correspondents, thus undermining Vauban's authority at long distance. Unfortunately for both Vauban and the French cause more generally, the fall of both Menin and Ath were very rapid, leaving him with little option but to comment on how fortunate the besiegers

[55] Rochas d'Aiglun, *Vauban*, vol. 2, p. 401, Vauban to Le Peletier de Souzy, Charleroi, 18 October 1693 and vol. 2, pp. 406–407, Vauban to Le Peletier de Souzy, Charleroi, 22 November.

[56] It is possible that had Vauban been present on the site, his more formal attack might have been accelerated due to the same weaknesses that Villars had noted.

were that the garrison's defenses were so weak and poorly-conducted. This rather feeble retort sounds just like Feuquières' criticism of his own Charleroi attacks, and this dismissal could just as easily be leveled against his own 1697 attack on the under-garrisoned and poorly-defended town of Ath. Vauban criticized these sieges not because they were unsuccessful, but because they were inefficient.

Further muddying the waters, even the most blatant Allied 'gaffe,' attacking the town of Douai in 1710 at its strongest point, is less damning than it at first appears.[57] Des Rocques' failure to provide posterity with an explanation of his decision leaves the field open to his opponents. At the time they attributed it, amazingly enough, to his failure to even bother reconnoitering what was purportedly the weakest side, instead assuming that a planned but not-yet-built hornwork displayed on published maps had already been constructed.[58] If true, this would be an unimaginable blunder for the head engineer—perhaps further research will turn up conclusive evidence. Regardless of the real reason for the decision, as it turned out, the attack on the western side of town had merit all the same. First, every map of the fortress indicated the inundations as being far more extensive on the eastern front. More importantly, French galleys from nearby Condé denied the Allies free movement up the Scarpe River, which would have brought the cannon right up to the front under attack.[59] Thus, an attack on the eastern side would have required the heavy siege cannon to have been transported from Lille across the Deule canal and then somehow manhandled past Fort Scarpe's guns (likely requiring this fortification to be captured first). As Vauban had indicated previously in his discussion on Valenciennes, access to

[57] Landsberg made a particularly harsh critique in *Nouvelle manière de fortifier les places*, p. 48. In a more general vein the Spanish veteran Santa Cruz de Marcenado went so far as to contend that chief engineers would not attack a fortress at an approach that previous engineers had targeted, lest they appear derivative! Alvar de Navia Osoris de Santa Cruz de Marcenado, *Reflexions militaires et politiques*, (Paris, 1738), vol. 10, p. 262.

[58] This accusation was made in Schulenburg, *Leben und Denkwürdigkeiten*, p. 458, 30 June; and also reported by the observer Johan Vegelin van Claerbergen in Ryksargyf Friesland (RA) familie Eysinga-Vegelin van Claerbergen, vol. 775, *Beschryvinge van myn Reys en veltogt de anno 1710*, f. 13, 4 July. It appears that the beginnings of a hornwork had been traced out, but it was not completed by the time the town was invested.

[59] BL Add MSS 61401, f. 25 Cardonnel to Walpole, 12 May 1710; and Murray (ed.), *Letters and Dispatches*, vol. 5, p. 21 Marlborough to Boyle, Douai, 12 May.

the site was a legitimate concern and could even be the determining factor.[60]

French observers were less critical of the Allies' engineers, conjecturing that they were motivated less by logistical anxiety than by relief concerns. The artillerist Armand de Mormes, sieur de Saint-Hilaire posited this theory for why the choice of attack was not as bad as it might seem at first glance:

> the enemy directed his attacks against the gates of Ocre and Esquerchin, although this place was the best fortified. They did this correctly assuming that the *maréchal* would not miss the opportunity of trying to relieve Douai by passing the Scarpe across from Arras, and would present himself in battle before them, and marching by the plains between them both. Thus they would cover their army on this side with good retrenchments erected on the most advantageous and favorable terrain, constructed with all the precautions and rules of the art. In this position their army covered their approaches that were behind them.[61]

Villars' efforts to lift the siege did indeed fail on exactly this front, despite his attempts to force a relief battle by attacking several posts along the Allied line. Whether these advantages were worth the additional time and effort spent attacking this front (in total 52 days of open trenches and 8,000 casualties) is a valid if unanswerable question, but the hypothesized concern about relief efforts certainly corresponds well with the Allies' tendency to favor observation forces over their siege armies. The facts that the biggest Allied 'blunder' in the theater could be justified by referring to Vauban's own successful conduct at Valenciennes, that it was recognized by the enemy as a reasonable choice, and that it was consonant with the Allied preference for privileging covering forces (even at the cost of weakening the siege effort) suggest that blanket condemnations and vague after-the-fact criticisms should be accepted only with hearty skepticism.

We further recall that the chief engineer had to have his attack plans approved by the commander and other generals in charge of the siege. Given Douai's 'self-evident' strengths, the engineers had

---

[60] For this reason, the chief of the Allied artillery had wanted to attack Menin by a different approach in order to facilitate their transports. AG Article 15 Section 3, Menin folder, Valory, Relation de ce qui s'est passé au siège de Menin, f. 4.

[61] Lecestre (ed.), *Mémoires de Saint-Hilaire*, vol. 5, p. 239. This conclusion may have been a widespread one among the French, for the provincial intendant of Hainaut and French Flanders also interpreted the Allied camp in this way. AG A[1] 2225, #162, Bernières to Voysin, 10 May.

surprisingly little difficulty convincing their generals of how weak its defenses were. In fact, even after they had broken ground against the strongest sector, we can still find Albemarle confident of this decision: "And I flatter myself that we will soon capture this place, because it is very weak and the least of all the places we have attacked this entire war."[62] It should come as somewhat of a surprise then that after the siege ended he conveniently forgot his own credence in this 'obviously' flawed assessment and performed an about-face in order to blame the engineers for underestimating its defenses:

> I cannot stop myself from telling you that our engineers are extremely ignorant of how to attack a fortress. It is certain that if we had attacked this place on any other front than where we did, we would have captured it in eight days. This is a truth that everyone must acknowledge, it being as clear as day; our conduct shouts out to our enemies our obvious incompetence and will force us to think twice before attacking another fortress.[63]

To this list of apparently hoodwinked officers we could add Marlborough's favorite William Cadogan and the Dutch quartermaster Yvoy heer van Binderen, while even the Captain-General himself vastly underestimated the length of the siege after a week of open trenches.[64] Similar criticisms of their engineers' underestimation of Béthune's works is also ironic given that it was the 'discredited' Des Rocques who exposed the flawed plan of attack of the chief engineer Hertel, who had been chosen as the director-general's successor.[65] Determining the best sector to attack was apparently not

[62] Veenendaal, Jr. (ed.), *Briefwisseling Heinsius*, vol. 10, p. 320 #623 Albemarle to Heinsius, Flines, 8 May. Even as late as mid-May he continued to believe the town would soon be theirs. Vol. 10, p. 343 #663 Albemarle to Heinsius, Flines, 15 May.

[63] Veenendaal, Jr. (ed.), *Briefwisseling Heinsius*, vol. 10, p. 500 #1004, Albemarle to Heinsius, Douai, 2 July.

[64] Cadogan's opinion reported in a Bernières report in A[1] 2225 #133; Veenendaal, Jr. (ed.), *Briefwisseling Heinsius*, vol. 10, p. 332 #650 Frederick Thomas d'Hangest-Genlis Yvoy, Lallaing, 12 May; and BL Add MSS 61401, f. 25, Cardonnel to Walpole, 12 May.

[65] See Jules Duvivier and J.W. Wijn, "Documents hollandais rélatifs à la guerre de succession d'Espagne dans le Nord de la France," *Revue du Nord*, vol. 39 (156), 1957, Mémoire comment on a trouvé la ville de Béthune en la reconnoissant pour en faire le siège, et les choses qu'il faut observer pour s'en rendre le maistre, 22 July. This example also points out the critical importance of intelligence, yet another factor that the engineers had little control over.

as straightforward as one might think, particularly for engineers seeking the most efficient solution, and even for those less knowledgeable in the fine points of siegecraft. Nor should this conclusion be surprising to anyone who has personally visited fortresses. From a viewer's ground level viewpoint, the specific design features of a fortress were not always easy to discern; this was the point after all, to hide the defenders and their works from the besieger's view—and the perspective can be just as confusing at the bottom of a ditch. Certainty would further recede when we recall that the various plans for a particular town, seeming to promise 'objective' measures, could be quite wrong in their details. For whatever reason, hornworks in particular were frequently substituted for ravelins or lunettes in published maps, despite their rarity on the ground. The imposing nature of a hornwork would certainly be enough to discourage an attack on that side. This difficulty of summing up the design of a fortress in a *coup d'oeil* is undoubtedly also the reason why almost every artistic representation of a siege or fortress that wishes to display more of the town than its skyline adopts a three-quarter elevation for visual interest. Viewing the fortress from a height in the foreground, the viewer (often times through a proxy, e.g. a general and his entourage or, more bucolically, a shepherd and his flock) is able to look down on the town as it spreads out before them. It is no surprise that the less 'artistic' and more 'scientific' aerial plan came relatively late, for it was a foreign view to the experience of most people.

In any case, it is also interesting to note that the French also became frustrated with their own engineers' conduct of the trenches at Douai. In their impatience Valory's critics censured him for being ignorant of a fortress he had just defended, assailing him for insisting that they delay their attack until he was able to examine the works in person.[66] It was his detractors, however, who appear ignorant, ignorant of the possibility that the fortifications could have changed in the past two years and of Vauban's caution to wait until they could view the works and take their measure of the garrison before giving a due date. In fact, the Allied garrison had indeed altered the fortress's defenses, erecting a number of additional earthen

[66] The siege commander Alègre confirmed Lefebvre d'Orval's report of the engineers' surprise at seeing such *nouveautez*, but his tone was more humble, concluding that "it is hardly appropriate for an officer of cavalry to comment on siegecraft." AG A[4] Carton 8 chemise 2, to Maine, 2 September.

works on the front of attack that sheltered the curtain wall from view—exactly why Vauban insisted on intensive on-site reconnaissance. If Valory committed a massive blunder regarding Douai, there is little indication of it when the Secretary of War Voysin reassured the engineer of the King's satisfaction with his conduct in a mid-September dispatch. However low Valory's standing had fallen, the Secretary also expressed his hope that the engineer would soon recover from his past exertions, as "it is up to you to determine the attacks [at the next siege of Le Quesnoy]."[67]

In criticizing the engineers, both Allied and French generals were rejecting not just their siege directors, but Vauban's own principles and practices as well. Given their continued acquiescence to the decisions of their scapegoats, we must conclude one of three things. Either the generals were just as incapable of judging the strength of a place as their engineers, in which case these criticisms are either unfounded or at least based on unrealistic expectations. The second possibility is that they left the reconnaissance of the fortresses entirely up to engineers that they deemed incompetent, something against the recommendations of James Duke of York. If this is true they must surely shoulder some responsibility for their apparently blind willingness to rely on the advice of experts who had 'proved' themselves utterly inept in siege after siege.[68] Or most likely, and what Vauban knew from long experience, the best point of attack was not always as obvious at the time as post-siege critics would have us believe. Impatient generals were forced to rely on the corps' expertise and only with hindsight criticized their technicians for their 'obvious mistakes.' Regardless of which option is true, none of these specific criticisms of the points of attack provide solid evidence of systemic technical stupidity. Even if the overworked engineers were mistaken in their initial choice of attacks (decisions which were largely irrevocable, as, once made, they could rarely be reversed), this would only show that they had difficulty attacking some of Vauban's strongest fortresses (such as Lille and Tournai) and Douai as well, one of the

[67] AG A[1] 2383, #220 Voysin to Valory, Fontainebleau, 13 September. At this siege there would also be disagreements between the various engineers and artillerists over the best approach to attack. AG A[1] 2384, #1, Villars, 16 September.

[68] If siege commanders were unable to reject their experts' misguided advice, this would be well worth an explanation, as it flies against the tendency of these same generals to ignore or overrule their engineers at every other stage of the siege.

more irregularly-fortified strongholds in the theater.[69] That their mistakes led to Vauban's crown jewel of Lille surrendering in four months instead of two or three hardly justifies the scorn that was heaped upon them—it does, as we shall see, give us a clue as to the real motivation for their criticisms. Trying to determine how best to attack a fortress required a careful consideration of many variables; had generals paid more attention to Vauban's emphasis on applying the general maxims to specific cases, or had they even spent more time interacting with the engineers, they might have recognized the difficulty of the decision, or at least been more sympathetic to the engineering challenges.

Closely related to the condemnation of misguided attacks was the most frequent complaint of all, that the engineers vastly underestimated the time required to capture fortresses. In the War of the Spanish Succession, complaints about underestimates peaked at the longer sieges late in the war, again suggesting that engineers had the most difficulty attacking the most imposing strongholds.[70] The Earl of Stair reflected the general disgust at Lille: "Our ingeniers tell us again the 19th wee shall have the town. If they don't keep their words I am for sending them to our friend the Emperor of Maricco for carrying on his siege of Ceuta."[71] The town only fell on the 23rd, but the threat of exile went unfulfilled. A year later, Cardonnel gloomily informed the English Secretary at War for the North: "Our siege [of Tournai] goes on in such a dark manner that I dare not give any guess when we shall be masters of the citadel."[72]

[69] Douai's advanced ditch presented a particular challenge for the Allies; Des Rocques explicitly mentions this in Veenendaal, Jr. (ed.), *Briefwisseling Heinsius*, vol. 10, p. 430, 12 June 1710. Landsberg also noted that "our engineers had never attacked this type of outwork before"; he called for more experienced engineers. As we have seen, there were few to be had. Landsberg, *Nouvelle manière*, p. 42. An advanced ditch at the approach against Béthune's St. Prix gate also forced the Allies to proceed by sap rather than storm. AG Article 15 Section 2, # 4bis, Journal du siège de Béthune, ff. 29, 30, 34, 37v.

[70] For an example of a complaint that does not identify the engineers as the source of the problem, see Veenendaal, Jr. (ed.), *Briefwisseling Heinsius*, vol. 1, p. 227 #390, Nassau-Saarbrücken to Heinsius, Kaisersweert, 23 May 1702.

[71] Royal Commission on Historical Manuscripts, *Report on the Manuscripts of the Earl of Mar and Kellie*, p. 465, Stair to the Earl of Mar, Rosselare, 15 October 1708. Spanish-held Ceuta (located across the straits from Gibraltar) withstood an epic blockade/siege by the Sultan Moulay Ismail from 1694 to 1720.

[72] BL Add MSS 61400, f. 190, Cardonnel to Robert Walpole, Tournai, 19 August 1709.

In this context, the besiegers were pleasantly surprised five days later when the citadel surrendered due to a lack of supplies, one of the strongest fortifications in the Low Countries holding out a mere thirty-four days of open trenches. Douai 1710 again provides the most striking example, where one Allied observer wrote that although he knew "we are often too hasty in computing the time that a town can hold out," most expected the town to last only ten to fourteen days of open trenches.[73] These projections soon proved overly optimistic, Albemarle informing Heinsius a month later of the opportunity they had lost: "Assuming our engineers want to take advantage of [the French abandoning their relief attempts] it would be of great use, but to my great dismay I must tell you that things could not be going more slowly than they are at this siege, because we still cannot judge when we will capture the town."[74] Early expectations were thirty-nine days too short, making this siege one of the Allies' most exasperating. The rest of the campaign would provide more occasions for frustration. Waiting for the next stronghold to fall, Marlborough's disgust grew as the engineers' uncertainties and caution confounded his plans for the rest of the campaign season. "Wee opened the trenches last night before Bethune. Our ingeniers have of late used themselves to go on, as they call it, with safety, [so] that I dare not name a time for our being masters of the town."[75] Of all their supposed faults, the engineers' inability to predict the end of sieges rankled generals most of all.

A more systematic examination of the issue is needed to test these claims. The common expectation (both then and today) that engineers could accurately predict the stages of a siege ignores a more complicated reality. The origin of this misconception derives from the acceptance of the rhetoric of siege history, in particular the widespread view of Vaubanian siegecraft as scientific, scientific in the sense of predictable. The difficulties related in chapter 5 should hopefully put to a rest this myth's underlying assumption that engineers' theoretical plans were slavishly applied on the ground. Convinced

[73] Royal Commission on Historical Manuscripts, *The Manuscripts of the Marquess Townshend,* (London, 1887), p. 64.

[74] Veenendaal, Jr. (ed.), *Briefwisseling Heinsius*, vol. 10, p. 408 #801, Albemarle to Heinsius, Douai, 6 June 1710.

[75] Snyder (ed.), *Marlborough-Godolphin Correspondence*, vol. 3, p. 1567 #1614, Marlborough to Godolphin, Villers Brulin, 24 July 1710.

of the systematic nature of the attack and uninterested in examining the matter in detail, historians have accepted this mistaken belief at face value and whittled the Vauban literature down to a wooden stereotype. The common claim that Vaubanian engineers could accurately predict the day a town would fall has even transformed into the assertion that "Vauban claimed to be able to predict [a siege's] course on a daily timetable."[76] Scholars have based such a position on a few of his writings that include timetables of siege stages and on claims by later 18th century authors, though these scholars have ignored similar claims of prognostication made by earlier military writers, witness the previously-cited claims that early seventeenth century siege durations could also be predicted within days.[77] A few scholars have noted that expecting accurate predictions was asking the impossible, though they undermine their admonitions by relying on the rhetoric of siege-as-science.[78] Vauban's own view was quite clear. His oft-cited 1672 treatise estimated that an average siege would require 41 days. He prefaced his calculation by stating that it was only an "instructive calculation and not a hard, fast rule," yet his timetable seems to have been taken much more literally than he ever intended.[79] The context of the schedule itself gives plenty of reasons for caution in accepting his figures as typical. His estimate's

[76] Dickinson, "Richards brothers," 77. For expectations of predictable progress by stage (or even by day): quote in Tallett, *War and Society in Early Modern Europe*, p. 511; see also Pollak, *Military Architecture*, p. xxxiv; and Lynn, "Vauban" entry, p. 459.

[77] Later French engineers such as Cormontaigne and Fourcroy de Ramecourt would extend Vauban's timetable to a much more precise, predictive measure, (though its validity was spurious at best). Duffy, *The Fortress in the Age of Vauban*, p. 153.

[78] Although Chandler's language elsewhere belies it, he does acknowledge at one point that Vauban's 48-day timetable was "only intended to be a rough guide to assist commanders estimate what they were up against." *Art of Warfare in the Age of Marlborough*, p. 246. Duffy also warned readers not to mistake Vauban's timetable with a prescription of how a siege should unfold, to little avail. *Fire and Stone*, pp. 101–104; and *The Fortress In The Age Of Vauban*, p. 153. Yet in *The Military Experience in the Age of Reason*, (New York, 1988), p. 294, he wrote that "if the besiegers made a serious trench attack on the fortress . . . . the progress of the rest of the siege could be predicted with reasonable confidence."

[79] This timetable comes from Vauban, *Mémoire pour servir*, pp. 191–193. In the text, forty-three days is written, but the days add up to only forty-one. Duffy mentions a similar timetable Vauban calculated for how long the fortress of Lille could hold out (*Fire and Stone*, pp. 103–104), while Faucherre and Prost mention a figure of 48 days (*Le triomphe de la méthode*, p. 53).

specific purpose was to calculate the supplies a garrison would need for a siege, since many previous fortresses had surrendered early due to a shortage of supplies.[80] He therefore made a number of significant assumptions that lengthened the hypothetical siege's length. First, he assumed that the fortress would have an adequate garrison and defend itself competently. Second, he assumed that the besieger would attack the strongest sector of the defenses, thus prolonging the siege and increasing the demand on the garrison's supplies. Third, the hypothetical fortress he based these estimates on was a regular-shaped, six-bastioned fortress with well-revetted demi-lunes and ditches, a covered way with good palisades and no other outworks. None of these assumptions could be taken for granted during any actual siege, but they satisfied Vauban's pedagogical need to illustrate planning for a worst-case scenario. Considering the purpose and assumptions of Vauban's timetable, it is clear that some historians have misinterpreted its predictive power. Our case study of Ath 1697 above reinforces this conclusion: predictions of its duration varied from twenty-five days (according to Vauban), to fifteen or more days (various Allied accounts), to only eight days according to many at the French Court—the town withstood fourteen days of open trenches.

The case of Douai 1710—where initial estimates were off by a factor of four—is difficult to interpret as anything but a major engineering gaffe, though our earlier discussion showed that the engineers were not the only ones to underestimate its duration.[81] The common failure to accurately estimate the lengths of attacks would not have surprised Vauban, for the master himself preached that such things were impossible to know beforehand, even when, at the pinnacle of his authority, he came closer than any to conducting a siege according to his specific wishes. Vauban was quite explicit about such claims at prognostication:

> You cannot estimate the time needed to reach the counterscarp, as this depends on its distance from the opening of the trenches, the vigor of the garrison, the quality of the terrain, the supply of materials and the number of workers. I have seen some sieges where we advanced

[80] Vauban, *Mémoire pour servir*, p. 187.

[81] Though hardly conclusive, the only case where a direct comparison can be made between the engineers' estimates and the commander's (Alègre during the siege of Le Quesnoy) shows the engineers' estimates more accurate than those of the commander. See Appendix E for details.

at a consistent rate, and others where we could not advance fifty paces in a night when we were near the place; it was the same at Montmédy [in 1657] where we could hardly advance sixty paces in eight days.

Vauban's opinion was only repeating the common sense of the previous generation as summarized by Mallet: "As it is difficult to find two towns situated in the same way, it is very difficult to conduct two sieges in the same manner, and ever more difficult to determine their duration."[82]

What is more, Vauban was in a rare position to practice what he preached, since his expertise and royal imprimatur granted him the ability to speak frankly with his superiors. One such example of Vauban's ability to dismiss unreasonable demands can be seen in his response to Louvois' repeated requests for an estimate of how much longer the siege of Luxembourg in 1684 would last:

> It is not possible to tell you when we will lodge ourselves on the grand counterscarp, because this depends on the difficulties and quality of the parallel and the resistance of the redoubts that we are attacking; when I see the day that I can predict the future with some appearance of certainty, I will be sure to tell you; but I am not an astrologer. There are some events of which God alone knows the outcome and its timeframe; it is up to men to do all that they can to succeed, as I am doing, God willing.

It was only twelve days later on 26 May that he felt confident enough to inform the director-general that "The time when this place will be captured is not something that a man of good sense would dare to guess at; but as far as I can conjecture, I expect that it will last seven or eight days longer, give or take two days."[83] The town fell eight days later on the third, within the wide five-to-ten-day margin of error he had given himself. Vauban's ability to put off the Secretary of War's demand for prognostication and his refusal to pin down a narrow date range when he did finally provide an estimate derived from his personal standing with the King and his blunt yet amiable relationship with Louvois, a rapport built up over decades of close

---

[82] Vauban quote in *Mémoire pour servir*, pp. 95–96. Alain Manesson Mallet, *Les travaux de Mars*, 3 vols. (Amsterdam, 1672), vol. 3, p. 225. Goulon states the same in his *Memoirs*, p. 18; likewise the marquis de Quincy's mid-18th century manual: *Maximes et instructions*, in *Histoire militaire*, vol. 7, p. 138.

[83] Rochas d'Aiglun, *Vauban*, vol. 2, pp. 234–235, Vauban to Louvois, Luxembourg, 14 May 1684; and p. 239, Vauban to Louvois, Luxembourg, 26 May 1684.

collaboration.[84] An undaunted attitude such as this was out of the question for lesser engineers who were interrogated by strong-willed commanders such as Marlborough or Eugene or Villars. Rarely given a chance to conduct the attack according to their own designs, they were unfairly condemned when a timetable they were forced to provide did not proceed according to their initial plan. It is hardly surprising that in such a hostile atmosphere the engineers became distrustful and defensive, an anxiety which prompted Marlborough's assurance to Des Rocques: "I ask you to immediately write me in confidence by return of this courier [how long you think the siege of Lille's citadel will last] and without reserve, since you can be assured that I will not show your letter to anyone."[85] Vauban was powerful enough to defer questions that he knew were impossible to answer without fear of consequences; his successors were not, and yet they were consequently criticized for failing to provide something that Vauban knew even he was unable to give: an accurate, precise estimate of a siege's length in the early stages of a siege.

A more systematic analysis of such estimates provides more evidence to moderate the criticisms targeted at the engineers. A selection of 83 specific length estimates is listed in Appendix E. The results indicate, first, that engineers underestimated the duration of their sieges by only three days on average, while the median estimate was exactly on target, though in only one of the actual cases was the exact date predicted in advance. These errors averaged a third of the length of each siege, with the median again lower at only a quarter of siege length—a frustrating delay for generals but hardly a sign of gross engineering incompetence given the relatively short time periods involved. The following chart illustrates the distribution of these estimates and the extent to which they group around the lower error values.

The distribution is skewed towards the negative values due to a few exceptional sieges like Douai 1710, but the overall trend suggests a moderately accurate estimation, which is likely due to the

[84] On the Louvois-Vauban relationship, see Blanchard, *Vauban*, pp. 136, 286; and Hebbert and Rothrock, *Soldier of France*, pp. 29–30. It can also be traced through the volumes of Rousset, *Histoire de Louvois*.

[85] Murray (ed.), *Letters and Dispatches*, vol. 4, p. 331, Marlborough to Des Rocques, Beerlegem, 2 December 1708. This letter suggests that the severe criticisms of Des Rocques and his engineers had made their way back to the chief engineer.

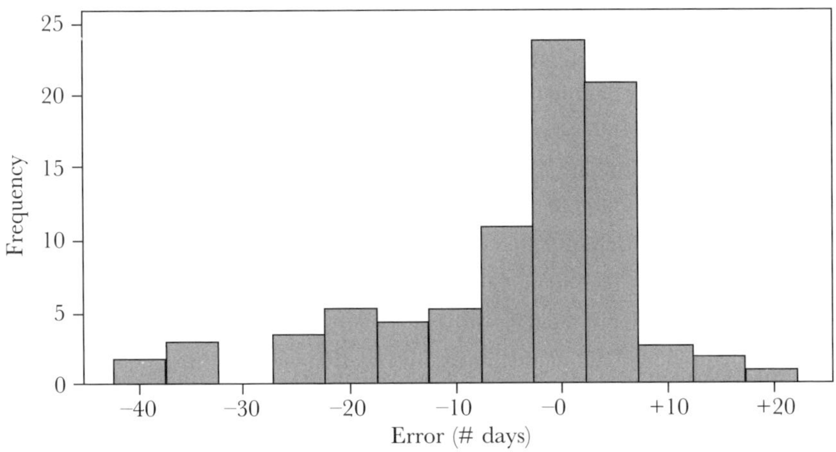

Figure 6.1 Distribution of Siege Length Estimate Errors

large number of estimates made toward the tail end of sieges. Looking at individual sieges, we see yet again the divide between early Allied sieges and the later ones amongst Vauban's *pré carré*. Contrary to the impression given by the generals, the predictions of sieges before 1708 usually *over*-estimated the length of the defense rather than *under*-estimating it. The engineers saw far fewer complaints of incompetence at these sieges, suggesting that the generals were not disappointed so much with the engineers' inability to accurately predict duration per se, but rather were only critical when sieges lasted longer than promised. We should also note that the accuracy of most estimates improved as each siege progressed, as at the Allied sieges of Venlo, Huy, Limburg, Menin, Lille, Douai and Aire, as well as with the French sieges of Douai and Le Quesnoy in 1712. That the besiegers were generally able to improve their accuracy as they approached the enemy walls is not very surprising, and hardly proves engineering ineptitude. Rather, it supports the logical conclusion that the longer and more difficult the siege, the greater the opportunities for things to go wrong and thus for initial estimates to be off, and, conversely, the closer to its conclusion, the less likely a siege would be thrown off by some major intervention or mishap. One could contend that the engineers suddenly lost their competency after 1706, but it seems more likely that they had understandable difficulty trying to predict the progress of siege attacks against some of Europe's strongest fortresses. Balanced against generic complaints, this data supports the reasonable conclusion that engineers were as accurate in their predictions as one might expect given the many unknowns of siegecraft.

2. *Who Was Right?*

It is difficult to determine conclusively whether the engineers consistently bungled the attacks, or whether alternatives suggested by non-engineers were a better option, because the sources rarely allow an adequate basis for sound judgment. But a more careful scouring of the sources does turn up some suggestive results. We discover several occasions where vigorous officers made clear mistakes in ignoring engineering counsel, examples which illustrate that sometimes there was no reasonable alternative to the cautious approach.[86] One Allied account of the siege of Aire informs the reader "At midnight we attacked the salient angle of the covered way across from the bastion which had been occupied by the besieged: this attack did not have the effect we had expected, we were repulsed with loss, which forced us to continue with the sap towards this angle in order to capture it."[87] Des Rocques justified his part in the botched storm by reminding the Duke that he had agreed to storm the covered way at *both* approaches at the same time in order to divide the garrison; he could not be held responsible when superiors ignored his recommendations and allowed the enemy to concentrate its forces.[88] On the French side, Broglie was sent to summon the Allied supply depot of Marchiennes to surrender right after Villars' victory at

[86] Blaming failed storms on the engineers, whether fair or not, did not lead detractors to question the viability of their preferred assault tactics. Across the Channel, Oliver Cromwell had earlier discovered the price of foregoing the patient approach in several of his Irish sieges, e.g. Duncannon fort, Waterford and Clonmel.

[87] AG Article 15 Section 2 §1 Aire, carton 1, #5 Relation du siége d'Aire en 1710 par un officier de l'armée des Alliés, 1 November. Marlborough confirms this in Murray (ed.), *Letters and Dispatches*, vol. 5, p. 205, Aire, 2 November.

[88] Veenendaal, Jr. (ed.), *Briefwisseling Heinsius*, vol. 11, p. 212 #387, to Heinsius, 2 October 1710; and BL Add MSS 61339, f. 187, to Marlborough, 28 October. We recall that Nassau-Saarbrücken had followed this advice at Kaisersweert. Des Rocques complained to Heinsius of an earlier instance of being ignored by the siege commander on p. 212, 2 October. For a French engineer's defense of his peers and a critique of infantry officers during the storming of Béthune, see AG Article 15 section 2, #4bis, Des Forges, Journal du siège de Béthune, ff. 40–42.

Denain. The garrison's recalcitrance led this harsh critic of the engineers to attempt two assaults against the town without any preparation—several hundred French soldiers were wounded and killed as a result of this futile effort. His fellow Lieutenant-General the duc de Guiche sounded much like an engineer in criticizing this brash behavior: "it is ridiculous that we have lost such men due to a lack of planning and precaution." Given a chance, the patient approach worked quickly, as the weakly-held town fell to a subsequent formal siege of brief duration.[89] The chevalier de Quincy highlighted another example at Freiburg in 1713 where a *maréchal de camp* foolishly attacked the covered way before the garrison's works had been sufficiently prepared by artillery and its countermines neutralized.[90] Had he been alive, Vauban would have shaken his head at the cost of such foolhardy actions; he had warned of exactly such setbacks—"precipitation in sieges never hastens its end, often retards it, and always makes it much bloodier."[91] Unfortunately for many of the rank-and-file, impulsive generals were slow to recognize this reality forty years after Vauban had made his first appeal. How much we can generalize from these examples is unclear, but the facts that so few self-justificatory engineering accounts survive and that the vast majority of extant sources were extremely unlikely to mention *anything* positive about the engineers make one suspect that a more varied evidential base would likely provide many more examples that one could use to counter the claims of impatient critics.

Accusations of utter incompetence are also difficult to believe given the impotence of the critics in dealing with the problem. It strains credulity to imagine that either the French or Allied engineers were nearly as bad as they were portrayed. If for no other reason, in siege after siege the same 'incompetents' were left in charge. In spite of English grumblings that inept Dutch engineers were blindly protected by interests within the States-General, the perpetual shortage of

[89] On Marchiennes' storms, see *Daily Courant*, 24 July 1712 O.S. #3364, from Tournai, 27 July. A report from Marchiennes, 25 July in the same edition gives 250 French casualties. From the French side, see Lecestre (ed.), *Mémoires du chevalier de Quincy*, vol. 3, p. 159. Guiche's quote is in AG A^4 Carton 8 chemise 1, #35, 28 July.

[90] Lecestre (ed.), *Mémoires du chevalier de Quincy*, vol. 3, pp. 255–256.

[91] Vauban, *Traité*, p. 263 Maxim XVII. He cites here the examples of Barcelona 1697 and Landau 1703.

engineers meant that the Allies had few alternatives.[92] Even the harshest detractors acknowledged that the engineering ranks were constantly short of capable members—'incompetents' at the top could hardly be replaced from below when their subordinates were censured just as harshly. The shortage of qualified engineers forced highly-critical commanders to set aside charges of gross incompetence and keep those deemed responsible for botched attacks in positions of authority.[93] The 1710 campaign illustrates this point most directly. Despite widespread condemnation from both British and Dutch generals, Des Rocques had managed to retain his post as director-general of fortifications. But after his widely-criticized attacks at Douai, he was initially refused the direction of Béthune as a veiled punishment.[94] The director of approaches Hertel replaced his superior but proved just as exasperating. A week after investing Béthune he had still not opened the trenches while Des Rocques provided evidence that he had significantly underestimated the strength of the town's works in his projected attack. Soon afterward Hertel returned to his original position and the formerly-disgraced director-general was rehabilitated and put back in charge of the siege.[95] The fact that it required the 'discredited' head engineer to illuminate Hertel's errors, that Des Rocques' critiques convinced the generals, and that

[92] For one such accusation, see BL Add MSS 61310, f. 3, Blood to Marlborough, Menin, 2 August 1706.

[93] The above-mentioned Holcroft Blood had wanted to conduct the siege of Menin, but the Dutch questioned his experience and feared the further disputes his command over a Dutch corps would engender. Later in this siege he would be proven wrong by Des Rocques regarding the necessity of an attack on an outwork. Veenendaal, Jr. (ed.), "De Mémoires van Sicco van Goslinga," p. 27.

[94] Veenendaal, Jr. (ed.), *Briefwisseling Heinsius*, vol.10, p. 546 #1095, Holland field deputy Willem Hooft to Heinsius, Aubigny, 16 July 1710; and Ryksargyf Friesland (RA), familie collectie Eysinga-Vegelin van Claerbergen 775, Johan Vegelin van Claerbergen (younger brother of the *Raad van State* field deputy Philip Frederik), *Beschryvinge van myn Reys en veltogt de anno 1710* , p. 29. The cover for Des Rocques was that he had to oversee the repairs to Douai's fortifications. ARA RvS 688–2, Vegelin van Claerbergen to *Raad van State*, Aubigny, 14 July; and Veenendaal, Jr. (ed.), *Briefwisseling Heinsius*, vol. 10, p. 519 #1044, Vegelin van Claerbergen to Heinsius, Douai, 7 July.

[95] Veenendaal, Jr. (ed.), *Briefwisseling Heinsius*, vol. 10, p. 558 #1114, Arnold Joost van Keppel, Earl of Albemarle to Heinsius, Villers Brulin, 21 July 1710. For Des Rocques' exposé on the true state of Béthune's fortifications, see p. 564 #1129, Des Rocques to Heinsius, Béthune, 24 July 1710. The attached intelligence report can be found in J. Duvivier and J.W. Wijn, "Documents hollandais rélatifs à la guerre de succession d'Espagne dans le Nord de la France," pp. 249ff.

he was reinstated, indicates the difficulties the Allies encountered finding skilled engineers and brings into question their independent judgment of siege matters. If directing a skilled attack was truly as self-evident to non-engineers as some suggested in their (after-the-fact) criticisms, then these inept men at the top should have been quickly cast aside and more talented ones from below (or even from outside their ranks, as William had done with Coehoorn) taken their place. After all, the chief engineer was subordinate to the siege commander, so messy disputes over seniority and precedence were not a concern, as they might have been had the engineers been fully integrated into the command hierarchy and given unquestioned authority.

Even the damning portrait of Allied engineers painted by their peer Landsberg is unconvincing when placed in the context of his work's endless series of hyper-critical assessments of almost all his non-German peers (including Vauban), as well as his many demonstrably false claims about the sieges in which he participated. In this broader light he appears more as a crank looking for employment by belittling his peers rather than the epitome of engineering efficiency and an unimpeachable witness to systemic ineptitude.[96] Disappointing Landsberg, the Dutch Republic would continue to rely on Des Rocques as director-general of fortifications well into the 1720s, and even promote him to general-major of the infantry in 1727. To those who criticized the corps' reliance on such incompetents, Heinsius could offer little assistance. Despite having ordered Goslinga to keep his eye out for "extraordinary" soldiers who might be cultivated as "the most necessary men in war," he threw up his hands several years later at the meager results of this effort: "I am frustrated to hear that our engineers are so incompetent, but I do not see any remedy. If we were able to get them from elsewhere, this would be good."[97] If the engineers truly were incompetent, the Allied commanders appear just as foolish as their bungling technicians for giving clear

---

[96] In many respects he repeated the maxims of other engineers, though presenting them as his own insights. To cite a striking example of his contrariness, he summarily dismissed Vauban's masterpiece of Neuf-Breisach as a "very poor imitatation of the Ancients." Landsberg, *Nouvelle manière de fortifier les places*, p. 14.

[97] Veenendaal, Jr. (ed.), *Briefwisseling Heinsius*, vol. 10, p. 510 #1025, Heinsius to Albemarle, The Hague, 5 July 1710. See also vol. 5, p. 581 #1127, Heinsius to Goslinga, The Hague, 6 October 1706.

incompetents such freedom in directing the trenches in the first place and their political leaders for refusing to pay to train better ones. There was in fact no easy solution to the 'engineering problem,' no matter what the officers of the cavalry and infantry claimed—they clearly could not do without them.

Given the lack of engineering accounts, favorable remarks from their critics must also go some way toward vindicating their performance. The Dutch field deputy Goslinga noted the larger structural issues the engineers were forced to deal with at Lille:

> It was decided yesterday to assault the tenaille and the hornwork; our directors were very apprehensive of its success. I was very much of their opinion, and to speak frankly, I despaired of success. Our infantry, which suffered extremely, is discouraged, and the size of the siege army is not nearly as large as is required for a siege of this importance. You will remember what I have already written to you on this. I told this to the Duke and to the Prince before the siege even began. Eugene agreed with me but did not want to press the Duke for a reinforcement.[98]

Even some of those normally critical of the engineering corps were honest enough to acknowledge on a few rare occasions the intense bias against the corps. Guiche, also a colonel of the Gardes françaises, defended the French corps to his patron and one of the corps' most partisan opponents, the *grand-maître* of the artillery: "it is certain that the engineers [at Douai in 1712] have committed faults, but it is also clear that there have been unfair complaints leveled against them."[99] We even find a few unguarded moments where generals admitted the unfairness of their earlier complaints, such as the ephemeral change of heart Villars experienced in 1712. At first quite critical of his technicians during the attack on Le Quesnoy, the self-assured marshal soon thereafter made a rare admission of fault:

> In truth, I cannot praise everyone too highly; my impatience made me think that we could accelerate the siege, but when I examined Valory's and de Vallière's explanations, both of whom are quite capable, I was convinced that they must be trusted and that it would be a mistake to force them to go faster than *raison* would allow.[100]

---

[98] Veenendaal, Jr. (ed.), *Briefwisseling Heinsius*, vol. 7, p. 531 #1167, Goslinga to Heinsius, Lille, 20 September. As previously mentioned, the Allies had a habit of strengthening the covering force at the expense of the siege army.

[99] AG A^4 Carton 8 chemise 2, 14 September 1712.

[100] AG A^1 2384, #174, Villars to Voysin, 30 September.

In spite of this unprecedented disclosure, Villars' conciliatory attitude towards the engineers quickly faded. As we have seen he promptly returned to his dismissive tone in successive sieges, and this uncharacteristically humble tone failed to find its way into his recollections of Le Quesnoy as written in his memoirs. These examples suggest that such charges were motivated by more than just concern over how closely the engineers approached Vauban's ideal attack.

We could examine the other criticisms mentioned at the beginning of this chapter and such analysis would only reinforce our conclusion that the vast majority of these condemnations tell us little about the engineers' competency. To the extent that these critiques were motivated by discrete events, most of the complaints provide little insight into the engineers and more into those making the complaints. The criticisms were either hypocritical, or came after generals forced actions which were explicitly disapproved of by the engineers and violated Vauban's precepts, or they were based on false assumptions, or else on wildly unrealistic expectations of the predictability and controllability of siegecraft. Undoubtedly individual engineers made mistakes in particular cases, yet this is in itself hardly damning to the entire corps or its proficiency—what is so surprising is that the generals' consistently singled out the engineers (as a group) as if they were the only ones to make mistakes. It is also baffling that despite the accusers' alleged certainty of engineering ineptitude, these critics did not feel confident enough to take matters into their own hands, but instead let the incompetents continue to plan the attacks at siege after siege. As Chapter 5 illustrated, the engineers judged incompetent were perpetually working under fire at the same time as they were under-staffed and under-compensated and under-appreciated, if not abused. In such conditions they were given the unenviable task of trying to rein in independent-minded non-engineers without the necessary command authority. To the conventional claims of engineering incompetence we must also consider this context of hostility and impotence in judging their conduct. How they were supposed to, for example, account for these many unknown variables in length projections demanded of them at the very beginning of sieges (forecasts which Vauban himself admitted were impossible to make), or why they were expected to easily see through a thinking enemy's attempts to conceal its inner fortifications and defensive strategies, remains a mystery. It is also certainly true that the engineers who followed Vauban and Coehoorn lacked the vast expertise of their predecessors, but this was an extremely high bar to set in the first place.

We also know that the chief engineers conducting these sieges had substantial experience with siegecraft before their moment in the sun, and we further recall that, to the extent that the corps was inexperienced, it was largely a result of the broader political refusal to support and institutionalize the engineers. This fundamental inequity is further highlighted by the fact that the engineers were even blamed for the failings of those personnel clearly beyond their control. The engineers were blamed for the artillery's mistakes, most strikingly when the Sun King's neophyte Secretary of War declared *Vauban* the cause of Charleroi's wasted powder, even referring to the engineer's purported "pleasure in seeing so many cannon firing and bombs flying in the air when this is not necessary."[101] On the Allied side, Marlborough evidenced his criticism of the engineers at Lille by pointing to the artillery's shortage of powder and shot, while Schulenburg bemoaned the miners at Tournai as if they were under the directors' direct command. The engineers were further blamed for the failings of the sappers who served under them—common soldiers (instead of the specialists the engineers had been requesting for decades) whom their own monarchs had refused to train and who could have been drilled in garrison or camp exercises had the generals truly considered it an important skill. Nor did commanders apparently take into consideration the role of delaying factors external to the siege itself, such as labor or supply shortages, not to mention the approach of relief forces and the requisite drawing down of siege regiments to reinforce the observation army. Du Mée acknowledged the error of their estimates at Lille in the dedication to his *Exact Journal*, but explained that these delays "beyond all expectations" were unpredictable and a function of the enemy's relief efforts,

---

[101] AG A[1] 1394, Barbesieux to the artillery lieutenant-general Vigny, Marly, 12 May 1697; see also a missive to the army intendant Desmadry on the same day. Taking Vauban's call to "burn more powder" out of context, the recently-appointed *secrétaire* specifically referred to the "habit" Vauban had developed in wasting powder, and even used the expenditure at Charleroi as evidence. The contrast with Vauban's own writings, including his complaints about the inefficiencies of the disobedient artillery during this same attack on Charleroi (cited earlier), suggests how little attention was being paid to Vauban's advice at the very top of the command hierarchy, and how in some highly placed circles, engineers were almost reflexively seen as the scapegoats whenever sieges went wrong.

including various observation army maneuvers, a minor field battle over a supply convoy at Wijnendael, and even a French diversion towards Brussels.[102] A siege was not a closed system that would allow predictive regularity, yet the engineers were criticized as if it was.

Excuses aside, the incompetency claims are even less persuasive when we recall that the great Vauban, whose skills could hardly be in doubt, found himself ignored and criticized only slightly less severely than his less-worthy successors and, importantly, criticized for *exactly the same mistakes*. With inconsistent support the engineers' numbers dwindled as they spilled their blood trying to manage the trenches as best they could. Such facts force us to draw several conclusions. First, generals not only had exceptionally high expectations for the management of sieges, but they also held the engineers responsible for the progress of the attacks regardless of the limited influence these technicians had over their actual conduct. Second, commanders nevertheless did surprisingly little to buttress the engineers' authority or improve their conditions of service despite their many complaints and despite opportunities to do so—the orders of a general would have gone far further than the appeals of engineers occasionally supported by a few field deputies and military administrators. To name a few ways generals could have improved the situation but chose not to: drill soldiers in sapping skills as both Vauban and La Vergne had been counseling for decades, adopt the engineers' advice during sieges, force all those associated with the siege artillery (infantry and artillery officers, as well as the conscripted infantrymen manning the guns) to follow the chief engineer's orders as Vauban had also been requesting for years, petition superiors to improve their pay and conditions of service, resist the urge to draw too many battalions from the siege army when a relief threat was near (or at least resist criticizing them for any resulting delays), and, least difficult of all, defend rather than accuse them for mistakes in the conduct of the approaches, particularly when those mistakes were made in opposition to the engineers' advice. As we turn to the next chapter, we shall learn that there is one critical reason for why the engineers were excoriated above all other branches and groups within the army: the engineers were seen as the one group most responsible

[102] De May, *An exact journal*, Dedication, p. 1. See also the comments in AG 1M 2365, Caligny, Relation de ce qui s'est passé au siège de Menin, f. 4.

for slowing down the pace of operations. As a perusal of all the citations above indicates, all the criticisms discussed above focused on time, a concern driven by the generals' constant obsession with the delays that sieges required. The paradigm siege was not only imperfectly applied, but its very foundation of efficiency was contested by non-engineers because it was seen to sacrifice time in order to save lives.

CHAPTER SEVEN

VIGOR: AN ALTERNATIVE PARADIGM

Many generals did more than just criticize. On rare occasions siege commanders intervened in the choice of approaches, but throughout the rest of the siege they called upon their expertise in field warfare to override many of the other engineering decisions. Looking closer, we see that all of the complaints about the engineers and all of the different interventions by different commanders in different armies were dedicated to the same goal: accelerate the tempo of campaigning. They simplified Vauban's precise balancing of casualties, delays and costs into a much more straightforward calculus of capturing the town as quickly as possible. To do so, they adopted techniques directly opposed both by Vauban himself and by his French and Allied heirs. Excessive safety and *industrie* were to be eschewed while speed and action were to take their place: the engineering cult of efficiency was opposed by the generals' cult of vigor. Putting Vauban back into his larger military context, we see a reservoir of hostility to the efficiency paradigm among both French and Allied generals not simply because they were jealous of Vauban's (or Coehoorn's) authority, or because these great engineers represented the status quo, but because the efficiency ideal that engineers strived to achieve appeared to protect human life at the cost of lost time.

The desire for immediate results and the corresponding impatience with delays is in some sense timeless. The cult of vigor—an early modern cult of the offensive—likely originated from a combination of time limitations and supply shortages.[1] The role of time was particularly important in the early modern world, when short campaign seasons and slow transportation technologies dictated significant delays

[1] This goes far beyond warfare, of course, and beyond the early modern period as well. For recognition of the shortage of time in the Middle Ages, see Murray, *Reason and Society in the Middle Ages*, pp. 105–107. Future research should explore competing explanations for the cult of vigor, including the impact of decisive action on morale. However, the only reason that participants in the War of the Spanish Succession explicitly gave, as we have seen, is the desire to avoid delays.

when coordinating armies and fleets across several theaters.[2] Thus strategic planning had to be done many months in advance and unforeseen delays might throw off coordination between theaters and allies, perhaps canceling military operations altogether. Political leaders were always anxious about when the army could start the campaign, since entering the field before your enemy opened up a whole variety of largely-uncontested targets.[3] Politicians similarly enquired when the army would finish up its current operation and move on to others—troop reinforcements might need to be sent to shore up another theater, or an approaching enemy corps might soon prevent the investment of another town or eliminate the possibility of a battle with numerical superiority, or peace negotiations might require a kick-start from a military victory. Conversely, delays might put a halt to successful post-battle pursuit, giving the enemy respite and time to recover from defeat, even allowing the enemy time to receive reinforcements.[4] Applied to siegecraft, the cult of vigor demanded a rapid siege. The historiography's emphasis on duration as a fundamental measure of sieges is not a coincidence, for this was by far the single most important variable to most military participants and observers. A siege might end too late in the season to start another one, or supplies in the surrounding region (or the contents of royal coffers) might be consumed before a siege's end was projected. Such delays allowed a strategic advantage to peter away. As a result, offensive wars would degenerate into less preferable contests between equally-matched powers, increasing the likelihood of requiring yet another year of war and all the costs associated with it.

---

[2] Scholars have examined the tyranny of distance in a number of works, most memorably in Fernand Braudel's *The Structures of Everyday Life: The Limits of the Possible*, trans. Siân Reynolds (Berkeley, 1992), pp. 415ff. In early modern European military history, the best recent example is Geoffrey Parker, *The Grand Strategy of Philip II*, (New Haven, 1998), chapter 2.

[3] John Lynn argues that Louis easily transitioned from a war-as-event strategy, i.e. his opening attempts to win wars quickly, to a war-as-process strategy wherein he readily accepted a war of attrition against allied forces after his initial efforts failed (*The Wars of Louis XIV*, pp. 375–376). Citing the example of the Dutch War, Lynn fails to mention, however, the two later cases of 1693 and 1706, where, as each war dragged on and became increasingly burdensome, Louis lost patience with war-as-process and sought to end it all by pushing for concerted efforts in every theater. Both of these attempts failed, *forcing* the Sun King to continue the wars of attrition.

[4] Points that contemporaries were well aware of, e.g. Feuquières, *Memoirs historical and military*, vol. 1, pp. 96, 98, 106.

Hence we find the desire for vigorous action to forestall the realization of such fears.

These seasonal limitations and the general lack of time appear to have encouraged military professionals to internalize a general impatience with inaction, though concerns over supply shortages and morale may also have played a role early on.[5] It is conceivable that this may have been in part an artifact of function: the engineers were focused on their sphere of expertise, after all, but the generals had to consider the larger operational issues beyond the siege lines. This hardly explains, however, why Vauban, whose strategic vision was unparalleled (e.g. his conceptualization of the *pré carré* and his constant traveling around the perimeter of France even in peacetime), would differ so strongly from his field *maréchaux* peers. What is certain is that those associated with the cavalry in particular cared (or knew) little of the intricacies of siegecraft.[6] As a result, every criticism of the engineers discussed in the previous chapter mentioned delays, while all but one dwelt *exclusively* on the time that was lost.[7] Not only do the generic complaints focus around the matter of delays, but their specific complaints also obsess over days lost. The engineers' purported failings might have resulted in greater casualties, but this is not what so incensed our officers. In fact, given their complaints, it would seem as if the out-of-practice engineers made mistakes that only postponed the advance of trenches. Flawed decisions to attack the strongest sectors received criticism only to the extent that it delayed the fall of the town. Only when engineers underestimated the length of a defense was condemnation rained down upon them. Their 'overly-cautious' method (a rather ironic

[5] Generals obviously varied in their adherence to the cult of vigor—some were barely-competent political appointees and others bureaucratic-minded administrators while yet others were proponents of the 'indirect' approach. But the commanders left in charge during the War of the Spanish Succession consistently relied upon the vigorous approach, and complained when they were forced to abandon it.

[6] Villars directly addressed this stereotype of the siege-ignorant cavalier Vauban had warned about, writing of himself that: "Although the marshal Villars had risen up through the cavalry, we saw in the beginning of these memoirs that he nonetheless had occasions to experience it from the infantry's perspective, and with this experience he was better equipped to competently judge the engineers' projects." Vogüé (ed.), *Mémoires du Maréchal de Villars publié d'après le manuscrit original*, vol. 2, p. 65 after his successful siege of Fort Kehl in 1703.

[7] Lecomte is the only Vauban scholar to explore this aspect in any detail. "Du service des ingénieurs militaires en France pendant le règne de Louis XIV," 28ff. Lynn briefly touches on the issue in *Giant*, pp. 576–577.

criticism considering the casualties they suffered) was frustrating to so many because it appeared to postpone the end of the siege.[8] So ingrained had this idea become among many (though by no means all) generals that even at the very beginning of a campaign season we still find concern over the need to achieve as much as possible in the precious few months left.[9] Whereas the engineering interpretation of efficiency weighted heavily the minimizing of casualties, to the extent that the top commanders were interested in efficiency, they saw it more simply in terms of minimizing the amount of time spent at the siege.

This dichotomy is evidenced in language itself, for general officers' discussions of sieges varied from their engineers in both content and form. Vauban's own treatment of these two categories set the standard for his subordinates: his improvements revolved primarily around saving lives. To the extent that he stressed the need to save time, he enlisted delays as a way to convince his audience (politicians such as Louvois and generals such as Burgundy) that his efficient siege would save time as well as lives.[10] Otherwise, engineers do not seem to have focused nearly as intently on time. The engineering siege journals, which were intended primarily for their peers, are empty of the dozens of length estimates generals demanded and dutifully reported in their correspondence. Although this metric could have offered a precise way to measure efficiency, engineers ignored it. Instead they dutifully reported the casualties from each day's trenchwork. More than one commentator emphasized how Ath was captured by Vauban with skill and art, rather than with mere unthinking force. Our journalist of the siege recognized the contrast between the cults of vigor and efficiency:

[8] Engineers like Pagan and Vauban explained that their attempts to improve the art of defending a place would not make them invincible, but would only buy time till a relief army or supply shortage or season change might force an end to the siege. Our critics of siege delays, however, almost never expressed concern that the outcome of the siege was in doubt—in many cases a relief army had already failed to lift the siege. The vigorous generals were interested only that the siege was taking time away from the rest of the campaign season.

[9] In fact, after the late May battle of Ramillies the Allies were extremely concerned to take advantage of such an early victory as quickly as possible—there was no time to lose.

[10] As with his instructions on how to capture the covered way. Vauban, *Mémoire pour servir*, p. 98.

> The capture of Ath was not the most vivacious [*éclatante*] action in this war. It was captured by the King's engineers. Mr. de Vauban is its hero, everything came from his capacity and experience, and this journal is an illustration of his art, from which the engineers alone can profit. You should not expect to find here events that would surprise and entertain the reader; it was captured not with weapons, but by the shifting of earth.[11]

This was obviously hyperbole, but by reinforcing the efficiency mindset and contrasting it with a cult of vigor it served its pedagogical point, a point that was likely lost on (or ridiculed by) many in the vigor camp. Such generals, on the other hand, cared little for technical discussions praising efficiency and minimal effort. In fact, minimal effort sounded suspiciously like laziness or, worse yet, cowardice. They focused instead on the end result as measured in chronological terms and, engineers were quick to note, whether the participants enthusiastically risked their lives or not. They kept close track of the deadlines and delays with no comparable interest in the body count, and suspected that those who called for less bloodshed were personally lacking the proper sacrificial spirit. They also made sure to tell their correspondents (often impatient political leaders) of every change in expectations of when the fortress would fall, and expressed frustration when such dates were pushed back. In contrast, they never complained of estimates of expected casualties (if they even asked for them)—time was foremost on their minds and loss of life was the unavoidable price. The contrast between the two types of sources could not be starker. If we had to oversimplify, the generals were focused on time while the engineers focused on lives.

The particular vocabulary generals used also illustrates their temporal focus. While the engineers praised efficiency, generals used the word "vigor" literally hundreds of times and its synonyms many more, and always the usage indicated the expectation that its application would lead to success. Defenders too were expected to hold out, preferably till an assault on the breach had been repulsed, but at the least they were expected to defend themselves with vigor. Honor demanded nothing less. Trenches were pushed forward with *vigor*, siege batteries played *vigorously* upon the town, while garrison sorties

[11] *Relation du siège d'Ath*, p. 1. Later he repeated that "it was captured only by cannon, shovel and pick . . . the King's troops had hardly a single occasion to distinguish themselves" (p. 32).

were *vigorously* repulsed. The future English King James II praised the French generals of the Fronde for their willingness to expose themselves in the name of "dispatch," and we have cited dozens of similar quotations. More generally, references to time (and fear of its loss) abound in their correspondence, of the need to press forward, to act quickly, to avoid further delay, to keep an eye on the approaching end of the campaign season. When the main force could enter the field, when a detachment would be able to reinforce the main army, when the siege army would be freed up for further tasks, when the army would be forced to go into winter quarters—correspondence orbited around these issues. This preference for vigorous behavior even appears to have been shared by much of the public at large. Both letters and news accounts (in French, English and Dutch) frequently reassured their readers that their sieges were being pushed with as much "vigor" as possible, and that the fall of the fortress was not far off.[12] Such appeals for patience were necessary, witness the tenor of the London public in mid-1710: "At our coffee houses [centers for disseminating the latest reports] we are very angry that the news talks of our beseigeing Douay; for their opinion is that we ought not to amuse ourselves in taking towns, but march directly to Paris."[13] Whether propelled by internal or external considerations, many generals' impatience with lost time (potential or real) and the need for action was the driving force behind the cult of vigor, and the shortage of time required that risks greater than sieges be undertaken.[14]

1. *The Appeal of Non-Sieges*

Before a siege even started commanders seeking vigor tried to avoid formal attacks altogether. If they could not force a field battle, they

[12] For example, *The Postman*, #1887, 20 May 1710 O.S.; the letter was dated 27 May. See also *Present State*, August 1710, p. 311.

[13] James Joel Cartwright, *The Wentworth papers, 1705–1739. Selected from the private and family correspondence of Thomas Wentworth, Lord Raby, created in 1711 Earl of Strafford*, (London, 1883), p. 90, Peter Wentworth to Lord Raby, London, 10 June 1710 O.S.

[14] We see this attitude reflected in Childs' discussion of how the 1697 campaign's "mathematical formalism encapsulated the unimaginative and conventional approach of contemporary military minds. Into this atmosphere, Marlborough was to burst like a ray of bright light." *The Nine Years' War*, p. 326.

pursued other positional tactics that promised faster results. Suggesting inevitable stasis, John Childs wrote of the Flanders theater: "Territory could only be gained by the systematic reduction of fortified towns and cities. . . . Every fort, however large or small, had to be besieged and captured."[15] In reality, however, only a small number of fortresses required such a formalized attack, even in a 'heavily-fortified' theater like the Low Countries. Time-conscious commanders had long tried to avoid sieges altogether, hoping to make an inexpensive gain with non-siege tactics.[16] With this goal in mind they called on a whole range of techniques to supplement or even replace the formal siege with a much smaller investment in time.

Even an incomplete sampling of such attempts in the Low Countries, found in Appendix D, shows that both sides looked to positional tactics other than sieges with great frequency.[17] In the blitzkrieg campaigns of 1668 and 1673, the French first attempted to capture the many fortresses by storm before settling for the more patient *siège en règle*.[18] Supporting the theory that time was of prime importance and casualty-avoidance only secondary, surprise attempts were the most frequent non-siege tactic used in Flanders during the war, at least sixteen attempts. Not only were such stratagems a low-risk option given the limited resources involved, but a millennium of campaign histories and war manuals stretching back into the Classical past gave the enterprising commander a multitude of ruses to attempt.[19] In addition to those actually attempted, there were likely several

[15] Childs, *Nine Years War*, p. 42.

[16] One mid-seventeenth century treatise noted that taking towns by storm (*vive force*) was the best method, though more formal sieges were the most frequent technique. Charles Common, *Traité universel et succinct des fortifications des places tout régulières qu'irrégulières, avec la manière de les attaquer et défendre*, (1647), pp. 29ff. De Ville represented the 'pure' engineering position by claiming that blockades were to be preferred over sieges because they spared men's lives. *Les fortifications du chevalier*, p. 316.

[17] As with brief sieges, ephemeral surprises (especially failed ones) are the most difficult to discover in the sources, since they required the fewest resources, lasted the shortest period of time, and usually occurred away from the main concentration of field forces.

[18] For example, Rousset, *Histoire de Louvois*, vol. 1, p. 135. Satterfield's *Princes, Posts and Partisans* provides an excellent overview of this alternative type of small-scale warfare.

[19] For recognition of the "infinity of astounding examples" to be found in prior history, see Guignard, *L'École de mars*, vol. 2, p. 464. Both Folard and Santa Cruz de Marcenado were particularly fond of mentioning Classical ruses. For modern discussion of the Classical legacy of military subterfuge, see Everett L. Wheeler, *Stratagem and the vocabulary of military trickery*, (Leiden, 1988).

times as many proposals for surprisals that were never carried out, submitted by everyone from local partisans to discontented townspeople to officers looking for patronage.[20] Unlike sieges and the other non-siege tactics, they did not require significant military or engineering expertise in their initial planning, beyond knowledge of a secret entrance or of a fifth column or bribable gate guard. Such projects were more likely to be approved as well, since they required far fewer resources than the larger manpower requirements of a blockade or the artillery a bombardment might require.[21] Surprise parties were also difficult for field forces to counter, for their slight logistical requirements made them more mobile than a siege army, more difficult for an intelligence network to detect on the march, and their intent and target more difficult to ascertain—even when spotted—due to a wider range of potential targets. Sieges, on the other hand, could be prevented much more easily because of their vast logistical requirements. Villars, for example, was able to disrupt the Allies' plans to besiege Arras by simply consuming all the fodder surrounding the place while the Allies were stuck before Douai in early 1710. Marlborough was similarly forced to abandon his designs on Ypres in 1706 since there was too little fodder to support a siege army. A successful surprisal could capture even the strongest town in a day or less, would leave the fortifications ready for immediate defense, the town's economic infrastructure intact, and would cost very little in terms of manpower, gunpowder and transportation costs.[22]

Surprises were an ever-present threat to even the strongest fortress—constant vigilance was required to defend against a bribed gate guard or a disaffected inhabitant with knowledge of the works. However much the surprise promised in theory though, its potential was rarely

[20] To cite just a few examples from a single year, we could mention proposals in 1710 to turn over Brussels to the French, as well as Condé, Calais and Boulogne to the Allies. On Brussels, see Veenendaal, Jr. (ed.), *Briefwisseling Heinsius*, vol. 10, p. 589 #1179, van den Bergh to Heinsius, Brussels, 31 July; for Calais and Boulogne, see Snyder (ed.), *Marlborough-Godolphin Correspondence*, vol. 3, p. 1436 #1471, Godolphin to Marlborough, 16 March and p. 1446 #1482, Godolphin to Marlborough, 25 March. For Condé, AG A[1] 2225, #5, Bernières to Voysin, 5 January.

[21] Such as we find, for example, in a design on Aire in Article 15 Section 2 §1 carton 1 #13.

[22] If a surprise was attempted near a concentration of the enemy, additional troops would be sent to reinforce the initial infiltration force, waiting for the infiltrators to open a predetermined gate.

achieved in the real world. Already by 1695 Vauban's subordinate Naudin had declared them largely *passé*, while in the 1720s the author Guignard confirmed that such indirect methods no longer succeeded as they had in the past, attributing this to the increased attention commanders were giving to matters of security, lest they be discredited when caught off-guard.[23] Despite many promising leads and a few notable successes in the War of the Spanish Succession, most attempts were unsuccessful. Their low success rate indicates that garrisons, often tipped off by their own intelligence networks, were able to implement adequate security measures in concert with the urban militias.[24] Thus attackers turned to other methods. Related to surprise attempts, storms also offered the possibility of a quick victory, though at a higher cost in resources and likely lives. We find several examples in the Flanders theater, but they were only successful against small garrisons defending isolated forts and entrenched posts.

Despite a relatively low success rate, both surprisals and storms remained popular because their rapid conclusion limited logistical demands and thereby made them low-risk, high-gain operations. Even the French, on the defensive in Flanders during practically the entire course of the war, could still conduct numerous attempts against enemy towns, with greatest success in 1708. A conventional judgment of the operational situation during these years showed French field armies consistently outmaneuvered and on their heels, yet these small-scale, low-risk tactics gave Louis the opportunity to go on the offensive all the same. As long as the entire length of the Low Countries frontier was contested (from coastal Spanish Flanders to the Meuse river in the east), neither side was able to provide enough troops to garrison adequately the many places spread along its breadth. Therefore there would always be towns vulnerable to such non-siege techniques. Dividing an army up into several smaller corps might provide more security for these towns, but there were still too many targets and these corps in turn would be exposed to a battle by a concentrated enemy field army. Lt.-Col. Blackadder perceived a

[23] Naudin, *L'ingénieur françois*, p. 273; Guignard, *L'École de mars*, vol. 2, p. 464. Possibly Guignard had in mind Villeroi's embarrassing capture in his winter quarters at Cremona in 1702, an Italian fiasco that merited a number of irreverent couplets.

[24] See, for example, the warning sent from the Council of State of the Spanish Netherlands to its towns to keep a strong guard on watch against any surprise attempts. BL Add MSS 61193, ff. 13ff., Brussels, 20 May 1710.

French preference for such tactics, and noted its concordance with Gallic perfidy: "we had a design upon Ypres lately, but it has mislucked. We must leave it to the French to take towns by trick and treachery; we never get any that way. We get all we win very honourably, with our blood and the sweat of our brow." He was only one of several Allied participants to bemoan their luck, though we should emphasize the self-serving nature of such a critique, witness their disappointment that their *own* attempts at subterfuge were unsuccessful.[25]

Less frequent, because of their greater demands, were bombardments and blockades. Attackers could bombard a place with mortars and red-hot shot from field pieces (often 8-pounders or smaller), setting fire to buildings and generally causing chaos within the town. Louis XIV and Louvois were fond of the technique, using it not only against declared enemies (setting Brussels on fire in 1695, for example), but also as a way to apply diplomatic pressure against allies of enemy states and even neutrals (e.g. against Genoa in 1684).[26] Vauban, on the other hand, opposed bombarding the towns themselves, and sought, usually unsuccessfully, to discourage their use. In a 1682 letter, for example, he spent several paragraphs explaining how a projected bombardment of Bruges would be far more expensive and time-consuming than they realized—its inefficiencies far outweighed any benefits.[27] In the Nine Years' War the Allies also resorted to such measures, English Tories being particularly enthusiastic about naval bombardments of coastal French towns (e.g. the port of Saint-Malo).[28] In the field they were most useful as a means to extort

---

[25] Crichton (ed.), *The Life and Diary of Lieut.-Col. J. Blackadder*, p. 391 letter to his wife, 1 June 1710 O.S.; Heinsius too fretted to Goslinga about their vulnerability to French "intrigues, finesses and ruses" while their main force was focused on Lille. Veenendaal, Jr. (ed.), *Briefwisseling Heinsius*, vol. 7, p. 493 #1070, The Hague, 30 August 1708. Also Marlborough in Snyder (ed.), *Marlborough-Godolphin Correspondence*, vol. 3, p. 1545 #1588, Douai, 7 July 1710. Millner also used similar terms in *A Compendious Journal*, p. 209. Satterfield's *Princes, Posts and Partisans* illustrates Louis XIV's long-standing dedication to this form of *petite guerre*.

[26] On Louis' use of the tactic, see Wolf, "Louis XIV, soldier-king," 215; and Lynn, *Wars of Louis XIV*, pp. 65, 173–174. See Naulet, *L'artillerie française*, pp. 243–248 for several case studies.

[27] Rochas d'Aiglun (ed.), *Vauban*, vol. 2, p. 218 Vauban to Louvois, Lille, 2 December.

[28] See John Stapleton, "The Blue-Water Dimension of King William's War: Amphibious Operations and Allied Strategy during the Nine Years' War, 1688–1697," in M.C. Fissel and D. Trim (eds.), *Amphibious Warfare 1000–1700: Commerce, State Formation and European Expansion* (Leiden, 2005).

contributions from walled towns or as a form of economic attrition rather than as a way to capture a town; Louis' infamous bombardment of Brussels in 1695 caused significant economic damage to the already-impoverished Spanish enemy, but did little to distract William III from his siege of Namur. In the Spanish Succession war, none of the five bombardment attempts resulted in surrender.

The blockade was the least desirable method for capturing a town, at least with the main field force. In most cases even a siege—though more bloody—was preferable to time-conscious commanders since it promised faster resolution than the much slower process of starving out a garrison.[29] As a result, blockades were quite rare in the Low Countries, the theater witnessing only four of note during the entire war. The tactic was especially used after one of the shorter, preferred techniques misfired—either an unsuccessful bombardment (Dendermonde 1706) or a stalled late-season siege that offered little prospect of success before the end of the campaign season would force the besiegers into winter quarters (as happened with both Guelders and Rheinberg 1702–1703). It could also be used in order to allow the main force to continue its otherwise-uncontested advance. Such subsidiary operations were conducted by small detachments of troops operating away from the main army's axis of advance. In the first year of the Flanders war the French advance saw almost all of Spain's (and Liège's) fortresses fall, yet the domineering Maas fortress of Maastricht and its Dutch defenders held fast. Wanting to pressure the Allies further north in Dutch Brabant and to the east in Guelders, they masked the garrison with a small corps of troops rather than redirect their offensive efforts to besieging this formidable fortress.[30] So too at Rheinberg do we find the Elector of Brandenburg, in command of eighteen battalions, eight regiments of horse and two of dragoons, quickly abandoning his still-born nine-day siege; he satisfied himself instead with establishing winter quarters

[29] Even Vauban preferred a siege of Besançon's citadel to the *longueur* of a blockade. Rochas d'Aiglun (ed.), *Vauban*, vol. 2, p. 97 at Breisach, 26 September 1673. In 1712 the French decided to make Bouchain the target of their final siege of the season—a blockade was ruled out because of the hardships the troops would suffer, as well as the difficulty of completely cutting the water-logged town off from outside support. AG A^4 Carton 8 chemise 2, Alègre to Maine, 18 September.

[30] The French had tried a similar strategy in the Dutch War, with Vauban eventually besieging it into submission. Satterfield, *Princes, Posts and Partisans*, p. 267.

in posts about the town. The town was eventually forced to surrender to this blockading force in early February.[31] Nor did a two-week Prussian bombardment of Guelders with twenty-nine mortars and forty cannon have the desired effect, and the town only surrendered to blockade in late December.[32] After Ramillies and an unsuccessful bombardment, an English detachment settled down to blockade Dendermonde while the main Allied force went on to besiege Oostende and then Menin. Marlborough informed Godolphin of Dendermonde's secondary importance: "I have sent Brigadier Cadogan to see if it be possible to shute it so up, that thay can put no more succors into itt. If it can be done without weakening the army to[o] much, I am desirous to do itt."[33] The Spanish governor refused to surrender, "whereupon my Lord Duke has orderd that the place be block'd up very close, till the troops of the Allies are more at leisure to attack it with greater vigour."[34] As the summer months passed, the fortress's expansive inundations evaporated in a drought, exposing the town's weak fortifications in the process.[35] With the waters at their lowest level in fifty years and the Allies now masters of Menin, the Duke's brother Charles Churchill converted the blockade into a formal siege thanks to reinforcements from the main army. It fell after only a week of open trenches.[36] Unwilling to tie down a large number of troops for a secondary objective, contemporaries turned to blockades only when more rapid techniques against a subsidiary target had failed.

Such non-siege techniques were beyond the scope of siege manuals that assumed garrisons impervious to lesser tactics (recall Vauban's assumptions for a 41-day siege), and were equally beyond the duties

[31] Vault and Pelet (eds.), *Mémoires militaires*, vol. 2, p. 615, Détail du siège de Rhinberg.

[32] Marquis de Quincy, *Histoire militaire du règne de Louis le Grand*, vol. 4, p. 43; Vault and Pelet (eds.), *Mémoires militaires*, vol. 3, p. 137; Wijn, *Het Staatsche Leger*, vol. 8 part 1, p. 384.

[33] Snyder (ed.), *Marlborough-Godolphin Correspondence*, vol. 1, p. 585 #597, Marlborough to Godolphin, Rosselare, 24 June 1706.

[34] *Daily Courant*, 8 July 1706 O.S. from the Army under the Duke of Marlborough at Harlebeck, 10 July.

[35] The seven weeks prior to the opening of siege trenches on 20 August saw no rain. Snyder (ed.), *Marlborough-Godolphin Correspondence*, vol. 2, p. 658 #667, Marlborough to Godolphin, Velaines, 9 September.

[36] Veenendaal, Jr. (ed.), *Briefwisseling Heinsius*, vol. 5, p. 451 #859, Albemarle to Heinsius, Helchin, 6 August.

of the engineers in general—they fell instead into the realm of 'small war,' a type of warfare admirably described by George Satterfield's recent *Princes, Posts and Partisans*. That the frequency of non-siege tactics followed their duration (surprisals being the most popular, then storms and bombardments, and only then the longer blockades) illustrates the widespread interest contemporaries had in avoiding delays. In most cases, commanders were even willing to suffer the higher casualties of a siege rather than a blockade if it meant shortening the length of the garrison's resistance and thereby freeing up the main force for further projects. Numerous alternatives existed to obviate the need for a potentially-lengthy Vaubanian siege or a blockade. To the chagrin of many, however, there was often no alternative to the lengthier siege if one wanted to continue the advance.

2. *Pressure to Accelerate*

The most decisive-minded generals, dissatisfied with the nature of the task before them and annoyed at the need for a siege, sought to force engineers to adopt more vigorous tactics whenever possible. The attention paid to time was explicit, as can be seen in letters sent to Amsterdam from France during the 1697 siege of Ath. They explained that "M. de Vauban, seeing all the avenues [to the town] occupied by three large French armies ready to unite in order to oppose any secours for the town, decided to conduct his attacks with a little more leisure [*un peu plus de tems*], in order to assure its success with less risk and fewer losses."[37] The King, remembering Vauban's earlier dismissal of the governor's facility, was disquieted by this leisurely pace and wrote somewhat disapprovingly from Versailles that "It does not appear that the governor of the Ath garrison merits an attack in strict form. This is why I am persuaded that you can without concern go faster than you believe; it is important to finish it promptly and to not lose any time." He similarly emphasized this point to the generals on site, expressing his hope that it would not last as long as Vauban predicted.[38] When framed

[37] *Gazette d'Amsterdam*, 3 June, Nouvelles extraordinaire, from Amsterdam, 3 June.
[38] AG A^1, 1394, 17 May, to Villeroi, from Marly. See also to Catinat, 22 May, from Versailles, printed in Rochas d'Aiglun, *Vauban*, vol. 2, p. 453.

as a choice between expending more powder and risking more blood, the new brash secretary of war was even more blunt. Four years after Vauban had appealed to "burn more powder and spill less blood" at the siege of Charleroi in 1693, Barbesieux criticized Vauban for his 'waste' of powder and inefficient cannonfire.[39] Even at Vauban's best-conducted siege, it seems, some hoped for it to proceed a little more quickly, efficiencies be damned. During Vauban's last siege, that of (Old) Breisach, the King showed a similar concern over the perceived delays. Louis' grandson Burgundy wrote to Chamillart from the siege: "I have received another letter whereby the King orders me to press the siege; I can assure you that we are doing all that we can and we are even going much faster than M. the *maréchal* Vauban would like."[40] The town fell quickly, but the tension between the King and his more-exacting expert is evident. A few years later, La Feuillade indicated his misunderstanding of Vauban's views when defending his projected attack on Turin. La Feuillade justified his planned attack *à la Coehoorn* by noting that critics such as Vauban had forgotten that Coehoorn's contrary method had managed to capture his own fortress of Namur.[41] Vauban had not forgotten Coehoorn's attack on Namur's citadel—how could he when over 2,000 men were cut down charging the covered way?—but he had long ago dismissed it as a model for others to follow.

Once the French returned to the offensive in Flanders in 1712, without Vauban, they abandoned their defensive strategy of delays and sought to shorten the lengths of their sieges. Chamillart's replacement Daniel-François Voysin requested daily reports on their progress and reminded Villars of their hopes that Douai would not be the final siege of the season. A few days later Voysin illustrated the bal-

[39] AG A[1], 1394, Barbesieux to Vigny, from Marly, 12 May 1697. See also to Desmadry, same date, where he notes disapprovingly of Vauban's galling tendency to act independently of the commanding general.

[40] Alfred Baudrillart and Léon Lecestre (eds.), *Lettres du duc de Bourgogne au roi d'Espagne Philippe V*, 3 vols. (Paris, 1912), vol. 1, p. 276 #166, Burgundy to Chamillart, Breisach, 3 September 1703. Consult also Chamillart's letter to the *maréchal* Tallard recommending he have Burgundy press Vauban to accelerate the siege. Vault and Pelet (eds.), *Mémoires militaires* vol. 3, p. 442, Versailles, 30 August.

[41] Esnault (ed.), *Chamillart*, vol. 2, p. 33 #164, La Feuillade to Chamillart, Turin, 20 September 1705. Interestingly, a decade earlier Vauban had accused La Feuillade's father, who served as a *maréchal de France* before his son, of criticizing the attacks on Philippsburg in a similar vein. See Vauban's correspondence with Louvois in Griffet (ed.), *Recueil lettres*, p. 105, 15 October 1688.

ancing act besiegers had to walk when he remarked that the King thought the siege could be conducted "with a bit more liveliness" as time was precious, but he also warned "not to expose the troops too carelessly." In a later letter he acknowledged the concerns Villars had raised about its slow progress, but counseled that although he did not fully understand their design either, he hoped the engineers would find ways to speed up the attack all the same.[42] Louis continued to push his army forward by then suggesting that Villars send troops to Le Quesnoy before Douai had even been captured, in order to preempt Eugene's army from positioning itself between the targeted town and their besieging force.[43] Villars heartily concurred with their desire for haste. In his memoirs he used every opportunity to highlight the contrast between the correctness of his own decisive judgment and the overly-cautious attitude of his engineers.[44] In his correspondence he wrote of pressing Valory and the siege commander Albergotti to accelerate their attack against Douai, accepting their excuses only begrudgingly. Recalling his successful rebellion against Vauban and the other engineers' advice regarding Fort Kehl, he notified Voysin that as the attack was going slowly, he would give Valory two more days and then "conduct the siege according to my taste."[45] Later in life he recalled of Valory's expectation of a fifty-day defense: "That was not my calculation, and I was used to leading the engineers a little bit faster than was their rule."[46] Elsewhere he reminded Valory of the need to follow the "general

---

[42] AG A[1] 2382 #49, Voysin to Villars, 19 August 1712; also #146, 25 August; as well as at the later siege of Le Quesnoy in vol. 2384 #45, 19 September. Louis expressed his desire to "manage" time to take advantage of Denain in Vault and Pelet (eds.), *Mémoires militaires*, vol. 11, p. 502, to Villars, Fontainebleau, 27 July.

[43] AG A[1] 2383, #32, to Villars, 2 September.

[44] See a reproach of Albergotti on the capture of Fort Manheim in Vogüé (ed.), *Mémoires du Maréchal de Villars publié d'après le manuscrit original*, vol. 3, p. 194. In 1705 Charles-Auguste Goyon-Matignon, comte de Gacé had assured Chamillart that he would work with Valory to "diligenter" their siege of Huy. AG A[1] 1835, #194, 30 May. For contrast, see the treatment of Vauban during the siege of Philippsbourg in Nicolas Catinat, *Mémoires et correspondance du maréchal de Catinat*, (Paris, 1819), vol. 1, p. 272, Jacques-Henri de Durfort, *maréchal* Duras to Louvois, 7 October 1688.

[45] AG A[1] 2382, quote in #88, Villars to Voysin, 22 August 1712; also #148, 26 August. Lefebvre d'Orval commented on Villars' desire to accelerate the siege "because he had other things to do." AG A[1] 2382, #102, from Valenciennes, 23 August.

[46] Anquetil, *Suite des mémoires rédigée par Anquetil*, vol. 2, p. 383. Broglie also considered Valory's attack far too cautious given the undersized garrison. AG A[1] 2382, #196, to Voysin, Douai, 29 August.

maxim for all military action": "the weakness of the garrison demands that we push forward our attack rather than delay it."[47] The town quickly fell and, with the campaign clock ticking, French troops were immediately sent to invest the next target. Villars encamped before Le Quesnoy and reassured Voysin that he was pressing the siege vigorously, explaining to the engineers (with whom, he claims, he was very content) that the King's orders insisted on a rapid capture. Made acutely aware of these time constraints, Valory promised to quicken his attack to please both his commander and Court.[48]

The 1713 campaign in Germany saw the *maréchal* exercizing his freedom of action once more. Describing his behavior in the third person, his approach was quite different from Vauban's quest for efficient preparation:

> He [Villars] forgot nothing in order to accelerate the preparations for the siege of Landau. The place was excellent, defended by a very strong garrison composed of the best Imperial troops. Thus people who like precautions had good reasons to object to the briskness of the marshal Villars, but as he followed the principal that vivacity is almost always necessary when it is not done carelessly, he only considered those precautions that were absolutely necessary.

After nine days of open trenches, he forced the commanding general to accelerate the siege, as not even an isolated redoubt well beyond the outworks had been captured. Lecturing Valory, he admonished: "one must not underestimate the enemy nor overestimate their strength, and, in judging its defense, even the enemy's greatest courage and skill is not enough to deter us from going faster."[49] Villars actively intervened in the details of siegecraft time and again in order to overcome the 'lethargy' of his engineers.

Allied political leaders took a less active role in the military affairs of the War of the Spanish Succession, leaving most of the operational details to Marlborough, Eugene and the Dutch generals. Nonetheless, political administrators reflected their interest in the calendar as well. On the Dutch side, Heinsius made clear his desire to see sieges end quickly.[50] In early 1710 the United Provinces' highest

---

[47] AG A[1] 2383, #53 Villars to Valory, 4 September 1712.

[48] AG A[1] 2383, #223 Villars, 14 September; and 2384, #22, Valory, 17 September.

[49] Vogüé (ed.), *Mémoires du Maréchal de Villars publié d'après le manuscrit original*, vol. 3, quotes on pp. 190–191 and 198.

[50] In 1704, van 't Hoff (ed.), *Marlborough-Heinsius Correspondence*, p. 144 #229,

representative body, the States-General, hoped to eliminate the delays by illustrating their keen interest in the daily progress; those conducting the siege were given "orders that an express be sent hither from the army every day."[51] The English Lord-Treasurer Sidney Godolphin for his part also considered how siege delays might affect the rest of the campaign.[52] Politicians and public debates back home, unable to coerce specific courses of action because of their distance from the scene, nevertheless focused their armies' attention on the passing days.

Allied politicians could count on their generals to apply consistent pressure on 'overly-cautious' engineers. From the very beginning, Marlborough wrote diplomatically-worded reminders to those conducting the sieges to push their attacks forward as quickly as possible.[53] As their targets shifted from *bicoques* to fortresses, this pressure only intensified. Goslinga reported on Marlborough's strident insistence that they accelerate the attack on Oostende; the field deputies responded by pointing to the severe shortages of supplies. This constant pressure could not be ignored forever though. Halfway through the siege, Goslinga noted in his diary entry of 3 July that "We did nothing this night except perfect the works and extend our parallel a little to our right. We could have advanced further, but our directors wanted to advance more carefully."[54] With cautious engineers overseeing the attacks, Salisch promised at Menin to "push the siege as much as will be possible."[55] A public account assuaged its readership's doubts by informing them that "the siege of Menin is pushed

Heinsius to Marlborough, The Hague, 21 October; and 148 #233, Heinsius to Marlborough, The Hague, 31 October. For 1706, see van 't Hoff (ed.), *Marlborough-Heinsius Correspondence*, p. 240 #398, Heinsius to Marlborough, The Hague, 28 June. In 1709: Veenendaal, Jr. (ed.), *Briefwisseling Heinsius*, vol. 9, p. 87 #176, Heinsius to Albemarle, The Hague, 24 July.

51 *London Gazette*, 25 May 1710 O.S., p. 2, from the Hague.

52 For example, Snyder (ed.), *Marlborough-Godolphin Correspondence*, vol. 3, p. 1464 #1501, Godolphin to Marlborough, St. James, 17 April 1710 O.S. For an appreciation of the defensive possibility of delay, see Snyder (ed.), *Marlborough-Godolphin Correspondence*, vol. 1, p. 540 #560, Godolphin to Marlborough, 10 May 1706 O.S.

53 For Venlo 1702, see Murray (ed.), *Letters and Dispatches*, vol. 1, p. 34, Marlborough to Coehoorn, Asch, 10 September. On the lengthy siege of Landau after the battle of Blenheim, see Marlborough's complaint in Murray (ed.), *Letters and Dispatches*, vol. 1, p. 513, to Henry St. John, Weissenburg.

54 Veenendaal, Jr., "De Mémoires van Sicco van Goslinga," 22; 23.

55 Veenendaal, Jr. (ed.), *Briefwisseling Heinsius*, vol. 5, p. 482 #931, Salisch to Heinsius, Menin, 21 August. An 18 August camp journal explained Marlborough's

on vigorously. Tis believed the enemy design to hold out as long as may be, tho the garrison run the risk of being made prisoner of war, that the Allies may have less time for undertaking another siege."[56] The siege of Lille two years later was a particularly frustrating experience for those in command. Schulenburg was particularly adamant that they step up their advance against its citadel as the defenders were quickly running out of powder: "Many would have thought that we would have already captured the covered way several days ago, but the enemies have profited much from the mistakes [*faux démarche*] that the engineers have committed."[57] Marlborough was particularly pessimistic regarding Vauban's most famous fortress: "I am sorry to tell you our situation is such, through the lateness of the season, the slowness with which the engineers have proceeded in their approaches, and the great difficulty we shall meet with in bringing up a further supply of ammunition, as to make us doubt of the success of the siege."[58] The attacks on the citadel prompted a chorus of complaints against the poor conduct and lack of vigor in the attacks both in the siege camp and back in England.[59] So too did the Duke stress to Des Rocques the need to push this attack as expeditiously as possible—after Du Mée's bungled storms on the town, the chief engineer had re-imposed control and insisted on slowing down the siege by sapping up the citadel glacis.[60]

At Tournai's countermined citadel, Schulenburg "wanted to storm the outwork the day before yesterday with the grenadiers, but because

going to the camp in now familiar terms: "in order to carry on the siege with all possible vigour." BL Add MSS 4742, f. 73b, while the Earl of Halifax wrote to Marlborough that everybody was "disatisfied with their delays." Snyder (ed.), *Marlborough-Godolphin Correspondence*, vol. 2, p. 634 note 8, 2 August. Heinsius also wished for a faster pace: Veenendaal, Jr. (ed.), *Briefwisseling Heinsius*, vol. 5, p. 373, to Goslinga, 5 July.

[56] *Daily Courant*, #1351, 13 August O.S., a letter dated 20 August from The Hague.

[57] BL Add MSS 61245, ff. 75b–76, Schulenburg to Marlborough, Lille, 1 October 1708; and f. 82, 14 October. For a French perspective, see Vault and Pelet (eds.), *Mémoires militaires*, vol. 8, p. 456, Senneterre to Chamillart, Douai, 28 September.

[58] Murray (ed.), *Letters and Dispatches*, vol. 4, p. 237 Marlborough to the Earl of Sunderland, Lannoy, 24 September 1708.

[59] Veenendaal, Jr. (ed.), *Briefwisseling Heinsius*, vol. 8, p. 106 #231, noted by Réné de Saunière de l'Hermitage (a Huguenot residing in England) to Heinsius, London, 30 November.

[60] Murray (ed.), *Letters and Dispatches* vol. 4, p. 293 Marlborough to Des Rocques, Rosselare, 8 November; and p. 331 from Beerlegem, 2 December.

of the difficulties and perils that several imagined this undertaking would meet, the resolution was taken to approach it underground and capture it foot by foot."[61] Albemarle complained from camp that "the attack on this citadel goes very slowly; order and good conduct are lacking, and I fear that the siege will be lengthy."[62] Later in the siege, Schulenburg explained the delays he experienced trying to pass the ditch: "all this would have been finished already if the engineers had not made so many detours, such as occupying the *place d'armes*, which had to be completely demolished to make room. These Messieurs are stretching out the length of this siege, no doubt so they don't have to conduct any others this year."[63] With Du Mée's death impatient generals lost a young engineer amenable to their concern for speed. Reflecting later on his passing, Goslinga concluded that his senior Des Rocques was "sage, well-versed in the craft, but a little slow and circumspect," whereas Du Mée was "very capable, lively, brave and determined to proceed as fast as possible when necessary. A little more age would have moderated his great fire and would have rendered him more capable in his art. . . . To tell the truth, he was worth more than Des Rocques."[64] Du Mée's many bloody assaults on Lille's counterscarp may have received criticism at the time, but such faults were forgotten quickly enough when contrasted with the 'plodding' pace of his senior, a tendency the ambitious Du Mée had used against his competition. The contrast between an older Des Rocques who had fully accepted the engineer's philosophy and the younger Du Mée who was not yet fully indoctrinated in the cult of efficiency shows that, even among those who supported the corps, the engineering mindset was considered too cautious and lacking in vigor.[65]

The first conquest of the 1710 campaign would be the most frustrating of all. Promising to use "all possible diligence in going on

[61] Schulenburg, *Leben und Denkwürdigkeiten*, p. 396, 18 August 1709.

[62] Veenendaal, Jr. (ed.), *Briefwisseling Heinsius*, vol. 9, p. 194 #406, Albemarle to Heinsius, Tournai, 24 August. As with most complaints, it is unclear where specific responsibility lies.

[63] Schulenburg, *Leben und Denkwürdigkeiten*, p. 466, August 1710.

[64] Veenendaal, Jr., "De Mémoires van Sicco van Goslinga," 23—he also noted Des Rocques' preference to do everything "methodically, without risking anything." Des Rocques was 42 in 1709, but the younger Du Mée's exact year of birth is unknown.

[65] A few did not make such distinctions, e.g. Schulenburg's criticism of Du Mée in BL Add MSS 61245, f. 99, to Marlborough, Tournai 11 August 1709.

with the siege," the Captain-General explained the larger picture to the *raadpensionaris*: "the French continue to march all their troops towardes this fruntier, but if M. de Rock could go so fast as my wishes, we should have [Douai] before thay could be together."[66] An important advantage had already been lost by the time the communiqué was written, however, for Villars' relief force had entered the field on the 25th—the Allied time advantage had dissipated. British quartermaster Cadogan conveyed to Lord Sunderland his hope of capturing the town by 12 June, "accounting in this calculation for accedent and the slowness of the temper as well as of the manner of attaquing of the ingineer De Roque, who has the principal direction."[67] The town finally fell at the end of June, but the extra month of delay allowed Villars' army to consume all the available fodder surrounding Arras, forcing Eugene and Marlborough to abandon their projected siege.[68] Instead they shifted westward to Béthune on the Lawe River. Here Des Rocques was secretly refused its direction as punishment for 'his mistakes' at Douai, but things did not improve much with the field promotion of the Dutchman Hertel to chief engineer. Albemarle grumbled about the new director's delay in opening the trenches (six days and counting), which he considered "a great blunder for the time we will lose, as the season advances rapidly while we do nothing."[69] Once the trenches were opened, the siege co-commander Fagel assured the Duke that he intended to push the trenches forward, assuming he could persuade the engineers.[70] Fed up with such delays, Marlborough groused to the Grand Pensioner: "It is certain wee are much longer ataking towns then wee aught to be, but your ingeniers will not be put out

---

[66] Van 't Hoff (ed.), *Marlborough-Heinsius Correspondence*, p. 489 #839 Marlborough to Heinsius, Douai, 8 May 1710. See also Murray (ed.), *Letters and Dispatches*, vol. 5, p. 13 Marlborough to Boyle, Douai, 1 May; and van 't Hoff, *Marlborough-Heinsius Correspondence*, p. 492 #843 Marlborough to Heinsius, Douai, 22 May.

[67] Public Record Office, State Papers 77/59, ff. 138–138b, Cadogan to Sunderland, Douai, 2 June 1710.

[68] ARA RvS 1897, Considérations sur les suites de la campagne et projet de la disposition pour les quartiers d'hyver, 10 August.

[69] Veenendaal, Jr. (ed.), *Briefwisseling Heinsius*, vol. 10, p. 558 #1114, Albemarle to Heinsius, Villers Brulin, 21 July.

[70] BL Add MSS 61181, f. 149, Fagel to Marlborough, Béthune, 26 July. Like other public accounts, Millner's compendium assured his readers that all measures had been used for the "expeditious carrying on of the Siege," and, later, that it maintained its brisk pace (*A Compendious Journal*, pp. 301–302).

of their methode, by which a strong place is not much longer taking then a week one."[71] After the town had finally fallen, Schulenburg (the other approach commander) averred that the town would have been captured ten or twelve days earlier if he had been given enough munitions, enough men, and if the plan of attack had not been changed several times.[72] Only with the campaign season winding down did the engineers themselves accept wholeheartedly the need to accelerate the attack, although even at Aire they did not work fast enough for some observers, who hoped that one last objective might be squeezed into what little remained of the campaign season.[73] The single siege at which Marlborough "had little to complain of" with the engineers was at Bouchain, where it had been pushed vigorously. After the town was captured, the Duke returned to his complaints, berating the engineers for underestimating how long it would require to repair the town's battered fortifications.[74] The next year would follow the same pattern: generals, field deputies and politicians calling for more vigor in the same breath as they criticized their engineers.[75]

3. *Waiting to Open the Trenches*

In addition to general pleas to accelerate sieges, there were a number of specific points in a formal siege when impatient commanders could force the engineers' hand. The Allied generals' opinion of the 'anti-Vauban,' Menno van Coehoorn is particularly instructive here. While French officers of the time might have idealized the Coehoorn-style siege as an escape from the plodding of a Vaubanian attack,

[71] Van 't Hoff, *Marlborough-Heinsius Correspondence*, p. 513 #884 Marlborough to Heinsius, Villers Brulin, 11 August 1710.

[72] Schulenburg, *Leben und Denkwürdigkeiten*, p. 472, 31 August; and a letter to Friedrich I Augustus of Poland, 31 August.

[73] Veenendaal, Jr. (ed.), *Briefwisseling Heinsius*, vol. 11, p. 38 #65, Vegelin van Claerbergen to Heinsius, Bevri, 11 August; and p. 197 #361, Dutch brigadier Isaac Cronström to Heinsius, Saint-Venant, 29 September.

[74] Murray (ed.), *Letters and Dispatches*, vol. 5, p. 485, Marlborough to Eugene, Bouchain, 9 September 1711. For Marlborough's complaint, see Snyder (ed.), *Marlborough-Godolphin Correspondence*, vol. 3, p. 1681 #1774, Marlborough to Godolphin, Bouchain, 15 October.

[75] ARA SG 9196, #46, Field Deputies to States-General, Le Quesnoy, 13 June 1712.

English and Dutch officers were more ambivalent about the Frisian's conduct. Coehoorn's acerbic personality gave plenty of opportunity for disputes to arise, but just as Allied commanders complained about his successors, so too did they complain about Coehoorn's method. For them, his tactics may have been an improvement over Vauban's predilection for formal techniques, but his sieges were still not fast enough. The lengthy preparations his tactics demanded appeared to jeopardize the potential time savings offered in the first place.

From the very beginning of the war Allied commanders had trouble restraining their eagerness. The delays in Coehoorn's preparations for an attack on Venlo—once he had been convinced that the siege was possible—concerned them as a result. One English participant in the theater predicted: "Couhorn is causing all thing be maide reddie and wont faire a shot untill all things are in good order, so that its thought when once they begin the town wont keek out fouer dayes."[76] Those in command wrote less approvingly of the engineer's efforts. Marlborough, anxious to clear the Meuse up to Maastricht in order to start the next campaign on enemy territory, expedited the English gunner/engineer Holcroft Blood to report on the siege's status. The resulting letter, though exonerating Coehoorn from charges of outright insubordination, included the unsettling suggestion that the Frisian may have shared Vauban's general concern for avoiding casualties even at the cost of lost time. As Blood recounted his discussion with the director-general: "he tould me he was sorry to see the time run away and fine wether spent in daring of nothing as much as any body, but to begin a ffoulish [foolish] thing and that that must cost severall lives mal a propos he could give no consent to."[77] Coehoorn brought the issue up with the Duke himself, explaining that the disorders they experienced when first opening the trenches were unavoidable when "one wants to speed up the siege [as] vigorously" as the commander did.[78] After the engineer had been brow-

[76] Royal Commission on Historical Manuscripts, *Report on the Manuscripts of the Earl of Mar and Kellie*, p. 225 Harry Erskine to the Earl of Mar, Breda, 30 August 1702.

[77] BL Add MSS 61306, f. 35 Holcroft Blood to Marlborough, August 1702. After Blood had left, Marlborough acknowledged the serious supply shortages Coehoorn had been complaining about. Snyder (ed.), *Marlborough-Godolphin Correspondence*, vol. 1, p. 100 #89, Marlborough to Godolphin, Everbeek, 10 August; and p. 109 #96, to Godolphin, Asch, 31 August.

[78] BL Add MSS 61187, f. 39, Venlo, 12 September.

beaten by his superiors into pressing the attack, the town lasted only eleven days of open trenches and a week of firing.[79] Marlborough may have been forced to acknowledge Coehoorn's justified complaints at Venlo, but this did not immunize the engineer from further criticisms along the same lines in later sieges, not did it slacken the generals' desire for haste. Prior to Roermond's investment, the Dutch general Jacob van Wassenaer-Obdam complained of the Republic's chief engineer: "Mr. de Coehoorn enrages us with his delays and the measures that must be made for him, because he never lets us know beforehand what he needs." A few weeks later he wrote despairingly:

> We are furious with M. de Coehoorn who always wants to act so methodically and will start nothing before having twice as much as is needed and everything complete, which makes us lose much time, but as the siege's direction is confided to him, we must manage him because he is *un esprit extraordinaire*; I do not know any remedy.[80]

Cardonnel, again mirroring the Duke's own opinion, informed the Tory politician Robert Harley back in England:

> I hope in my next to send you the like good news of Roermond, where we reckon Mr. Cohorn is more nice than wise. He is losing time there as he did before Venlo, and will not begin till he has everything ready to a tittle, though half the preparations might do the busyness; for we reckon Stevensweert must be the strongest of the two.[81]

Similar complaints were made at Coehoorn's siege of Bonn in early 1703. Marlborough grumbled to Heinsius that "Wee shall not bee

---

[79] Veenendaal, Jr. (ed.), *Briefwisseling Heinsius*, vol. 1, p. 456 #935, Wassenaar-Obdam to Heinsius, Roermond, 2 October. With this evidence, we might even consider revising our opinion of the quintessential Coehoornian attack, the bloody storm on Namur's citadel in 1695. It is conceivable that it was pressed on him by William in order to end the siege quickly, or, alternatively, in later sieges Coehoorn applied the lesson learned from the Terra Nova experience to follow more closely Vauban's admonitions.

[80] Veenendaal, Jr. (ed.), *Briefwisseling Heinsius*, vol. 1, p. 440 #898, Wassenaar-Obdam to Heinsius, Venlo, 24 September 1702; and p. 462 #949, from Roermond, 5 October.

[81] Quoted in Wijn, *Het Staatsche Leger*, vol. 8 part 1, p. 171. Coehoorn's "farther demands of stores," wrote Marlborough, "gives mee the spleen." Van 't Hoff (ed.), *Marlborough-Heinsius Correspondence*, p. 25 #47, to Heinsius, Everbeek, 21 August. On the Duke's desire after Roermond and Stevensweert to make use of the remaining good weather, see Murray (ed.), *Letters and Dispatches*, vol. 1, p. 43 to the States-General, Zutendaal, 8 October.

able to make a farther step till M. de Coehorn has all that he expects; which gives mee a good deall of concern, for he complains very much."[82] Just as Vauban had been forced to accelerate the attack on Breisach in 1703, by early May the Duke had finally convinced his recalcitrant engineer to open the trenches before all their supplies arrived "for fear of losing to[o] much time."[83]

Surprisingly then, even the brusque Coehoorn, renowned for his supposed love of bombardments and storms, was too methodical and systematic for Allied generals in a hurry. That such grumbling surfaced at practically every siege directed by Coehoorn in the Spanish Succession, most of which lasted only a few weeks, illustrates how powerful and almost instinctual was the commanding officers' distaste for delays. It also illustrates how generals refused to believe the engineering explanation that time spent preparing for a siege would shorten its overall length, despite several examples where initial lost time was quickly made up in later stages of the siege (Venlo, Oostende, Menin . . .).[84] While we can in some cases blame the Dutch logistical system for not providing the chief engineer with all the supplies he demanded sooner, Allied commanders were certain that Coehoorn wasted time demanding far more guns and ammunition than were actually needed to capture the towns. He may have improved upon the stereotyped Vaubanian attack in his superiors' eyes, but he did not go far enough, was still unwilling to take the necessary shortcuts the operational situation demanded. He was still too wedded to the engineers' insistence on preparations and planning. Coming from generals who did all they could to avoid sieges in the first place, such impatience is not surprising.[85]

Coehoorn's exit did little to lessen their impatience. A journal from Marlborough's army noted how the Duke and his Dutch con-

[82] Van 't Hoff (ed.), *Marlborough-Heinsius Correspondence*, p. 61 #101, Köln, 21 April. See also Snyder (ed.), *Marlborough-Godolphin Correspondence*, vol. 1, p. 165 #157, Marlborough to Godolphin, Köln, 9 April.

[83] Van 't Hoff (ed.), *Marlborough-Heinsius Correspondence*, p. 65 #106, Marlborough to Heinsius, Bonn, 4 May.

[84] The generals, of course, could respond that their pressure to accelerate the pace is what caused the town to be captured more quickly than if the engineers had had their druthers.

[85] More surprising, we even find Barbesieux making the same accusation of Vauban, ordering an artillery commissaire to keep an eye on Vauban's powder usage, "which is several times more than what is needed." AG A[1], 1394, to Desmadry from Marly, 12 May 1697.

federates met "in order to concert measures for carrying on that siege [of Menin] with all possible vigour."[86] Attending the siege of Ath from afar, Marlborough was already looking toward the future; so impatient was he to see the place captured that he mistakenly reported the trenches opening a night earlier than they actually did.[87] Later that same day he corrected himself in a letter to Secretary of State (and future nemesis) Robert Harley, reporting that "the trenches will be opened this night, in order to begin to work on the batteries in the morning, from whence they may fire on Thursday or Friday, after which I hope it will not be many days before I shall be able to give you an account of being master of the counterscarp, beyond which the French have not yet defended any place in these parts."[88] Several years later, Marlborough illustrated how entrenched a practice delaying the trenches remained, as he wrote hopefully to the Lord-Treasurer of their attack on Douai: "And that we might not continue intierly idel, though it be against the rule to open the trenches before we have our canon, we think of doing it tomorrow, or the next day, at farthest, for if it were possible we would faine be masters of this town in this month of May."[89] The trenches were opened three days later. Des Rocques also acknowledged such pressures by including the following point in his *mémoire* on preparations for Béthune: "Beg the generals to not give the order to open the trenches until everything is ready at the site, and particularly the artillery, because a shortage of it causes many difficulties in the approaches."[90]

French technicians also struggled to resist premature trenchwork. Vauban's own correspondence combined the appeal for efficiency with an indication of the pressures he faced to start the siege at Old Breisach:

[86] BL Add MSS 4742, f. 65, Helchin, 5 August 1706.

[87] Snyder (ed.), *Marlborough-Godolphin Correspondence*, vol. 2, p. 673, Grandmetz, 20 September.

[88] Murray (ed.), *Letters and Dispatches*, vol. 3, p. 141, Grandmetz, 20 September.

[89] Snyder (ed.), *Marlborough-Godolphin Correspondence*, vol. 3, p. 1468 #1505, Marlborough to Godolphin, Flines, 1 May 1710. See also Van 't Hoff (ed.), *Marlborough-Heinsius Correspondence*, p. 487 #836, Marlborough to Heinsius, Flines, 1 May; and Snyder (ed.), *Marlborough-Godolphin Correspondence*, vol. 3, p. 1469 #1506.

[90] Jules Duvivier and J.W. Wijn, "Documents hollandais rélatifs à la guerre de succession d'Espagne dans le Nord de la France," Mémoire comment on a trouvé la ville de Béthune, p. 250.

> It appears that His Majesty is resolutely determined to besiege Fribourg; he even orders Monseigneur [Burgundy] to press the conclusion of this siege; on this point I must tell you that we opened the trenches five or six days before the siege lines were finished against custom and only with a view toward accelerating it, and that it was pushed with so much speed that the batteries did not have enough time to have their effect with all the success we should expect.[91]

As we have seen already, these pressures had not abated a decade later, with Villars advocating an early start: "I am pressing vigorously [*vivement*] to open the trenches [at Bouchain] . . . two days saved will be of great benefit when the rains arrive."[92] So dominant was the cult of vigor that engineers used it to support their calls for greater efficiency. In their manuals Vauban and La Vergne warned that premature haste would only delay the end of a siege, and the cannonier Vallière also enlisted it in his efforts to convince others of the benefits of complete preparation:

> The general of artillery should oppose as much as possible the opening of the trenches until the artillery park has at least half of everything needed for the siege, and the last convoys are en route. This will eliminate all the disorders that can arrive, and it is better to postpone the trenches several days and have everything ready for a vigorous [*vivement*] attack, than to open it too early, when any number of problems will prolong the siege.[93]

In both Allied and French camps, such advice was ignored wherever possible. Vigor demanded that the trenches be started as soon as possible, and as close as possible too, far closer in fact than Vauban's guidelines suggested.

4. *Capturing the Covered Way*

In addition to opening the approaches at the earliest opportunity, impatient besiegers pushed their sieges even faster by foregoing the sap wherever possible. The most frustrating stage of a Vaubanian siege for vigorous officers was the slow trench advance towards the

---

[91] Rochas d'Aiglun, *Vauban*, vol. 2, p. 527 to Chamillart, Biesheim, 5 September.
[92] AG A[1] 2383, #223, to Voysin, 14 September 1712.
[93] Vallière, *Memoires d'artillerie*, pp. 10–11. Exactly such a thing happened at Douai. AG A[1] 2382, #102 Lefebvre d'Orval, Valenciennes, 23 August.

fortifications—much of this delay could be avoided if they could jump forward to the next phase, the capture of the covered way. As Vauban had identified in his early treatise, this was the most important tactical challenge, since he argued that the casualties incurred capturing the covered way by storm accounted for three-quarters of a besieger's casualties over the entire length of a siege.[94] All three of his tactical 'innovations' consequently improved the odds of seizing this work, while two of them (trench cavaliers and ricochet fire) were solely dedicated to making this task easier and safer. With these and other techniques, industry was now to replace the brute force of a storm whenever possible.[95] In his tome dedicated to Burgundy Vauban provided his mature judgment that capturing the counterscarp with the use of ricochet fire and trench cavaliers was preferable to bloody assault whenever possible.[96] Vauban's first choice was clear to all by the end of his life, as the non-engineer marquis de Feuquières explained half-way through the Spanish Succession:

> I am not very copious in my Maxims on this subject [of attacking the covered way], because I am entirely averse to the ancient Practice of attacking Counterscarps by Storm, as being an Operation very destructive to the Assailants. The manner of forcing the Besieged to abandon their Counterscarp, and Covert-way, practiced by Mr. de Vauban, is certainly best, and most effectual, at the same time that it is least fatal to the Men.[97]

Noticeably, here is where the Vauban method was most strenuously resisted by both Allied and French commanders. La Vergne enumerated the various choices available to besiegers:

> There are several ways of lodging on covered ways, which is always one of the most deadly tasks of a siege; some want to attack it in broad daylight without concern for the casualties, others want to attack it at night, others to sap regardless of how long it takes, and others

[94] Vauban, *Mémoire pour servir*, p. 98.

[95] Vauban, *Traité*, p. 130. Vauban did not categorically reject general storms of the covered way (see for example his advocacy for one at Valenciennes in 1677), but he condemned unprepared and unnecessary attacks when *industrie* would suffice. For 1677, see Rousset, *Histoire de Louvois*, vol. 2, p. 285ff.

[96] Vauban, *Traité*, p. 263, maxim XIV. In those rare cases where the effectiveness of ricochet fire was neutralized by terrain, he gave additional advice on the best way to weaken the covered way before storming. Vauban, *Traité*, pp. 87–89, pp. 128–130. See also his juxtapositioning of sapping and storming in Rochas d'Aiglun (ed.), *Vauban*, vol. 2, pp. 229–230.

[97] Feuquières, *Memoirs Historical and Military*, vol. 2, p. 286; also 289.

> want to blow up the angles of the covered way with mines and, without knowing for certain if they have been successful, lodge in the debris.[98]

To these traditional methods, the model of Ath 1697 offered a "new manner" of capturing the work: supplementing the advance of the sap with ricochet fire.[99] Although this technique appealed to those seeking the ideal combination of speed and safety, siege commanders consistently chose the quickest, most-direct route—foregoing the recommendation of Vauban and successive chief engineers to sap up to the palisades, instead sending a 'forlorn hope' of grenadiers charging up the glacis *à la Coehoorn*. These troops would rush past the shattered remains of palisades into the covered way itself, which ideally had already been softened up with artillery fire and hand-grenades. Scores of fusiliers would follow, as well as a group of workers directed by engineers to construct a secure lodgment. Additional waves of reserve troops would be thrown into the fray as needed, in all several thousand men dedicated to establishing posts on the work. With lodgments established (usually at the salient angles), the besiegers could then begin the task of sapping to the right and left, clearing the branches of the covered way and starting descents underneath it towards the ditch. Against major fortresses, the casualties resulting from this massive effort numbered a thousand men or more. Vauban, struggling against his own Court's admiration of William's vigorous storm of Namur's Terra Nova fort in 1695, alerted the director-general of fortifications to the danger of its siren song: "It seems that you consider this action a feat of bravery, while I deem it one of the most spectacular follies ever made when attacking a place. And if God desires the enemy to always act this way, we would be only too happy."[100] God must have been part French, for in siege after siege the Allied commanders of Spanish Succession attacks chose this bloody option rather than accept the delays that were required for the slower process of advancing *à la sap*. When they justified it, it was in terms familiar to modern commanders, with the counterfactual argument that we would take more losses exposing our men while sapping than what we will suffer in the

[98] Jacob de La Vergne, *Nouveau exercice du gabion*, p. 58.

[99] Mesgrigny noted the novelty of this approach in a letter to Barbesieux during the siege. AG A[1] vol. 1400 #286, 31 May 1697.

[100] Rochas d'Aiglun (ed.), *Vauban*, vol. 2, p. 439, Vauban to Le Peletier de Souzy, Brest, 25 September.

much shorter time needed to assault.[101] More efficient-minded individuals were skeptical of such arguments.

The Allied commanders were so enamored with covered way assaults that it practically became their doctrine. Charged with the defense of Kaisersweert, the French commander and son of *le grand* Colbert, Jules-Armand Colbert, marquis de Blainville appreciated Vauban's arguments and somewhat incredulously reported to the *maréchal* Boufflers on the enemy's tactics:

> After seeing the manner with which the enemies have attacked us up till this point in time, the slowness and extreme caution that they have shown in all their actions, and their recent attempts to sap towards our angles [of the covered way], I would never have believed that they would decide to attack the covered way by storm, and to risk losing a good part of their troops in order to establish themselves on a work that they could have captured by sap in two or three days without losing hardly anyone; this, however, is exactly what they did yesterday between seven and eight in the evening.[102]

Blainville misinterpreted the Allies' slow pace as a sign of their caution. At the Dutch attack at least, they had expected to storm the counterscarp after only a week of trenchwork, but delays at the Prussian approach forced them to defer their assault in order to help put their ally's works in order. It was only after a six-week postponement that the besiegers were able to launch storms at both approaches against the divided garrison. The town surrendered within a week, but the storm had cost the Allies 2,800 casualties (500 dead among the Dutch alone).[103] After this bloody start, the next several Flanders campaigns (1702–1705) settled down into a more comfortable

---

[101] G.T. Coehoorn, *Het leven van Menno baron van Coehoorn*, p. 80. See also Timothy Kubik's discussion of Renaissance military thought on artillery fire compared with the 20th century U.S. Army's view. Kubik, "Is Machiavelli's Canon Spiked? Practical Reading in Military History," *The Journal of Military History*, 61(1) January 1997, pp. 16–17.

[102] Wijn, *Het Staatsche Leger*, vol. 8 part 1, p. 688, Blainville to Boufflers, Kaisersweert, 10 June 1702.

[103] For the Dutch attack, see in particular Veenendaal, Jr. (ed.), *Briefwisseling Heinsius*, vol. 1, p. 138 #221 28 April. For the storm's casualties, p. 285 #497 Geldermalsen to Heinsius, 14 June from Kaisersweert mentions 2,800 casualties in total; while Wijn, *Het Staatsche Leger*, vol. 8 part 1, p. 691 Lijste van dooden en gequesten van de troupes van Den Staat enumerates 2,101 casualties for the Dutch. French losses were initially reported at 700, but they were quickly revised downward to only 350. Wijn, *Het Staatsche Leger*, vol. 8 part 1, p. 690, Blainville to Boufflers, Kaisersweert, 14 June 1702.

rhythm. The Allies conducted a series of short sieges against weak fortresses that did little to tax their resources or challenge their besieging skills. The garrisons were small, their powder supply often limited; the small volume of fire they offered made the slow pace of sapping *à la Vauban* unnecessary. Most of these French garrisons wisely surrendered only a day or two before a general storm was imminent. If they did not, they were to be taken at bayonet point. The rare instance of this was at Venlo's Fort Saint-Michel, where:

> the design of the attack was only to drive the enemy from the covert-way, that they might not disturb the workmen in making their lodgment; however, if they found them give way with precipitation, we were to jump into the covert-way, and pursue them, let the consequence be what it would. We all thought these were very rash orders, contrary both to the rules of war, and the design of the thing.

Their risk-taking paid off here, for the assault party forced the few defenders in the covered way to retreat, pursued them into the fort itself and accepted its surrender all in the same day.[104] Storming small forts such as these presented little challenge and undoubtedly validated the generals' decision to use brute force tactics.

The situation began to change in 1706. After Ramillies, the Allied army eventually reached the coastal fortress of Oostende. Des Rocques' initial plan for its siege indicated his intent to postpone opening the trenches until all the supplies were ready, and then to push back the decision whether to capture the covered way by storm or sap depending on the effectiveness of their own artillery fire as well as the garrison's dispositions.[105] Ten days later Goslinga apologized to the Grand Pensionary for the unexpected length of the siege, three days after the trenches had been opened, but a frustrating eleven days since the town was invested. Despite early delays due to a shortage of fascines and the high water table, it quickly became obvious that the garrison was performing far below expectations. By 3 July the siege commander Ouwerkerk was already discussing the upcoming attack on the counterscarp, which took place two days later.[106]

[104] Chandler, *Robert Parker and Comte de Mérode-Westerloo*, pp. 21–22.

[105] ARA SG 9188, #86, Disposition des points principaux pour faire le siege d'Ostende, 20 June.

[106] Veenendaal, Jr. (ed.), *Briefwisseling Heinsius*, vol. 5, p. 357 #674, Goslinga to Heinsius, Oostende, 1 July; and p. 371 #694, Ouwerkerk to Heinsius, Oostende, 3 July.

Resistance was quite light, and the grenadiers quickly chased the defenders away and lodged on the covered way with only 54 casualties.[107] The very next day the town capitulated, to the relief of those who had expected a much longer defense.[108] Thus far, the 1706 campaign looked like a continuation of previous sieges and a confirmation of anti-Vauban tactics.

Menin was their next objective, described as one of the strongest fortresses in the entire theater; it presents a clear example of the conditional nature of the engineers' authority over the siege, as well as providing a flashback to Kaisersweert. The governor of Menin had expected that the besiegers' attention to their parallel and saps indicated their "desire to capture the covered way *à la façon de Vauban*, and not attacking it by *vive force*," but this was reading too much unity into the enemy's behavior.[109] The Dutch engineers concerned themselves with perfecting the trenches, but the threat posed by a gathering French relief force (more apparent than real, as Marlborough noted), a growing shortage of fodder and their success at Oostende encouraged the generals and field deputies to halt Des Rocques' sapping and demand a storm.[110] The assault took place on the night of 17 August, resulting in several lodgments along the covered way. The cost, however, was far greater than what Oostende had given them hope for: 1,300 casualties instead of the hundred or fewer experienced in most previous assaults.[111] They reckoned the storm

---

[107] *Europische mercurius*, July 1706, p. 86; and BL Add MSS 61180, f. 5, Ouwerkerk to Marlborough, Oostende, 5 July. According to the semi-official Dutch monthly, the besiegers suffered 438 casualties in all. *Europische mercurius*, July 1706, p. 87.

[108] Veenendaal, Jr. (ed.), *Briefwisseling Heinsius*, vol. 5, p. 382 #720, Heinsius to Goslinga, The Hague, 10 July.

[109] AG A[1] 1939, #203 Caraman to Chamillart, Douai, 27 August 1706. Much more frequent than explicit mention of an attack *à la façon de Vauban* is French discussion (and only French from the documents I have consulted) of an attack "à la Coehoorn," in opposition to the slower techniques of Vauban.

[110] Veenendaal, Jr., "De Mémoires van Sicco van Goslinga," 30. In a letter Goslinga showed his annoyance that they had "lost" four days sapping and they would still have to storm the counterscarp. Veenendaal, Jr. (ed.), *Briefwisseling Heinsius*, vol. 5, p. 476 #915, Menin, 18 August.

[111] On the storm itself, see Veenendaal, Jr., "De Mémoires van Sicco van Goslinga," 30. Wijn, *Het Staatsche Leger*, vol. 8 part 2, p. 758, Lijste van de dooden en gequetse van den 17en tot den 18en augustij enumerates 1,305 casualties; Goslinga also mentioned 200 dead and 1,100 wounded. A French journal suggested 2,000 casualties and noted that most of these losses were inflicted by fire from the outworks and town ramparts, as most of the defenders in the covered way itself abandoned their posts. AG 1M 2365, Caligny, Relation de ce qui s'est passé au siège de Menin, f. 4v.

saved them eight days of sapping, though it is doubtful their calculation took into consideration the effects the continual bombardment would have had on the garrison had they persevered with the sap for a few more days. In the season's last siege, Ath, the Dutch commander Ouwerkerk spoke frequently of the expected upcoming storm, while Du Mée reassured both him and Marlborough that they would decide on sap or storm depending on the countenance of the garrison, "but always with the goal of losing the least amount of time possible."[112] In the end, the undersized garrison abandoned the covered way to the besieger's saps at a cost of only 150 Allied casualties. Looking back on the 1706 campaign, one finds a mixed picture of Allied engineers following Vauban's adage of adapting their tactics to the resistance of each fortress with a preference for the sap, and of their generals pushing for storms with sporadic success.

Though storms usually proved worth the risks against *bicoques*, the higher casualties seen at Kaisersweert and Menin would soon be repeated again and again when the Allies collided with the main fortresses of Vauban's *pré carré*. Initial efforts by the engineers to sap quickly gave way once again to the generals' insistence on shock tactics. The Allied siege of Lille is a particularly instructive example. Here Goslinga described a council of war wherein the chief engineer expressed his desire to continue sapping towards the covered way rather than storm it outright: "We wanted to assault the counterscarp, but Des Rocques believed it was necessary to lodge on it by the sap. This method requires more time, but it will save us many brave men if it succeeds. We have decided with Prince [Eugene] to give him several days to attempt it." This deadline was not extended, for after a few days of digging the generals soon tired of such trenchwork and on the night of 7 September stormed the covered way. Goslinga described the results:

> Yesterday at sunset we assaulted the counterscarp; the success did not meet expectations, the heavy fire added to the darkness put many of our workers in disorder, and of the 4,000 we could only gather together 1,600 to make the lodgments. Our poor men and regiments however were obliged to withstand the enemy fire and could only lodge at a few angles.[113]

---

[112] BL Add MSS 61310, f. 74, Du Mée to Marlborough, Ath, 27 September.

[113] Veenendaal, Jr. (ed.), *Briefwisseling Heinsius*, vol. 7, p. 497 #1085, Goslinga to Heinsius, Lille, 3 September; and p. 509 #1110, Goslinga to Heinsius, Lille, 8 September.

These few lodgments cost the confederates 2,667 casualties, the butcher bill approaching that of Kaisersweert four years earlier.[114] With little progress to show for their efforts, the generals, using the peculiar logic of a vigorous siege, demanded further assaults, overruling yet again the concerns of Des Rocques. These high casualties became the pattern from 1706 onward. At every siege hereafter (with the three exceptions of Lille's citadel, Tournai's citadel, and Douai in 1710) the engineers preferred sapping while the generals demanded assaults.[115]

In this vigorous context it is not surprising that the Duke was careful to avoid accusations of cowardice in 1709 by attributing sapping to the engineers when outsiders might otherwise blame him for Tournai's slow pace: "the enemy now make very little fire, and we have carried on our several approaches within a few paces of the counterscarp, but our engineers being apprehensive of the enemy's mines, we now proceed by the sap in order to save our men."[116] Sapping here was a sensible precaution, though, for the complex system of permanent countermines was untouched by the besieging batteries and thus would have posed a significant threat to storm parties. Only by sapping and sending out miners to probe underground could these hazards be neutralized. At the less imposing fortress of Charleroi in 1693 Vauban considered the lives that would be saved by ferreting out enemy mines well worth a delay of three or four days.[117] So serious could this threat be—garrison mines exploding underneath lodgments or batteries constructed on the covered way—that a good system of countermines was one of the very few defenses that could prevent the Allies from using their favorite method of storm.[118] In fact, both Low Countries fortresses equipped with prepared

---

[114] Wijn, *Het Staatsche Leger* vol. 8 part 2, p. 788.

[115] Des Rocques expressed the desire to speed up the siege of Aire (due to the late season) by storming the covered way. It is unclear if this was his own idea or whether he wrote of it under pressure. Veenendaal, Jr. (ed.), *Briefwisseling Heinsius*, vol. 11, p. 212 #387, Des Rocques to Heinsius, Aire, 2 October 1710.

[116] Murray (ed.), *Letters and Dispatches*, vol. 4, p. 547 Marlborough to Boyle, Tournai, 15 July 1709. For a similar example careful to identify the engineers as the cause of the delays, see Van 't Hoff (ed.), *Marlborough-Heinsius Correspondence*, p. 454 #772, Orchies, 12 August: "for our inginers have great respect for the myns."

[117] As reported in an account from Namur in the Public Record Office, State Papers 8/14, f. 202, 1 October 1693. My thanks to Dr. John Stapleton for bringing this collection on Charleroi to my attention.

[118] Another defense against storms was the use of inundations or an advanced wet ditch at the foot of the glacis—the besiegers would have to build bridges across

countermine galleries forced the besiegers to dig underground to ferret them out: Lille's citadel, and Tournai's citadel. Tournai's was particularly intimidating: "We fear the taking of [it citadel] will take us up thirty days from this time, for there is not a foot of ground that is not undermined and casemated, which will oblige us for the preservation of our people to approach it à la sapp, which indeed is very slow but sure," as a Colonel Revett informed a correspondent back in England.[119] But rather than acknowledge such losses as an unavoidable necessity, impatient commanders dodged responsibility, refused to acknowledge the inevitable and focused instead on the shortcomings of the engineers. We even find an occasional complaint that the engineers botched the storms, rather ironic given that the engineers had discouraged their use in the first place and as they had far less control over this aspect of the attack than any other.[120] If the generals leading the assaults were not to blame, we must wonder why commanders knowingly and repeatedly ordered engineers whom they deemed 'incompetent' to conduct bloody storms at siege after siege.

The French were also captivated with shock tactics, Villars using the technique again and again in 1712.[121] The *ingénieur en chef* Valory perfectly encapsulated the engineering preference for greater certainty and the reluctance to spill unnecessary blood as he resisted pressure to storm Douai's counterscarp after only a week of trenchwork. With Vauban's warnings no doubt ringing in his ears, the

this moat before starting up the glacis to reach the crest of the covered way. Douai 1710 and one of the approaches at Aire are the only examples from Flanders where the attackers were unable to fully drain the front of attack and thus were forced to sap instead.

[119] Royal Commission on Historical Manuscripts, *Report on the Manuscripts of Mrs. Frankland-Russell-Astley of Chequers Court, Bucks*, (London, 1900), p. 198, Revett to Pollhill, Tournai, 1 August 1709.

[120] See, for example, Marlborough's odd complaint that the engineers compelled them to conduct dangerous night-time assaults at Lille against their inclinations—Vauban declared himself against nocturnal assaults. Murray (ed.), *Letters and Dispatches*, vol. 4, p. 252, Marlborough to the Earl of Sunderland, Ronck, 4 October 1708. Why the commanders, who forced many of these assaults in the first place, would then feel constrained to follow the "incompetent" engineers' recommendations for a plan that they opposed is unclear.

[121] The French temperament has often been seen as ideally suited to vigorous actions. Christopher Duffy, *The Military Experience in the Age of Reason*, (New York, 1988), pp. 19–21; and John Lynn, *Bayonets of the Republic: Motivation and Tactics in the Army of Revolutionary France, 1791–94*, (Boulder, CO, 1996), pp. 185–193.

engineer alerted Villars of the inevitable casualties and confusion that would result from a premature storm across open ground. He concluded his missive to the marshal by echoing Vauban's appeal for patience, promising that "several days more will make the assault's success certain and our loss infinitely smaller."[122] The Allied garrison had inundated the sector under attack, and after diverting this water away from the trenches the covered way was finally stormed on 7 September. Valory was somewhat mollified that they had only one hundred wounded and around twenty to thirty killed, "which is not many when one attacks the covered way by storm [*vive force*]."[123] The garrison beat the chamade the next day. Bouchain and Le Quesnoy quickly followed suit. An attack on one of Le Quesnoy's demi-lunes had less success than expected at a cost of almost one hundred casualties; one French officer admitted that it was "too precipitous" an attack.[124] Alègre identified how Bouchain's countermines prevented them from storming its counterscarp: "the place although small is all the same very strong, fortified regularly enough, reinforced with earth and the enemies have a quantity of mines which will oblige us to approach the works with a little more ceremony."[125] A besieging engineer's journal indicated more precisely that the defender's fougasses only forced them to sap at the left attack. Eight companies of grenadiers stormed the covered way on the right attack and established themselves there at a cost of one hundred wounded and eighteen killed; in contrast, the lodgment established by sapping inflicted only twelve to fifteen casualties at the left attack.[126] Despite such evidence, Villars remained skeptical of his engineers and at

[122] AG A[1] 2382, #89, Valory to Villars, Douai, 21 August. See similar sentiments in an engineer's journal in AG Article 15 Section 2 §1, Douai folder, Précis du journal du siege de Douay fait en 1712, ff. 2, 8–9.

[123] AG A[1] 2382, #215, to Voysin, 31 August. Valory also pointed out to his political master that in the rush to capture the town, they had not been given time to reinforce the fascine bridges across the advanced wet ditch, which quickly collapsed and stranded the van of grenadiers. AG A[1] 2383, #124, to Voysin, Lallaing, 8 September.

[124] AG A[1] 2383, #47 Contades to Voysin, Lewarde, 3 September.

[125] AG A[1] Carton 8 chemise 2, Alègre to duc du Maine, Bouchain, 13 October. On Bouchain 1712 more generally, see Lecestre (ed.), *Mémoires du chevalier de Quincy*, vol. 3, p. 198.

[126] AG Article 15 Section 2 §1, Douai folder, Siége de Bouchain, ff. 22–24. Villars' memoirs also claimed 100 casualties. Anquetil, *Suite des mémoires rédigée par Anquetil*, p. 391. For Le Quesnoy, Villars admitted fewer than 200 casualties attacking the counterscarp (p. 388).

Landau the next year he justified his intention to embrace Coehoorn-style tactics in the following terms:

> The sentiments of the *maréchal* Villars, in the first days of the siege, had been to attack the covered way from a little further away than is normally practiced. The reason was that with this covered way completely countermined, it was certain that the very wet terrain would force the enemies to charge the mines three days before they expected it to be attacked. The engineers opposed this plan, thinking it rash and too perilous. However, the rest of the siege would show that beyond the loss of time, which is very precious in wartime because thirty days of open trenches were needed to capture the detached outworks, the loss of men was also very considerable during this period of time which a premature attack would not have cost, and it was recognized by the end that the *maréchal* Villars' opinions were proved correct.[127]

Again, rather than acknowledge the necessity of the sap's delays, Villars was willing to risk an assault on a covered way only briefly prepared by artillery fire, despite the cautionary example of Marchiennes the previous year. Whether his plan would have succeeded or not is impossible to know; instead they conducted a conventional trench attack and still stormed the covered way at a cost of 2,000 casualties. Christopher Duffy has disapproved of "such an un-French and costly method of reducing the stronghold," yet it was in many ways a typically French tactic—Vauban had complained in 1672 of just this predilection, was unable prevent its application at some of the attacks under his own command, and the chevalier de Quincy testified to its continued popularity forty-five years later.[128] At yet another 1713 siege the impetuous Villars refused his chief engineer's cautious recommendation to delay the storm of Freiberg's covered way for two or three days. In this case Villars' vigor was especially untimely, for just as his grenadiers reached the crest of the covered way they met a force of 600 garrison soldiers sallying out—the resulting combat saw perhaps 700 French troops killed or wounded.[129] Despite such costs, this storming mentality remained popular with many Allied and French commanders.

[127] Vogüé (ed.), *Mémoires du Maréchal de Villars publié d'après le manuscrit original*, vol. 3, p. 199.

[128] Duffy, *The Fortress in the Age of Vauban*, p. 47. Elsewhere (p. 65), Duffy argues the opposite, that French commanders preferred Coehoorn's speedy tactics over Vauban's more patient approach. See also Hebbert and Rothrock, *Soldier of France*, p. 54.

[129] Augoyat, *Aperçu historique*, vol. 1, pp. 430–431.

The historiography explains the supposed rarity of the attack à la Coehoorn in terms of skill: it was purportedly easier to learn Vauban's mechanical attack than to become skilled at knowing exactly where, when and how to storm the covered way, an achievement that demanded the gift of *coup d'oeil*.[130] Not only was the Coehoornian assault in fact the norm, but, as Vauban preached, memorizing his maxims was of little worth without the ability to apply them to a particular context, a skill that required years of practice. General assaults were in fact far easier for field commanders to understand and control—the ability to quickly recognize the advantages of terrain and dispose large bodies of troops for an advance is exactly what made good battlefield commanders and officers.[131] The unattractive alternative was to rely on distrusted engineers who, to their eyes at least, maddeningly whiled away the campaign season measuring angles and turning dirt. Implementing Vauban's attack in siege after siege further required a large and well-trained group of engineers overseeing acquiescent generals, gunners and workers. Coehoorn's method required far less, allowing commanders to minimize their reliance on engineers by returning to a more-traditional, better-understood, less-systematic technique, the same tactics that had spurred Vauban to reform siegecraft in the first place.

Engineers who strove for Vauban's efficiency found themselves with the almost impossible task of living up to Vauban's reputation. Their problems were many, not least that they could not act quickly enough for time-conscious generals in a hurry. Notably, Coehoorn was also criticized for the exact same reason—even his more vigorous style of siegecraft was still deemed too ponderous. With such high expectations, the lesser technicians that followed these two bright lights could hardly hope to win the favor of their superiors. They lacked powerful royal patrons. They were understaffed in the trenches and dependent on the other branches for all their supplies and manpower: civilian and military administrators for their powder, shot, tools, fascines and gabions; infantry generals for their workmen;

[130] On the rarity of Coehoorn's *coup d'oeil*, see Duffy, *The Fortress in the Age of Vauban*, p. 96; Lynn, *Giant*, pp. 576–577. Langins describes how later 18th century French engineers attempted to coopt *coup d'oeil* as a skill uniquely suited to engineers. *Conserving the Enlightenment*, p. 179.

[131] We recall Villars' assertion that the head of the trenches was most suited to the "commander's eye." AG A[1] 2382, #124 Villars to Voysin, 24 August 1712.

artillery officers for their fire support. They were given a limited amount of time to draw up plans that were changed more often than not at the whim of the commander or his generals in the trenches. Those truly in charge of the siege, frustrated with the season slipping away and less concerned with the cost in lives, gave the engineers only a few days of sapping against some of the continent's strongest fortresses before declaring them incompetent, shelving their plans, and forcing them to conduct storms in many ways antithetical to their training. Given all these obstacles, it is not surprising then that the technicians were overworked, their estimates at times inaccurate, their conduct of the trenches less than perfect, and their morale low.

As the commanders' complaints suggest, much of the detailed planning of the attack—trenchworks, batteries, gabions and the like—had to be left largely to the technicians, the engineers and artillery officers. However, the overall results of a siege were determined in large part by several critical decisions, and at these pivotal points field generals relied on the vigor of brute force rather than patient *industrie*. Unlike the cult of efficiency, this cult of vigor centered around the belief that individual lives were disposable *when necessary to gain an immediate, tactical objective*, and that such sacrifices were the only way to achieve true decision. Thus their single-minded focus on the avoidance of delays rather than the engineer's careful balancing of delays, costs and casualties. Although Vauban may have changed the engineering corps' methods over the course of his lifetime, the underlying assumptions of the field officers that commanded them remained the same. The most trusted Allied and French generals still sacrificed thousands of men in their quest to accelerate the attack.

CHAPTER EIGHT

SUCCEEDING WITH VIGOR

Despite the claims of engineering incompetence, despite the numerous complaints about delays, and despite the explicit rejection of many of Vauban's sound techniques, besieging forces in the Low Countries were still surprisingly effective. The resulting sieges were often poorly conducted in a technical sense, straying far from the efficient siege Vauban conducted at Ath in 1697 and even further from the perfect siege he had envisioned in his writings. But they were effective, and for many generals and politicians this was sufficient. Attacking armies captured their Low Countries targets 88% of the time and every major siege that was undertaken was successfully concluded; in the non-Flanders theaters the rate was a slightly lower 82%. Furthermore, two-thirds of the Flanders sieges lasted no more than a month, while 40% of these places could hardly resist two weeks. Apparently Vauban's efficient attack was not the only way to capture places in a relatively short period of time. The competing histories of the period provide few explanations for this unexpected result: the Marlborough biographers (themselves immersed in the cult of vigor) present sieges as a time-consuming betrayal of the Duke's battle-seeking strategy, while the Vauban literature portrays rejection of his scientific techniques as the highest form of ineptitude. Why, then, were besiegers so successful when their commanders rejected the underlying impetus for Vauban's method and forced engineers to use exactly those bloody techniques that he had railed against? There were of course many factors that dictated the exact outcome of any particular fortress's defense: the relative strength of the attackers and the defenders (the size of the forces available to them, their morale, their leadership, their equipment, their logistics, the competency of their technicians, etc.), the design and maintenance of the fortifications under attack as well as the terrain in which they were situated, the weather, the role of relief efforts, and even chance occurrences. The length and resolve of each garrison's defense depended on a unique combination of such factors—and we have already discussed the importance of weak fortifications for the many

Spanish *bicoques*—but we should emphasize the common traits that all these sieges shared and which practically guaranteed (statistically speaking) their success—the brute force techniques the besiegers used to conduct the attacks. Three factors were of critical importance: the besiegers' trenchworks of course, but just as importantly, their reliance on overwhelming firepower, coupled with their political masters' abilities to continually resupply their armies with the men and materiel needed to maintain this vigorous approach in siege after siege.[1] In short, attackers supplemented Vauban's imperfectly-implemented attack with overwhelming power, enabling them to succeed while ignoring Vauban's appeal for finesse and efficiency.

1. *Vauban's Trench Tactics*

As Chapter 6 suggested, the military engineers and Vauban's methods deserve at least some of the credit for the success of besiegers. Even the most biased commanding generals needed their engineers' advice on and supervision of the trenches, as they simply could not ignore their expertise altogether, nor did they replace them with more vigorous-minded candidates from outside the corps. Certainly Vauban's trench parallels were a pivotal part of practically every siege. How closely besiegers followed Vauban's specific advice to "always make three parallels; locate and construct them well, making them as wide as is necessary" is not always clear, though there is plentiful evidence that in the War of the Spanish Succession both Allied and French attackers made use of trenches dug parallel to the front under attack—contemporary plans of almost any siege of this

[1] The impotence of relief forces also merits further analysis. We should further note that there is very little evidence of the importance of pre-siege blockades by partisan forces that George Satterfield emphasizes in his discussion of the Dutch War (*Princes, Posts and Partisans*). There are several reasons to discount this claim, in the last of Louis's wars at least. Most of the targeted towns were chosen only a month or less before they were invested, while the defending side communicated freely with its frontier fortresses and even reinforced threatened garrisons up until the investment itself. As we shall see in this chapter, the advantage of weakening the defenders with a pre-siege blockade would rarely have been decisive in any case, since almost every fortress surrendered long before it ran out of supplies, while garrison commanders never mention this factor when explaining their need to surrender. Partisan forces were more active in the Spanish Succession war attempting to surprise strongholds away from the main area of operations.

period will illustrate these works.[2] In most cases, the system of parallels worked exactly as Vauban had intended. The majority of both Allied and French siegeworks included three parallels enveloping the fronts under attack. The garrison commander Pierre-Paul Riquet, comte de Caraman attested to their efficiency while he was defending Menin:

> Our garrison was strong and had good morale, but the enemies had a parallel 140 toises [900 feet] from the palisades which was guarded by more troops than we had in our entire garrison, and this parallel was supported by 110 cannon and more than 200 mortars of all sizes, so that any sorties were impossible, since their batteries were spaced along the length of the parallel, we therefore had nothing else to do but keep up a continual fire to slow down their trenches.[3]

One of the Allied approach commanders corroborated the defenders' impotence, and noted that their side had little fear of garrison sallies now that their parallel had a "great front."[4] With parallels widely adopted, external sorties (that is, sallies against the trenchworks beyond the glacis) were almost always repulsed with losses, while the most successful could expect little more than to see most of their defenders return safely after filling in several yards of trenches, with perhaps a few captured gabions as trophies. Only at the Allied siege of Douai was a sortie against early trenchworks successful—and only because the sally targeted the gap between the two approaches' un-joined parallels. While the exact layout of such trenchworks may have varied from Vauban's theoretical strictures, by and large Spanish Succession parallels followed the broad outlines of the Vaubanian attack.

---

[2] Quote from Vauban in his *Traité*, p. 261 maxim 5. Writing of the 1706 Flanders campaign, Duffy noted that "The Allies were now using Vauban's system of parallels as a matter of course" in *The Fortress in the Age of Vauban*, p. 37. For visual examples, see the contemporary published maps by Eugene Fricx and Anna Beek, as well as Pelet's accompanying *Atlas des Mémoires militaires relatifs à la succession d'Espagne sous Louis XIV*, (Paris, 1862). Some of these images are viewable online courtesy of the University of Michigan Digital Library, at http://images.umdl.umich.edu/m/misc/ (search the Miscellaneous collection under "ostwald").

[3] AG A[1] 1939, #203, Caraman to Chamillart, Douai, 27 August 1706; confirmed by the garrison's engineers in AG 1M 2365, Caligny, Relation de ce qui s'est passé au siège de Menin, f. 3v.

[4] BL Add MSS 61162, ff. 99–99b, George Hamilton, first Earl of Orkney to Marlborough, Menin, 15 July.

There are a few instances where their application did not necessarily follow all of Vauban's maxims. In a small number of cases (e.g. the Seven Fountains attack at Tournai and the Hartevelt attack at Guelder), besiegers attacked along a narrow front, making it impossible to envelop the front under attack. The Allies also tended to attack with multiple approaches on different sectors, violating Vauban's sixth maxim.[5] In most of these cases there was little alternative, as inundations or waterways divided up the approaches into isolated narrow attacks. Separate attacks had the additional advantage of allowing different nationalities to command their own approaches. The Allies occasionally broke other of Vauban's rules, though one could certainly point to Ath 1697 for precedence. The Allies used a false attack at the siege of Ghent in 1708; the French did the same at Bouchain in 1712. Despite Vauban's fifth maxim, the sieges of Ghent and Bouchain each included only a second and a third parallel, while the trenches at Saint-Venant were pushed forward so quickly that only a third parallel was constructed, this impatience resulting in 140 casualties in a single day of trench digging.[6] Vauban did allow some flexibility even in his late system, but the trenchworks at these Allied approaches did not follow his admonitions as faithfully as he undoubtedly would have wished. As Vauban admitted in 1704, even implementing his 'simple' parallels was surprisingly difficult. These exceptions encourage us to conclude that parallels were frequently used and served their purposes adequately, although they did not conform to the exact forms Vauban demanded and thus were undoubtedly less efficient than engineers might have wished.

Trench cavaliers also saw use in the War of the Spanish Succession. It is less clear whether it was a concern for Vaubanian efficiency that led to their use, for natural terrain and poorly-designed fortifications could create the same effect, offering inspiration to those unfamiliar with Vauban's writings and earlier precedents. At the siege of Venlo, we find an English correspondent reporting of the besiegers that: "They have made their works so heigh that they can see into the midle of the market place."[7] At Menin, Valory wrote in the garri-

[5] A point made by Duffy in *The Fortress in the Age of Vauban*, p. 43. These conclusions are necessarily based on the assumption that engineers did not omit their mistakes from their plans.

[6] *Relation de la campagne de l'année 1710*, p. 141.

[7] Royal Commission on Historical Manuscripts, *Report on the Manuscripts of the Earl of Mar and Kellie*, Harry Erskine to the Earl of Mar, Breda, 30 August 1702.

son's journal that the enemy batteries were placed on high ground "which gave them a height advantage that exposed the base of the face of the Capuchins bastion" well before they had even reached the covered way. Within a day sixty breaching cannon had reduced its face to rubble.[8] Du Mée, encamped before Lille, wrote of "increasing the work on the right to look into the covered way before the grand breach."[9] At Douai 1710, the garrison commander François-Zénobie-Philippe comte d'Albergotti explained late in the siege that although his men still held the traverses of the covered way, they "would be chased out very quickly by the height of [the enemy's] lodgments, as is their method."[10] At the next siege, Béthune's château was breached down to the very base of its walls long before the besiegers reached the covered way.[11] Height was height, regardless of how one gained the elevation, so the mere use of elevated batteries cannot be considered a legacy of Vauban without more evidence of his direct influence, especially considering the long lineage of the technique. In short, Vauban's trenchworks clearly played an important role in the conduct of the attack, but more important in the eyes of contemporaries was the dominance of artillery.

2. *Overwhelming Firepower*

At the same time as officers relied on Vauban's parallels, they still refused to follow his other suggestions at pivotal moments in the siege. Whereas Vauban encouraged his pupils to conduct their attacks

---

[8] AG Article 15 Section 3 Menin, Relation de ce qui s'est passé au siège de Menin . . ., f. 5, 9 August 1706. This is confirmed by an Allied account which claimed that two paces of the bastion were "entirely ruined" by their batteries on the 10th. BL Add MSS 4742, Journal of the siege of Menin, Helchin, 9 August.

[9] Sieur de May [i.e. Du Mée], *An exact journal of the siege of Lisle*, (1708), p. 12. Earlier, see p. 4, 29 August 1708. And BL Add MSS 61312, f. 50, Des Rocques to Marlborough, Lille, 22 August 1708. For confirmation of the breach, see Royal Commission on Historical Manuscripts, *Report on the Manuscripts of the Earl of Mar and Kellie*, p. 464, Lt. William Nodding to the Earl of Mar, Lille, 20 September 1708.

[10] AG A[1] 2215, #223, Albergotti, Douai, 19 June. Also #51, Albergotti, Douai, 9 May; and AG A[1] 2225, #183, Albergotti to Bernières, Douai, 14 May; and #222, Albergotti to Villars, Douai, 5 June.

[11] AG Article 15 Section 2 Béthune, #4bis, Des Forges, Journal du siège de Béthune, f. 21v.

with *industrie*, generals most often sought to solve the problem of slow-paced sieges with the application of overwhelming firepower. With large artillery trains even a strong and well-stocked fortress could fall surprisingly quickly without excessive reliance on Vauban's techniques. Besiegers massing gunpowder cannon overcame the numerous offensive weaknesses discussed in Chapter 5 surprisingly easily, particularly when compared with the long sieges in the previous century. The string of formidable fortresses the Allies captured between 1708 and 1712 clearly indicate the potency of a revitalized attack. To explain this trend contemporaries emphasized the predominance of besieging artillery, and a more concrete assessment of the artillery used by besiegers reinforces these anecdotal judgments. The ways in which these pieces were used, however, often deviated from Vauban's techniques, just as Vauban's perfect parallels were often abbreviated. The sieges of the Spanish Succession war illustrate how besiegers could ignore many of Vauban's specific recommendations and still conduct successful sieges against even the strongest of fortresses. By late in Louis' reign, firepower played a critical role in making well-fortified towns rapidly untenable.

Unlike modern historians, military contemporaries in the Spanish Succession did not acknowledge Vauban's three tactics as the critical element of the early 18th century siege. Trench parallels, his most widely-disseminated technique, were always mentioned matter-of-factly in siege accounts. Trenches certainly protected soldiers, but without artillery to suppress the enemy's fire and breach their works, the attackers had little hope of success. This use of gunpowder facilitated the rapid advance of the trenches, and it was speed after all that most interested contemporary observers. Revolutionizing siegecraft in the 15th and 16th centuries, gunpowder weapons once again slowly shifted the equilibrium between the attack and defense over the course of the 1600s, authorities such as Pagan detecting the trend early in the century.[12] Until the 1670s, however, the artillery served as only a junior partner in French siegecraft. David Parrott described

[12] Pagan, *Les fortifications de Monsieur le comte de Pagan*, p. 60; and later in the century, Claude-François Milliet Dechales, *L'art de fortifier, de defendre, et d'attaquer les places, suivant les methodes francoises, hollandoises, italiennes & espagnoles . . .*, (Paris, 1684), p. 384. The Dutch had increased their siege parks in the 1590s, though few exceeded 100 guns. See the figures in Kuypers, *Geschiedenis der Nederlandsche artillerie*, vol. 2, pp. 226–234, 292–294.

conditions right before Vauban's service began: "The role of artillery in [sieges of Richelieu's France] was relatively limited. Few of the French sieges of this period were brought to a conclusion because artillery bombardment had rendered the defences untenable."[13] By 1700, on the other hand, the one factor receiving the most comment was not Vauban's three tactics nor his reliance on efficient methods, but the domination of the besieger's artillery.[14] In his twilight years Vauban acknowledged that "there is nothing more important in a siege than good use of the artillery," for "it is with the good use of cannon and bombs that places are captured and sieges shortened."[15] Most other commentators focused not so much on their proper *use* as their *quantity*. Feuquières put the superiority of the artillery within a historical perspective. By 1709 governors could no longer rely upon un-revetted redoubts beyond the covered way to effectively defend medieval walls as they had only a few decades previously:

> As it was not usual at that time [circa 1654] to convey a great train of artillery before places that were besieged, M. de Schulemberg [governor of Arras] believed, with sufficient reason, that their works could not easily be destroyed by the enemy's cannon, and that he should be in a condition to defend them for a considerable time, by his strong garrison, whose fire would oblige the enemy to make their approaches to his works with circumspection and by opening trenches; by which means, he should greatly retard them in their advances to the body of the place, which was but indifferent. This was a judicious manner of thinking at that time, but such a conduct would now be subject to great inconveniences; because works of that nature which are separated from the place would soon be destroyed by the artillery and the bombs to such a degree that the men within the place could not long be in any state of security, and these demolished works would be converted into sure lodgments for the enemy, and afford commodious situations for the batteries.[16]

---

[13] Parrott, *Richelieu's Army*, p. 70.

[14] Earlier historians of the French engineering corps also recognized the importance of artillery, at the same time contrasting Vauban's skillful attack with how little talent was needed to conduct the Coehoornian attack, the reverse of what later historians argue when describing the purported dominance of the Vaubanian attack. See, for example, Allent, *Histoire du Corps Impérial du Génie*, pp. 407, 416, and 589.

[15] Vauban, *Traité*, p. 238; also p. 239. Given Vauban's interest in efficiency and recalling his earlier scuffles with the artillery, his stress on the proper *use* of the guns should be emphasized.

[16] Feuquières, *Memoirs Historical and Military*, vol. 2, p. 271.

The Spanish veteran Santa Cruz de Marcenado gave four reasons for why "nothing is more advantageous in a siege than to have a lot of artillery": their counter-battery fire would quickly dismount the garrison's guns, the quicker outcome facilitated by numerous cannon would limit the threat of relief efforts, assure the place's capture before the campaign season ended, and would save powder and cannonballs as well, since many breaching cannon firing together formed a breach much more efficiently than a smaller battery of guns firing the same number of rounds. He opined elsewhere that "Today it is less with the large number of troops than the quantity of artillery that we force places to surrender," going so far as to contend that even a small besieging force could overcome a more numerous garrison, as long as they had superior firepower.[17] Vauban's erstwhile assistant Goulon similarly assumed the superiority of enemy fire, recommending that garrisons not attempt to duel head-on with the besieger's artillery, but instead to frequently shift their pieces in a game of cat-and-mouse with the enemy counter-batteries.[18] The growth of artillery arsenals, on both sea and land, upset the established offensive-defensive balance and forced engineers to search for ways to return the balance.[19]

Contemporaries discussing the specifics of the various Low Countries sieges were struck by the dominant role of siege artillery. Engineers rarely received mention, yet the correspondence expressed again and again that guns were the key. Against the weakest of fortresses, the combination of voluminous cannon and mortar fire might be enough to force surrender soon after the besiegers began firing. The semi-official French monthly *Le Mercure galant* prepared its readers for the rapid fall of Venlo by reminding them that "today one attacks places

---

[17] Santa Cruz de Marcenado, *Reflexions militaires*, vol. 9, pp. 3–4; and vol. 10, pp. 292–293.

[18] Goulon, *Memoirs of Monsieur Goulon*, p. 10.

[19] For a discussion of the increasing armaments of naval vessels in the 16th century and their impact, see John Guilmartin, Jr., *Gunpowder and galleys: changing technology and Mediterranean warfare at sea in the sixteenth century*, 2nd edition (Annapolis, 2003), pp. 269ff. For later eighteenth century recognition of this same fact of offensive artillery dominance, see Langins, *Conserving the Enlightenment*, pp. 102–103. Luh's *Ancien Regime Warfare and the Military Revolution*, pp. 107–109, also incorrectly locates the watershed period later during the War of the Austrian Succession, while his discussion of the Spanish Succession Low Countries sieges (pp. 105–109) is riddled with unrepresentative examples and errors.

with very large armies and with a great amount of artillery."[20] Millner's *Compendious Journal* noted that the siege commander Walrad von Nassau-Saarbrücken initially thought the sixty-four cannon, twenty-four mortars and howitzers and numerous small mortars were "not sufficient thereof for the quick dispatch of the siege" and ordered more heavy cannon (p. 30). The batteries targeting Roermond's "thin Brick walls" took only a day to create a practicable breach and force the garrison to surrender in order to preempt a general storm.[21] Liège's Fort Chartreuse fell quickly as well, the Duke of Marlborough noting that: "our mortars having started to fire on the Chartreuse this morning, the bombs had such an effect that at 2 o'clock in the afternoon the garrison asked to capitulate."[22] At the beginning of the 1703 campaign Coehoorn promised to capture Bonn in short order with his large artillery train.[23] Back on the Meuse, Huy's small forts swiftly collapsed in turn, an Allied account emphasizing yet again the role of the guns:

> On Tuesday last at 8 o'clock at night our bombs began to play upon a fort called St. Joseph. On Wednesday about 3 o'clock in the morning our bombs and cannon played so briskly upon it that they were forced to hang out the white flag in order to capitulate. . . . Our batteries continued firing so vigorously upon the two other forts that they were obliged to capitulate upon the same terms and accordingly marched out this morning about 9 into the castle. We have nothing now but the castle. We are firing upon it from seven or eight batteries and it cannot hold out I believe above a day or two.[24]

Limburg in the Ardennes also fell quickly, the besieging cannon making quick work of the earthen fortifications that the commander had been too slow to abandon.[25] Emphasizing the improvements the Allies had made to Huy's Fort Joseph, Villeroi noted during his siege in

---

[20] *Le Mercure Galant*, octobre 1702, pp. 332–333, Relation de ce qui s'est passé à la déffense de Venloo, depuis le 29 aoust jusqu'au 24 septembre.

[21] Millner, *A Compendious Journal*, p. 37.

[22] Murray (ed.), *Letters and Dispatches*, vol. 1, p. 50, to the States-General, Liège, 29 October 1702.

[23] Vault (ed.), *Mémoires militaires*, vol. 3, p. 12, Louis to Boufflers, Versailles, 16 April 1703.

[24] Royal Commission on Historical Manuscripts, *The Manuscripts of the Duke of Beaufort, K.G., The Earl of Donoughmore, and Others*, (London, 1891), p. 96, T. Welby to the Earl of Coventry, Huy, 12 August 1703.

[25] *The Daily Courant*, #445, 20 September 1703 O.S.; Allent, *Histoire du Corps Imperial du Génie*, p. 416.

1705 that "only the artillery can finish such enterprises."[26] In response Louis approved of his marshal's efforts to acquire another ten 24-pounders for the attack, noting that "it is the artillery which contributes the most to sieges."[27] After Villeroi recaptured the works at Huy, the Allies returned from the Moselle and endeavored yet again to capture this strategic location. Although the French garrison's ensuing surrender was considered a day premature, the overall brevity of the defense in this, the third attack on the beleaguered place, followed earlier trends: "we battered the château in quite an extraordinary manner and the artillery had a surprising effect in very little time."[28] Albemarle described soon afterward the impression his tour of the battered citadel made: "Nothing is more astounding than to see these poor walls which have been so terribly shattered in four weeks of attacks, from both the [earlier] French [siege] and our own."[29] Reflecting upon his own wartime experience, Landsberg confirmed the effects of constant shelling on such *bicoques*:

> I had already participated in five or six sieges before that of Lille, including Venloo, Roermonde, the citadel and chartreuse of Liège, Bonn, Huy, Limbourg, but as these sieges were not difficult because the enemy found itself vigorously attacked with well-served artillery, these towns could not defend themselves as they do at present. We saw their garrisons terrorized by the artillery, they asked to capitulate before a breach had even been made or their covered ways captured, so that we never experienced any difficulties in our approaches and there was never any need to fill a ditch or make a gallery.[30]

For the first half of the war, the large Allied siege trains rapidly brought the walls of the Spanish *bicoques* crumbling down around their garrisons' ears.

Nor were the mightiest strongholds immune from the massing of siege cannon. The stronger fortresses of Vauban's *pré carré* would

[26] AG A[1] 1835, #210, Vignamont, 2 June.
[27] AG A[1] 1835, #247 Louis to Villeroi, Versailles, 8 June.
[28] Cited in Jean-Pierre Rorive, *La guerre de siège sous Louis XIV en Europe et à Huy*, (Brussels, 1997), p. 250.
[29] Veenendaal, Jr. (ed.), *Briefwisseling Heinsius*, vol. 4, p. 241 Albemarle to Heinsius, Latinne, 13 July 1705.
[30] Landsberg, *Nouvelle manière de fortifier les places*, p. 34. Indicative of Landsberg's hypercritical personality, he is simply wrong here in one important respect, for even the besieging commanders' correspondence indicates that almost every French garrison in Flanders beat the chamade when Allied general storms on the main breaches in their enceinte were only a day or two away.

require far more effort as Landsberg noted, although an eventual collapse was usually just as inevitable. In 1706 the defenses of once-mighty fortresses were quickly reduced to rubble. At Oostende, thirty-two 24-pounders and thirty-six mortars and howitzers covered the creeping Allied trenches: "We continued to make a great fire all day. . . . The French were so shocked that they allowed our people to approach to the edge of the counterscarp without firing a shot."[31] A French account described the enemy's firepower at Menin in similar terms:

> The 9th, at three o'clock in the morning, the enemy's guns began to fire from the parallel with forty cannon and forty mortars; and by ten o'clock, there was a breach eight-toises [fifty feet] wide in the Capuchins bastion between the two demi-lunes, the place's masonry being poor. They fired many bombs and an infinity of double grenades and several carcasses [incendiary bombs] which set fire to several houses in the town, and which blew up one of the mines that we had constructed under the glacis; this fire became so terrible from afternoon until night that all our best officers and soldiers who had seen the sieges of Kaisersweert and Landau were amazed at the volume of fire. We could no longer distribute our food, our brasseries and butcher shops were burned, and our soldier carpenters who were laboring on works in our covered way could not continue due to the heavy bombardment and abandoned their labor; our cannoniers and bombardiers could no longer fire, their platforms being absolutely ruined.[32]

The French commander of Ath later in the year noted the same pattern, explaining how the enemy's batteries battered their bastions with "a vigor beyond imagination."[33] Reading reports of the siege, Vauban could not help but contrast his efficient siege of Ath in 1697 with the Allies' brute force attack in 1706. He recognized the reality of overwhelming firepower, but was still unable to abandon his efficiency-mindedness, for although he believed that the Allies had blundered by attacking Ath at its strongest point (as they had at

[31] Veenendaal, Jr. (ed.), "De Mémoires van Sicco van Goslinga," p. 23.

[32] AG A[1] 1939, #204 Journal du siége de Menin, 9 August 1706, printed in Vault and Pelet (eds.), *Mémoires militaires*, vol. 6, p. 535. AG A[1] 1988, #57 Relation ou journal du siège de Menin, p. 11 gives seventy 24-pounders plus smaller pieces, and forty-four mortars. Goslinga confirms this in Veenendaal, Jr. (ed.), "De Mémoires van Sicco van Goslinga," 29. The garrisons of Oostende and Menin both stressed the defective materials used in their respective fortifications, which quickly crumbled under cannonfire.

[33] AG A[1], 1939 #401, Saint-Pierre to Chamillart, 24 September.

Menin), he still predicted, correctly, that "with the fury of so many batteries and the hail of stones that are flying continually, they will quickly become masters of this place."[34] The year 1708 saw the Allies expend almost all their munitions in their successful attempt to capture Vauban's masterpiece of Lille. Louis' trusted military advisor Jules-Louis Bolé, the marquis de Chamlay had predicted the result back in 1706 when discussing potential relief operations in case Vauban's crown jewel came under attack after Ramillies. He identified the challenges such firepower now presented: "Even a very considerable garrison would likely be unable to save Lille if the enemies were to attack it with an artillery train as large as what they usually employ in their sieges."[35] While Boufflers' defense of Lille was the longest of the theater, his four-month resistance is far less impressive when we recall that the well-provisioned garrison actually held the town for only two months, while the citadel resisted for another two months.[36] The less-imposing fortress of Aire held out for just as long as each of these did, while Douai's mediocre fortifications persevered a similar length of time with fewer troops manning a much more extensive circumference. Given the 'impregnable' nature of Vauban's masterpiece and the Allied mistakes made in its attacks, Lille's relatively swift capture suggests that the siege attack had indeed come far since the late sixteenth century.

Lille's neighbors suffered a similar fate. At Tournai, one garrison witness described how they were "attacked by the most frightful artillery ever seen before a place."[37] With three separate attacks, the enemy intended to "throw many bombs in the works in order to

[34] Rochas d'Aiglun, *Vauban*, vol. 2, p. 594, Vauban to Chamillart, 3 October. In an earlier letter to his good friend Mesgrigny, he noted that if Ath's garrison had had its full complement of 4,500 men instead of only 2,000, the inefficiencies of the Allies' brute force attacks would have been readily apparent. Emile Ouverleaux, *Mesgrigny, ingénieur militaire, lieutenant-général des armées du Roi, 1630–1720*, (Brussels, 1928), p. 44, 30 September.

[35] Vault and Pelet (eds.), *Mémoires militaires*, vol. 6, p. 513 #315, Mémoire de M. de Chamlay, 25 July.

[36] Most histories put the garrison at 12,000 men or more, but Maurice Sautai's exhaustive study of the siege suggests that Marshal Boufflers could only muster half that number. *Le siège de la ville et de la citadelle de Lille en 1708*, p. 59.

[37] AG A[1] 2159, #408, Doles (lieutenant du roi of Tournai), 29 July 1709. The field deputy in charge of the artillery confirmed that the besiegers had "all the artillery that anyone could ask for in a siege." Veenendaal, Jr. (ed.), *Briefwisseling Heinsius*, vol. 8, p. 668 #1365, Vegelin van Claerbergen to Heinsius, Tournai, 30 June.

fatigue the garrison which they believed was undermanned and in a poor state."[38] Quickly silencing the garrison's fire, the threat of garrison countermines forced the besiegers to settle down to discovering them under the glacis with the sap.[39] The next year at Béthune Des Forges' engineering journal noted the effectiveness of the attacking artillery: the overwhelming weight of shot—the besiegers fired one hundred shots for every round fired by the garrison—dismounted most of the garrison's guns after only a day of firing, while the indirect fire was extremely effective even against men and batteries sheltered from sight by intervening breastworks.[40] The Tory-leaning *Post Boy* newspaper reported in 1711 how the Allied batteries at Bouchain "very much favour[ed] the carrying on our approaches" by dismounting the garrison's cannon, ruining their defenses and annoying the troops in the works.[41] Recounting the 1712 siege of Le Quesnoy, the chevalier de Quincy noted that "Prince Eugene pushed the siege with all possible vigor. He had a large train, with which he pounded the place into dust: the defenses were thus soon ruined."[42] An Allied participant in the siege acknowledged their reliance on firepower: "we hope to oblige him to capitulate soon, for we propose to make a great and continual fire from the opening of the trenches."[43] The same fate was expected of Eugene's next target: "The town of Landrecies is also well fortified; but the works are very small: thus it is to be presumed that it would have quickly been reduced to dust by the numerous artillery that they would have battered [*foudroyer*] it with."[44] This hypothesis would not be tested, however, as the Prince lifted the siege after Villars overran the entrenched camp at Denain and captured his supply depot at Marchiennes.

[38] AG A[1] 2151, #144, d'Hermanville, 4 July 1709.

[39] Murray (ed.), *Letters and Dispatches*, vol. 4, p. 547 Marlborough to Boyle, Tournai, 15 July 1709. Also: *Relation de la campagne de Tannières* . . ., (La Haye, 1709), p. 36.

[40] AG Article 15, section 2, Béthune, #4bis, Journal du siège de Béthune, ff. 19v, 24, and 61.

[41] *The Post Boy* of 25 August 1711 O.S., reporting from the Camp before Bouchain, 31 August.

[42] Lecestre (ed.), *Mémoires du chevalier de Quincy*, vol. 3, p. 127. The chevalier also noted that the garrison commander Labadie had been unjustly thrown into the Bastille for the town's poor defense; his subordinates defended their commandant's honor, swearing that "the numerous Allied guns had made so many breaches in the place that he had been forced to submit to the Allies' conditions." Vol. 3, p. 128.

[43] *The Daily Courant*, 14 June 1712 O.S., from the Camp before Haspre, 16 June.

[44] Lecestre (ed.), *Mémoires du chevalier de Quincy*, vol. 3, p. 135.

Louis' armies likewise made relatively short work of fortresses with their artillery arm once they regained the initiative in 1712. The French commander before Douai assured Voysin that "the siege will not be very long, Monseigneur, as long as we are careful to have all the necessary artillery and munitions."[45] At the next siege Alègre hoped to quickly extricate his troops from mud that rose up to their knees, for "after a full day of the cannon firing, I think, Monseigneur, it will be easy to capture [Bouchain]."[46] One French officer extended artillery's labor-saving benefits further by arguing that large batteries could actually substitute for brainpower, suggesting that eighty pieces would quickly force many German fortresses to surrender without the need of *les premiers ingenieurs*.[47] Contemporaries predicated speedy success on the availability of well-supplied siege trains, and even the strongest of fortresses succumbed far more rapidly than they had fifty or one hundred years earlier.

A more systematic measure of siege artillery reinforces contemporary commentary, for it was the number of siege guns and stocked magazine that formed the basis of the new-found offensive supremacy. The general techniques of attacking a fortress with artillery—e.g. the division of labor between different caliber weapons—were already established back in the 15th century, while ballistic performance (i.e. range and rate of fire) was nearly static across the early modern period.[48] Although the numbers of pieces in use has received almost no exploration compared to the more technical changes in

[45] AG A[1] 2382, #8, Alègre to Voysin, Douai, 16 August 1712.

[46] AG A[4] Carton 8 2, Alègre to Maine, Bouchain, 13 October 1712.

[47] AG A[1] 1944, #211 *Maréchal de camp* François de Granges de Surgeres, marquis de Puyguion (or Puiguyon) to Voysin, Lille, 22 July 1706.

[48] For the division of labor, see Salamagne, "L'attaque des places-fortes au XVe siècle," pp. 65–113. On the need to maximize the number of breaching cannons firing at once and on the need to have them cross their fire, see Errard Bar-le-duc, *Fortification demonstrée*, pp. 5–6 and De Ville, *Les fortifications*, p. 309. More generally, see Guilmartin, *Gunpowder and galleys*, Conclusion; Bert Hall, *Weapons and Warfare in Renaissance Europe*, pp. 153–156; Chandler, *The Art of Warfare in the Age of Marlborough*, pp. 176–193; and Lynn, *Giant*, pp. 500–509. On the general performance of Dutch cannon circa 1700, consult Olaf van Nimwegen, "Kannonen en houwitsers. De Staatse veldartillerie in de eerste helft van de achttiende eeuw," *Armamentaria* 32 (1997), pp. 50–66. The period's technological advances in artillery tended to decrease the cost and increase the precision of manufacture, further standardize the arsenal's variety of calibers, and increase the mobility of the guns by decreasing the powder charges needed (and thereby the weight of the gun) as well as improving the carriages.

gun construction and gunpowder manufacture, histories of 16th and early 17th sieges highlight both the very small proportion of large-caliber pieces as well as the small number of pieces used overall, measured usually in the dozens.[49] The examples from this early period are sketchy, but the numbers given range anywhere from one dozen up to one hundred guns in a few extraordinarily-large trains. De Ville, writing in the first quarter of the seventeenth century, complained of the inadequate siege trains used in his lifetime: against the strongest fortresses he would recommend thirty pieces instead of the usual eighteen or twenty-four![50] No surprise then that Francis I's Spanish engineer Pedro Navarro could write that a sixteenth century city could "expect to have more guns than an army can carry with it; whenever you can present more guns to the enemy than he can range against you, it is impossible for him to defeat you."[51]

By the turn of the eighteenth century the situation had changed dramatically, with siege trains consistently numbering well over one hundred pieces (up to 300 pieces in the largest sieges), while a large percentage of each train consisted of the heavy breaching guns, the now-standard 24-pounders and occasionally even larger calibers.[52] At the beginning of the Dutch War circa 1672 French besiegers usually employed no more than fifty pieces of cannon and Vauban can be found asking for twenty-five cannon for a siege, but by the siege

[49] See Eltis, *The Military Revolution in Sixteenth-Century Europe*, pp. 81–85; Pepper and Adams, *Firearms and Fortifications*, pp. 11–15, pp. 166–168; Wood, *The King's Army*, pp. 153–162, pp. 178–183, pp. 272–273; Simon Pepper, "Castles and cannon in the Naples campaign of 1494–95," in D. Abulafia (ed.), *The French Descent into Renaissance Italy, 1494–95. Antecedents and Effects*, (Aldershot, 1995), pp. 286ff.; Parrott, *Richelieu's Army*, pp. 65–71; and Thomas Arnold, *The Renaissance at War*, (London, 2001), p. 33. In *The Army of Flanders*, pp. 18–19, Parker mentioned an increase in siege trains over the course of the Dutch Revolt, but saw little corresponding improvement in the attack as a result.

[50] *Les fortifications*, pp. 281–282. Earlier in the sixteenth century the duc de Guise had argued similarly that considerable places required trains of thirty-two pieces (twenty-four for breaching), and some 8,000–10,000 rounds of ammunition. Michaud and Poujoulat (eds.), *Mémoires de François de Lorraine, duc d'Aumale et de Guise*, (Paris, 1839), p. 251. Flamand also recommended at least twelve cannons and two culverins with 6,000 rounds for "une forte place." *La guide des fortifications*, p. 195.

[51] Quoted in J.R. Hale, "The Early Development of the Bastion," 477. Pepper and Adams argue for the relatively small number of 16th century garrison guns in *Firearms and Fortifications*, p. 14.

[52] This varied somewhat by theater: the Iberian peninsula was infamous for its rocky terrain that made transporting large numbers of heavy siege guns almost impossible in much of the theater.

of Cambrai in 1677 this number had jumped to 106 pieces (including the smaller pieces taken from the field train). By the 1690s French siege armies were amassing 125 pieces against the strongest fortresses.[53] Saint-Rémy's treatise written before the turn of the century included several examples of French siege trains from the Nine Years' War, ranging from 110 to 196 pieces of cannon.[54] It is in this context that the efficiency of Ath must be seen. Though the artillery park contained 73 cannon, Vauban used no more than 41 cannon and 22 mortars at any one time, while the early batteries saw only 36 cannon in action. The recent work of Frédéric Naulet has tallied the steady rise and then slight decline of French artillery inventories over the course of Louis XIV's reign—from 5,000 pieces of all calibers in 1688 to a peak of 7,444 in 1697, declining to 6,006 by 1711.[55] A portion of these were dedicated to sieges, the size of these artillery trains increasing along with the pool from which they were drawn. On the Allied side, William of Orange's siege of Namur in 1695 witnessed 300 cannon of all calibers.[56] Projected siege trains for Allied sieges in the Spanish Succession fall within the same range, medium-sized trains including fifty-six cannon (a similar number as the French), but those for larger attacks ranged from 80 cannon up to 135 pieces.[57]

Appendix F provides more systematic data on siege artillery used in these Flanders sieges. From it we see that besiegers in the Spanish Succession opened fire against the largest fortresses with more guns in multiple batteries than most earlier 17th century siege parks pos-

[53] Naulet, *L'artillerie française*, pp. 220–225. Vauban's siege of Luxembourg used only forty-two cannon and twenty-one mortars and *pierriers*. Rochas d'Aiglun, *Vauban*, vol. 2, p. 229.

[54] Saint-Rémy, *Mémoires d'artillerie*, vol. 2, p. 253 suggests 110 guns; a table on vol. 2, pp. 269–270 includes three sieges with 130, 196 and 149 guns each. The number of heavy breaching cannon (24-and 33-pounders) in these examples ranged from 35% to 45% of the totals. Michel Decker, "Louvois, l'artillerie et les sieges," *Histoire, économie et société* 15(1) 1996, p. 92 puts names to these numbers, identifying 130 cannon for the siege of Mons in 1691, 196 for Namur in 1692, and 149 for Charleroi in 1693.

[55] Naulet, *L'artillerie française*, p. 320, Figure 7.

[56] 137 cannon and 44 mortars opened fire against the citadel on 22 August. Georges Baurin, "Les sièges de Namur de 1692 et 1695," in F. and Ph. Jacquet-Ladrier (eds.), *Assiégeants et assiégés au coeur de l'Europe, Namur 1688–1698*, (Brussels, 1991), p. 101.

[57] See the various projects and lystes in ARA, Collectie van der Hoop 120. A project of 1701 called for 135 cannon (eighty 24-pdrs), a 1702 project totaled eighty pieces (fifty 24-pdrs), a 1705 project numbered ninety-two pieces (again eighty 24-pdrs), while a 1707 list counted 100 pieces (eighty 24-pdrs).

sessed in their entire train—attackers opening fire with 100, even 120 pieces in a few cases. Predictably, siege trains were smallest when facing the *bicoques* besieged in the first half of the war.[58]

Increasing train sizes gave besiegers fire superiority over their opponents. As long as the batteries waited to fire together (a point emphasized by Vauban), their numerical superiority against defenders was assured. The Sun King had perhaps 1,000 pieces at the beginning of the War of the Spanish Succession to defend all of northern France, though the dozen front-line fortresses averaged only thirty guns each, while some of these were unavailable for service due to shortages of carriages, cannonballs and a general lack of maintenance.[59] Even when besiegers could amass only a relatively small number of cannon, these were still usually enough to outnumber the garrison's pieces, as we see at Bouchain in 1711. In the rare case where a garrison was fortunate enough to have a large number of guns, the advantage quickly shifted to the attackers all the same. The Allied garrison recently installed at Le Quesnoy was well-armed, for Alègre and Villars both wrote that they had never seen a garrison with so many guns. They went so far as to make the almost unheard of admission that their trenches were being razed by the defender's intense fire—"Never was a place so well supplied with cannon, mortars and all the necessities that go with it."[60] Besieging

---

[58] For example, a siege of Dendermonde called for only twenty-six 24-pounders, six 12-pdrs, 15 mortars and 10 howitzers. Compare this with BL Add MSS 61339, Mémoire qu'on pourroit avoir en vue de faire, concernant l'attaque de la ville de Mons, f. 45, a 1706 document which noted that besieging Mons would require "an artillery far more numerous than what we have needed for the sieges conducted thus far."

[59] Naulet, *L'artillerie française*, pp. 149ff., 159–160. This count of one thousand guns includes the 700 at the arsenal of Douai, most of which were evacuated before the town was invested in 1710. For an example of how few pieces second-line fortresses might have, consult an *état* from the beginning of the 1706 campaign regarding Ath in AG A[1] 1936 #263; and also Saint-Hilaire's discussion of the shuffling required to provision it when a siege became imminent, in A[1] 1939 #92.

[60] Léon Lecestre (ed.), *Mémoires du chevalier de Quincy*, vol. 3, p. 187. See also AG A[4] Carton 8 2, Alègre to Maine, 23 September 1712; AG A[1] 2384, #81, the engineer Naudin to Voysin, 23 September; and AG A[4] Carton 8 2, Villars to Maine, 27 September. After the town had been captured, a French *état* enumerated 56 Allied cannon (22 still in working condition), 46 French guns captured after the earlier conquest and 47 mortars of various calibers. See Vault and Pelet (eds.), *Mémoires militaires*, vol. 1, p. 527 #160, État de l'artillerie et des munitions trouvées dans le Quesnoy, 4 October. See also the chevalier de Quincy's comments in his memoirs, vol. 3, p. 196.

breastworks could easily be reinforced, however, and even the most heavily-armed garrisons were quickly silenced once the besiegers were able to respond with their own batteries. Villars explained how they overcame Le Quesnoy's many pieces:

> [Our] batteries began the 25th at daybreak: there were sixty 24-pounders, thirty mortars and several smaller pieces that fired ricochet. The enemies had more than one hundred 24- and 36-pounders on the ramparts; but as the besiegers had all the terrain that they could want on which to place their cannon, and the besieged on the contrary were forced to squeeze theirs together in a small space, from the first day of firing we gained the advantage, and the 26th at noon, two-thirds of the place's batteries had been demolished.[61]

Within a few days the majority of the garrison's pieces were put out of service and by the 29th almost all were dismounted.[62] Despite an encouraging start, the well-endowed garrison held out for only sixteen days of open trenches and ten of cannon-fire.[63] As with breaching batteries, suppressing fire depended as much on outflanking as outgunning the enemy. As illustrated by the description of Menin cited several pages earlier, cannon were most easily silenced by smashing their carriages or destroying their firing platforms; garrisons did not have the space and rarely the reserves to make up for those they lost as a siege dragged on.[64] As the knowledgeable Saint-Rémy explained it:

> [the besiegers] firing bombs into the places where they will have placed their cannon, the garrison will be forced to abandon them, having no other space where they can shield themselves because the most advantageous locations for batteries in a fortress are always known by the defenders when it is constructed before the attack, and when these are destroyed, it is impossible for the besieged to find others that are as well-positioned to damage the attackers; this diminishes the effectiveness of their fire.[65]

---

[61] Anquetil, *Suite des mémoires rédigée par Anquetil*, pp. 387–388. Villars likely exaggerated slightly—a French *état* shows the garrison as having only fifty 24-pounders, still a large number for a garrison.

[62] AG Article 15 Section 2 §1 Douai folder, #19 Siège du Quesnoy en 1712, f. 17.

[63] AG A[4] Carton 8 2, Hautefort to duc du Maine, 29 September 1712; Lecestre (ed.), *Mémoires du chevalier du Quincy*, vol. 3, p. 196. The French also attacked Bouchain later in the year under similar circumstances. AG A[1] 2386, #32 to Villars, 17 October.

[64] Naulet, *L'artillerie française*, p. 160.

[65] Saint-Rémy, *Mémoires*, vol. 2, p. 268. A director-general of Swedish fortifications

Except at the most isolated of sites, there was more room to site the besiegers' batteries than the garrisons could find on their ramparts. Besiegers relied on a large number of multi-gun batteries to provide them with the firepower necessary to silence their foe's cannon and cover their approaches.

A final illustration of the prominence besieging artillery had acquired by 1700 is the inevitability of the siege's final stage. Before the reign of Louis XIV, the most difficult stage of a siege was blasting a wide enough breach in the main wall and suppressing the murderous crossfire of its retrenchment. By the time of the War of the Spanish Succession, such concerns were almost irrelevant in Flanders. Millner summarized the Allied Flanders campaigns: "of all the cities and towns taken throughout the whole war on our side thereof, none stood a storm, but at the last extream surrender'd on reasonable conditions."[66] Though some imputed this to spineless defenders, garrisons usually had little choice, for the besiegers' guns quickly opened up the defensive works of even the strongest places. This would lead 18th century contemporaries throughout Europe to declare that a fortress's fate was sealed once its covered way was taken.

After supporting the capture of the covered way, the final tasks for siege cannon were to breach the outworks and enceinte. The process of punching gaping holes in masonry and brick-faced earth had progressed significantly since the 16th century despite few changes in ballistic performance. Reflecting the change, histories of this early period give much attention to the lengthy process of mining underground—a preferred tactic where high water tables did not preclude their use.[67] As the number of breaching pieces increased and their supply of ammunition became more assured over the course of the 17th century, besiegers could increasingly rely on breaching by cannon rather than mine. Following earlier authors, Vauban discussed

---

later in the 18th century, Jean Bernard Virgin, also identified this as a critical offensive advantage. Virgin, *La défense des places*, Au Lecteur, pp. iii–iv.

[66] Millner, *A Compendious Journal*, p. xiv. He goes on to note that "several castles, citadels and forts, etc. were taken by storm." In this we should include Venlo's Fort St. Michel.

[67] For example, Israel, *The Dutch Republic and the Hispanic World*, p. 101; Parrott, *Richelieu's Army*, pp. 70–71; and the case study by Kenneth Wiggins, *Anatomy of a Siege: King John's Castle, Limerick, 1642*, (Woodbridge, 2001). In the 18th century miners were still needed to dig galleries descending down underneath the covered way that would exit out onto the floor of the ditch.

in his 1672 treatise the medieval task of attaching a miner along the face of a work, though he recommended a "newer, faster and at the same time more certain" method of using cannon to accelerate the task. As with trench parallels and ricochet fire, he felt compelled to emphasize the efficiency of breaching batteries by describing the several ways in which they economized on both labor and lives:

> 1. It takes much longer to breach a work with the miner than with cannon, because you must cross the ditch to reach the foot of the wall before you can see any effect. 2. The blind sheltering this lodgment is further at constant risk of being destroyed by the bombs and infinity of rocks that the enemy hurls down on it from above. 3. The enemy cannon in flanking batteries threaten the same. 4. The fire from above will either kill the miner inside or force him to flee. On the other hand, breaching batteries will provide several advantages. 1. Using the guns you can create a cavity for the miner before you even cross the ditch. 2. The miner will have nothing to fear from rocks or bombs and even from flanking batteries as he will start his work lodged five or six feet inside the wall, where he will be able to push away from the opening any bombs that the enemy might drop on him. 3. If there are countermines within the wall, the cannon will expose them, which would be difficult to do otherwise. 4. A final advantage of breaching artillery is that after the mine has been blown, you can then use these same guns to fire into the breach.[68]

This technique indicates the direction breaching tactics took over the course of the century: by its end the miner had became only an adjunct to the breaching power of the siege artillery. At the end of his career thirty years later, Vauban reprised his method, but in general he now advocated abandoning the use of miners and mines altogether when attempting to breach most works:

> I rarely use mines in normal attacks; not that I renounce them, far from it, but I prefer cannon because their effect is nowhere near as uncertain. With cannon you make a breach where you want, when you want, and how you want; mines cannot do this with the same certainty.[69]

---

[68] Vauban, *Mémoire pour servir*, pp. 118–119. Lazard contends that Vauban himself started the shift away from mines and towards breaching batteries. *Vauban*, p. 465.

[69] Vauban, *Traité*, quote on 163; pp. 186–188 for his comparison of the old and new methods, which varies little from his earlier work. La Vergne similarly preferred cannon over mines. *Nouveau exercice du gabion*, pp. 63 and 81. See also Allard, "Notice sur Vauban," *Le Spectateur militaire* 18 (1835), p. 440; and more generally Jean-François Pernot, "Vauban et la guerre des mines" *Histoire et Défense. Sapes et contre-sapes: la guerre des mines à l'époque moderne et contemporaire* 32 (1995), pp. 73–104.

Feuquières' summary of warfare in Louis' reign vouched for the widespread use of this technique:

> The custom of waiting for the effect of mines in the body of the place has not been observed by those governors who sustained a siege since I engaged in the Service; because the besiegers did not find it necessary to set on the miners at the body of a place to compel it to surrender. The faceing of bastions are always so demolished by the cannon and their inside is shattered to such a degree by the bombs that it is impossible for an attack to be sustained there when it has been carried on near enough to be made against the bastions, and especially since it is usual in the most obstinate sieges for the lodgments along the covert-way and before the breaches to be always so extensive and solid that it would be imprudent in a governour to expose his garrison to a destruction that might be completed in half an hour, and particularly when his bastion had not been previously intrenched in the gorge.[70]

Once again, Spanish Succession sieges confirm the impressionistic reporting. Significant breaches were formed at several Allied sieges before the enemy had even reached the covered way (not only *bicoques*, but also at Menin and Lille and Béthune)—whether through the use of trench cavaliers or batteries planted on heights beyond the covered way. Otherwise, the base of the outworks and curtain walls were only visible once batteries were raised on the covered way. In either situation, by the time passages reached the other side of the disputed ditch, the breaching guns would have created enough of a breach in the work to allow a storm.[71]

As a last resort, a particularly defiant garrison could build a retrenchment behind the breach where they were expecting an assault.

---

[70] Feuquières, *Memoirs Historical and Military*, vol. 2, p. 293. The Marquis de Quincy noted that such retrenchments across the gorge of a bastion "could only serve to resist one or perhaps several assaults, and to gain the time necessary to make a capitulation." *Histoire militaire*, vol. 7, p. 176. For other contemporary views that reach the same conclusion, see Vallière, *Memoires d'artillerie a l'usage du sieur St. Vallier[e]*, p. 15 (which assumed breaching cannon would require a week of firing); Feuquières, *Memoirs historical and military*, vol. 2, pp. 296ff.; Santa Cruz de Marcenado, *Reflexions militaires*, vol. 9, p. 4; as well as Guignard, *L'Ecole de mars*, vol. 1, pp. 313–315; vol. 2, p. 452. Duffy gives details of the process in *Siege Warfare*, pp. 98–99.

[71] For one example where the Allies did attach a miner to the wall, see Veenendaal, Jr. (ed.), *Briefwisseling Heinsius*, vol. 5, p. 482 #931, Salisch to Heinsius, Menin, 21 August 1706. A garrison account noted that this technique was used "where they could not make a breach wide enough" by other means. AG 1M 2365, Caligny, Relation de ce qui s'est passé au siège de Menin, f. 5.

Piling up earth, wood, stone and anything else available at hand, the defenders thereby hoped to provide the besiegers with yet another obstacle between them and the town. Historians of the 16th and early 17th centuries point to many examples where assault parties failed to penetrate such debris-strewn breaches. Duffy cites eleven famous fortress defenses of the 16th century whose garrisons were able to turn back assaults on the breach: storming troops were slaughtered as they scrambled up the slope of rubble and into the killing ground beyond it.[72] In the 1590s Sir Roger Williams recalled his defense of the town of Sluis, which saw five assaults repulsed 'at push of pike.' He went further, however, and even claimed that "the furie of all batteries are past at the first, I meane within two dayes: yea, commonly in one. For the defendants knowing the place, and the successe of the furie, will re-enforce their breaches and re-entrench themselues in such sort, that the assailants can doe small hurt with their approaches."[73] By the Sun King's reign, however, garrisons were much less likely to hold out in such a 'last-ditch' effort; constructing such defenses were usually intended only to encourage the attackers to grant an honorable surrender.[74] With a heavily-armed attacking force, retrenchments behind a breach would be difficult to construct under constant shelling and just as difficult to defend with the besiegers targeting the area behind the breach with indirect fire. This late in a siege, few garrison commanders had the manpower, the munitions, the time, the materials, or the support of the townspeople, much less the inclination to undertake such an immense task with such significant risks. Vauban's siege of Ath in 1697 exemplified

[72] Duffy, *Siege Warfare*, p. 15; see also Eltis, *The Military Revolution in Sixteenth-Century Europe*, p. 88. The eight bloody assaults at La Rochelle in 1573 described by James Wood (*The King's Army*) indicate just how defensible such retrenchments could be in the 16th century, although several of the initial assaults were repulsed due to casemates in the ditches—Arnold's *The Renaissance at War*, p. 38 provides a good visualization of how such casemates might have appeared. A more complete dataset would inform us how representative this defense of the breach was.

[73] Evans (ed.), *The Works of Sir Roger Williams*, quote on p. 91, see also p. 48, where he mentions how thirty cannon and eight culverins fired some 4,000 rounds, a battery of previously unseen fury.

[74] This was true despite Louis' 1706 order that garrison commanders should only surrender after an assault on the breach. Historians have assumed far too much about the period's siege warfare from this one ordinance, particularly regarding the stoutness of French defenders. Louis usually overruled this blanket ordinance by giving besieged commanders specific orders to save their garrison before it risked being slaughtered or captured in a general storm.

the artillery's newfound dominance—the retrenchment across the breach was quickly destroyed and the garrison forced to capitulate.[75]

Turning to the few examples from the Low Countries theater in the Spanish Succession, participants in the attack on Lille mentioned how the defenders had "dexterusly underminded and fortified not only behind [the breach] but likewise have laid a vast pille of dry wood and other combustibles behind the breach to set fier to when they can defend it no longer."[76] A day of artillery fire destroyed this ad hoc defense, as Du Mée related: "This night we finished the new batteries, but began not to fire from them till Sunday morning, where they did such good execution all that day and the next, that at four of the clock in the afternoon the garrison beat the chamade."[77] Vauban's cousin DupuyVauban stacked logs and wooden beams across the wide breach in Béthune's walls, but the approach commander Schulenburg assured Marlborough that they would easily overcome this obstacle, assuming the necessary munitions were provided.[78] At the 1711 siege of Bouchain, Goslinga may have been uncertain of whether the garrison would attempt to defend the breach or not, but the end result was predictably clear:

> The siege has advanced to a point where the enemy, if they want to have an honorable capitulation, must beat the chamade today or tomorrow, but the presence of the [relief] army will animate them according to all appearances to hold out to the last. It appears that they are preparing for this because they have retrenched a part of the bastion where the breach is already fairly large. Perhaps they will hold out until an assault, which would be the first example during the course of this war, but in either case, they will be forced to surrender in the end, having no relief to hope for.[79]

The town capitulated a few days later, and even a retrenchment across the breach was not enough to save its defenders from becoming

---

75 *A Journal of the Siege of Ath*, p. 138. La Vergne's new fortification design, regardless of its practicality, was prompted by the recognition that storms on the breach were inevitably successful. La Vergne, *Nouvelle fortification imprenable par force d'armes*, Dédicatoire.

76 Royal Commission on Historical Manuscripts, *Report on the Manuscripts of the Earl of Mar and Kellie*, p. 464, Lt. William Nodding to the Earl of Mar, Lille, 20 September 1708.

77 De May, *An exact journal of the siege of Lille*, pp. 16–17.

78 BL Add MSS 61245, f. 133, Schulenburg to Marlborough, Béthune, 21 August 1710.

79 Veenendaal, Jr. (ed.), *Briefwisseling Heinsius*, vol. 12, p. 323 #548, Goslinga to Heinsius, Bouchain, 10 September.

(treacherously, according to the commander) prisoners of war. So effective had the bombardment been that, reviewing the shattered works, the newly-appointed Dutch governor Frederik Sirtema van Grovestins reported that if the besieged "had not surrendered as they did, we might in two days more have entered the town, and us'd them at discretion."[80] As siege trains and their supply convoys grew larger and powerful enough to first suppress the garrison's fire and then breach twenty-feet or more of earth faced with brick or masonry, there was little hope that a makeshift wall of rubble, earth and wood would last long enough to present a serious obstacle to the attackers.

A more systematic examination of the length of siege stages reinforces the cumulative impact of the artillery in the later phases for it is in the final stage of most sieges that we find the most radical deviation from Vaubanian theory. Whereas Vauban projected the period from the capture of the covered way up till the hypothetical siege's conclusion at twenty-five days, the twenty-two Flanders sieges required only two to ten days from the covered way to capitulation (a mean of ten days and median of five days). In other words, most towns surrendered two to three *weeks* sooner than might be expected from theory. More appropriately, when we standardize these numbers we find that almost every siege's final phase (excepting the best defenses of Lille, Douai and Aire) was significantly shorter as a percentage of the overall length of the siege than what Vauban's estimate suggested. In his hypothetical siege Vauban expected the final stage to account for 60% of its entire length, whereas most actual sieges in Flanders required more days from investment to open trenches than his estimate, and far fewer days than he estimated after the counterscarp had fallen—the percentage of the sieges spent pushing through the outworks and forcing the garrison to capitulate averaged only 25% of the overall duration. In some cases this was a function of towns not meeting Vauban's hypothetical assumptions, but this trend of truncated sieges held true even with many of the stronger fortresses in the *pré carré*, where additional outworks should have further increased the time needed to advance from the counterscarp to the curtain wall. The result of this overwhelming offensive advantage was that sieges ended more abruptly as besiegers smashed

[80] *The Post Boy* of 11 September 1711 O.S., from the camp near Bouchain, 17 September.

through defensive works and quickly reached the final line of defense in less than half the time one might have expected. Unlike the long siege/blockades of the earlier 17th century, where fortresses may have capitulated due to starvation or mining, the vast majority of garrisons in the Spanish Succession surrendered when their last layer of defense had been breached and a general storm was only a few days away at most.[81]

There are a few exceptions, of course, and their details are worth noting. In addition to Tournai's citadel, which surrendered due to a shortage of victuals, the small force defending Venlo also surrendered early, thinking a general storm was about to commence when the besiegers began firing a *feu de joie* for the capture of Landau in Germany—the fear of a general storm was enough to win their surrender.[82] Menin was also widely condemned for surrendering prematurely. However, an impartial observer and civilian field deputy unsteeped in the cult of vigor revealed that there was in fact a diversity of opinions in the army over whether the governor had surrendered too early, some claiming that he could not have held out beyond another twenty-four hours.[83] The commander of Ath committed the most heinous crime of all, surrendering early without even acquiring honorable terms; the 800 remaining troops (minus the Swiss soldiers who joined the Allies' ranks) marched off as prisoners of war. Here too, however, we find the Dutch field deputy Cuper concluding from his survey of the battered fortifications that "we understood well that the governor was correct to save himself [by surrendering]."[84] With these cases, however, it is important to distinguish weak resistance from a premature surrender—the defense

[81] The time needed to cross the ditch varied obviously: Marlborough complained of their difficulties early on, but a garrison engineer's account of the siege of Béthune noted that Allied pontoon bridges were set up across their final wet ditch in only two *hours*. AG Article 15 section 2, Béthune, #4bis, Journal du siège de Béthune, f. 65.

[82] Royal Commission on Historical Manuscripts, *The Manuscripts of the Earl Cowper*, vol. 3, p. 16 Pope to Coke, Sutendaal, 25 September 1702.

[83] Veenendaal, Sr. (ed.), *Het Dagboek van Gisbert Cuper*, p. 242, to van den Bergh, 9 June.

[84] From the French side, see Ouverleaux, *Mesgrigny*, p. 44, Vauban to Mesgrigny, 16 October 1706; AG A[1] 1940, #46 Chamillart to d'Antin, 11 October. For Allied comments and criticisms, see quote from Veenendaal, Sr. (ed.), *Het Dagboek van Gisbert Cuper*, p. 159; and Veenendaal, Jr. (ed.), *Briefwisseling Heinsius*, vol. 5, p. 575 #1117, Salisch to Heinsius, Leuze, 4 October.

may have been incompetently waged, as Vauban often argued, but criticism of the defenders should not be extended to their decision to surrender when they did. The besiegers' correspondence makes clear in all these cases their intent to conduct a general storm within a few days of when the garrison actually did surrender.[85] Thus, with the exception of the citadel of Tournai, 'premature' meant a matter of days, a week in the most egregious case. And if these defending governors were as incompetent as some claimed, this was an even better reason to try to save the garrison by surrendering early, rather than risk a worse fate in a botched storm defense. The critiques of incompetent defenders, then, are convincing only regarding those few fortresses that could have reasonably resisted far longer, or else if we accept the cult of vigor's postulate that even a few days were critically important to the outcome of the campaign and even the war.

Overall, offensive firepower had come to dominate early 18th century siegecraft in every phase, from the opening of the trenches to the beating of the chamade. The effect of massed firepower on both body and mind was exponentially more effective when targeting the many undermanned Flanders garrisons who were under direct orders to save their battalions from capture, who were defending poorly-maintained fortresses against bombardments the likes of which few garrisons had ever seen before, who had no hope of assistance from impotent relief forces, and who were only days away from a general storm that, the besiegers promised, would lead to the slaughter of the garrison.

All participants relied heavily on gunpowder, but how besiegers actually used their artillery to achieve success is far more varied than a reading of the secondary literature, which stresses the supremacy

[85] Liège 1702: Murray (ed.), *Letters and Dispatches*, vol. 1, p. 47. Stevensweert 1702: Vol. 1, p. 42. Limburg 1703: Snyder (ed.), *Marlborough-Godolphin*, vol. 1, p. 243. Huy 1703: Murray (ed.), *Letters and Dispatches*, vol. 1, p. 169. Huy 1705: AG A[1] 1835, #256. Diest 1705: AG A[1] 1838, #331. Dendermonde 1706: Veenendaal, Jr. (ed.), *Briefwisseling Heinsius*, vol. 5, p. 451 #859. Lille town 1708: Schulenburg, *Leben und Denkwürdigkeiten*, p. 357. Tournai's town 1709: PRO SP 77/58, f. 166. Douai 1710: ARA RvS 687–2, Orange to Raad van State, 27 June 1710. Bouchain 1711: *Post Boy*, #2549, 11 September 1711 O.S. Le Quesnoy 1712 (Allied siege): ARA SG 9196, #71. Marchiennes 1712: AG A[1] 2381, #2. Bouchain 1712: AG A[1] 2385, #49. Le Quesnoy (French siege): AG A[4] Carton 8 2, 2 Alègre to Maine, 6 October 1712.

of Vauban's doctrine, or an analysis of Ath 1697 would suggest. The artillery corps' autonomy from the engineers almost guaranteed such an outcome, and the trend was further encouraged by generals looking to speed up sieges. Time pressures led both Allied and French armies to use artillery in ways that Vauban had explicitly rejected, while at the same time they refused to follow other prescriptions of his geared towards improving its efficiency. In this, they could invoke the name of the period's other great engineer, Coehoorn. The Frisian's method of attack relied heavily on bringing enormous numbers of guns to bear on the enemy, although this aspect does not receive as much attention in the wider historiography as his predilection for storms.[86] This straightforward use of overwhelming firepower deviated from Vauban in many significant respects, loosely summarized by the replacement of Vauban's finesse with Coehoorn's brute force. Sheer weight of shot allowed less-skillful besiegers with large artillery parks to supplement or even replace the precision firing demanded by Vauban's ricochet.

Generals diverged from Vauban's plans by insisting on the massive use of artillery at the earliest possible opportunity. Impatient for the trenches to be opened, they were just as eager to hear the sound of their cannon firing. Vauban had opposed constructing batteries early in the siege, concluding by the end of his career that batteries located in the first parallel served only to expend shot and burn powder uselessly.[87] Not only would fire against the works be ineffective at such extreme range, but early counter-battery guns dedicated to targeting the defenses' cannon were unnecessary, as enemy fire only became dangerous at a distance of one thousand feet or so (i.e. the distance of the theoretical second parallel from the counterscarp).[88] The gunner Saint-Rémy joined Vauban in warning against firing too early, admonishing his artillery officers to resist the inevitable

---

[86] For recognition of his reliance on overwhelming firepower, see Allent, *Histoire du corps impérial du génie*, pp. 316–317; Wijn, *Het Staatsche Leger*, vol. 8 part 2, p. 121; and Duffy, *The Fortress in the Age of Vauban*, p. 30. For works that focus solely on Coehoorn's assaults (largely from a more modern, French perspective), see: Chandler, *The Art of Warfare in the Age of Marlborough*, p. 280; Pujo, *Vauban*, p. 280; Lynn, "Vauban," p. 58; and Blanchard, *Vauban*, pp. 503, 510.

[87] Vauban, *Traité*, pp. 107–108; and Deidier, *L'ingénieur parfait françois*, pp. 220–221.

[88] Vauban, *Traité*, p. 89. Later in the century, Lazare Carnot argued that one cannonball fired at short range was as deadly as five hundred fired from long range. Duffy, *Fire and Stone*, p. 124.

pressure, for "the troops quickly become impatient if they do not hear the cannon of batteries firing from the very beginning of the siege."[89] At the ideal siege of Ath in 1697, the first parallel was established 270 toises from the outworks, i.e. too far away to open fire, since at this distance it would "only serve to imbolden the Enemy, and burn Powder to no purpose" (p. 102). Once again we note the close attention paid to efficiency—saving powder and effort whenever possible.

One might think that Vauban's success at Ath would have resolved the matter, but it did not. The pursuit of improved efficiency was quickly pushed aside by other commanders keen to end their own sieges, men who were convinced that such preparation was wasted time. In the case of Vauban's attack on Breisach, his young pupil Burgundy dared to criticize the master of siegecraft, opining to his mentor Paul, duc de Beauvillier that:

> I do not see the *maréchal* Vauban preparing the cannon to fire as soon as possible after the trenches have been opened, which will cost us many men, because the garrison's cavalier batteries will plunge [into our trenches]. I will attempt to bring the cannon into the batteries as soon as possible; or if he insists on not firing the cannon until they are all in battery, at least use mortars to attempt to dismount first the cavalier batteries that will cause us the most damage.[90]

With the trenches opened on the 23rd, Vauban shrugged off the royal counsel and only on the 27th did the first battery on solid ground open fire, 100 toises [640 feet] from the covered way. It was only on 31 August that all of the batteries fired together, the relatively-small number of thirty-two cannon and twenty-six mortars, which quickly silenced the garrison's guns.[91] Vauban had dismissed Burgundy's plan because it would have required delays while waiting for the poorly-sited battery platforms to be constructed and the guns hauled into position. With a tone of vindication he noted on 27 August that "Up till now our affairs are progressing so well and the trench breaks the rules [*pèche*] by advancing much faster than

[89] Saint-Rémy, *Mémoires*, vol. 2, p. 264.

[90] Marquis de Vogüé (ed.), *Le duc de Bourgogne et le duc de Beauvillier: Lettres inédites, 1700–1708*, (Paris, 1900), p. 198, Breisach, 20 August 1703. Villars similarly pressed for the batteries to open fire at Freiburg in 1713. Vogüé (ed.), *Mémoires du Maréchal de Villars publié d'après le manuscrit original*, vol. 3, p. 224.

[91] Vault and Pelet (eds.), *Mémoires militaires* vol. 3, pp. 428–431.

it should. I have never seen a place of Breisach's merit and reputation approached so quickly in four days of trenches."[92] Such progress came cheaply and their cannon had played almost no role thus far, as he explained that "the enemies are well served by their cannon, at least they make lots of noise, but do very little work. We are not yet close enough for them to hurt us as they need to."[93] Burgundy's concerns ultimately proved to be misplaced, for the first several days of the garrison's firing caused fewer than a dozen casualties each day.[94] It was only as they approached nearer the town that they were forced to adopt the sap in order to shield themselves from the town's fire. In spite of Burgundy's early misgivings, after two weeks of trenches the town surrendered following Vauban's plan at a cost of only 830 casualties.[95] More generally, Vauban noted in his 1704 treatise that elevating batteries above the trenches was a good idea but this required more time than was usually available, for "the French are not given to patience."[96] The struggle over such inefficiencies continued in 1712. The French besiegers located their batteries against Douai too far away to damage the works, requiring a delay while they were resited; at Le Quesnoy, on the other hand, Vallière won support for his plan to open fire against the heavily-armed garrison only when all of the cannon were ready to fire *en masse*.[97]

Allied commanders were similarly skeptical of withholding their fire for so long, a tactic that appeared, to them at least, to only lengthen the duration of the siege. Marlborough exhibited this opinion from the first campaign, explaining that he was "very impatient to hear of the canons being arrived at Venlo, which place was

---

[92] Rochas d'Aiglun, *Vauban*, vol. 2, p. 522 Vauban to Chamillart, Biesheim, 27 August 1703. Confirmed by Vogüé (ed.), *Le duc de Bourgogne et le duc de Beauvillier*, vol. 1, p. 274 Burgundy to Chamillart, Breisach, 23 August.

[93] Rochas d'Aiglun, *Vauban*, vol. 2, p. 524 Vauban to Chamillart, Breisach, 27 August 1703.

[94] Vault and Pelet (eds.), *Mémoires militaires*, vol. 3, p. 902, Vauban to Chamillart, Brisach, 24 August 1703; vol. 3, p. 904 Vauban to Chamillart, Biesheim, 25 August 1703.

[95] Casualty figures from Vault and Pelet (eds.), *Mémoires militaires*, vol. 3, p. 459. The heir to the French throne also acknowledged the rapid fall: Baudrillart (ed.), *Lettres du duc de Bourgogne au roi d'Espagne Philippe V*, vol. 2, p. 167 #18, Burgundy to Maintenon, Breisach, 8 September 1703.

[96] Vauban, *Traité de l'attaque*, p. 109.

[97] For Douai 1712: AG A[1] 2382, #55 Contades to Voysin, 20 August. At Le Quesnoy, Villars noted his grudging assent to the delays. AG A[1] 2384, #60, to Voysin, 21 September.

invested last Munday, but [we] can make noe great progress till thay have their artillery."[98] After dragging his feet at Venlo, Coehoorn promised to have the batteries ready to fire against Roermond within two days of the trenches being opened, and was happy to see the fortress capitulate after only a day of breaching fire.[99] At Liège, they opened the trenches and opened fire on the same day.[100] At Bonn the next year, the Duke admitted to Heinsius that "for fear of losing to much time I have prevailled with M. Coehorn to open the trenches, which wee did last night with very litle lose. I hope everything will be here by the time our batteries will be ready." Marlborough continued to exhibit concern for how soon the batteries would begin firing at other sieges as well, particularly against the weaker places whose defenses would quickly disintegrate under such fire.[101]

Vauban's recommended technique of ricochet fire also found minimal use despite its potential. Notwithstanding his delight with the technique, the efficiency of such fire did not overawe many others. He complained in his last treatise of the inefficiencies that resulted from the failure to embrace this approach: "I know well that the reputation of ricochet fire is poorly established, because people do not know its usefulness [*mérite*]."[102] The neglect of ricochet fire was not a function of Vauban's secrecy: the technique was illustrated in the *Mercure galant's* journal of Ath at the end of the previous war. It would have been quite easy for any besieger to adopt as simple a change as decreasing the amount of powder and resiting the battery; there was no corresponding demand for structural changes in technology, organization or logistics, areas much less amenable to rapid modification. Nevertheless, I have found only a handful of ref-

[98] Snyder (ed.), *Marlborough-Godolphin Correspondence*, vol. 1, p. 104 #92, Marlborough to Godolphin, Asch, 20 August 1702.

[99] Veenendaal, Jr. (ed.), *Briefwisseling Heinsius*, vol. 1, p. 456 #935, Wassenaar-Obdam to Heinsius, Roermond, 2 October 1702; Murray (ed.), *Letters and Dispatches*, vol. 1, p. 45, bulletin from Zutendaal, 9 October; Veenendaal, Jr. (ed.), *Briefwisseling Heinsius*, vol. 1, p. 463 #949, Wassenaar-Obdam to Heinsius, Roermond, 5 October.

[100] Murray (ed.), *Letters and Dispatches*, vol. 1, p. 49, bulletin from Liège, 23 October 1702.

[101] For Bonn, Van 't Hoff (ed.), *Marlborough-Heinsius Correspondence*, p. 65 #106, 4 May 1703. At Huy in 1703, Murray (ed.), *Letters and Dispatches*, vol. 1, p. 162, Marlborough to Coehoorn, Val Notre Dame, 20 August O.S. At Limburg 1703, see Snyder (ed.), *Marlborough-Godolphin Correspondence*, vol. 1, p. 242 #239, Marlborough to Godolphin, Verviers, 9 September O.S.

[102] Vauban, *Traité*, p. 116.

erences to the tactic in the dozens of siege accounts and thousands of letters I have consulted, whereas his other two tactics are mentioned with far greater frequency. In the Allied sources we find rare references to their use. Schulenburg recommended its use during their attack on one of Lille's hornworks.[103] Two years later, the defending engineer Des Forges noted from inside Béthune that the Allies' single battery dedicated to ricochet fire was particularly effective against their covered way.[104] Overall, though, contemporary maps of the Allied sieges show their batteries consistently targeting the faces of the work with direct (perpendicular) or oblique fire rather than using ricochet fire, which would fire laterally along the covered way and other firing platforms as Vauban intended.[105] While the Allies knew of the tactic and used it on occasion, they did not privilege it as Vauban demanded—multiple batteries with small numbers of guns firing ricochet would remain a unique feature of Ath 1697. They had not yet fully accepted the superiority of ricochet fire (or embraced the cardinal virtue of efficiency), for they relied primarily on bombardment with more traditional *pierriers* and mortars. Ricochet fire found little appreciation among the Allies.

French sources mention their use of ricochet fire slightly more often. The defenders at Douai 1710, for example, used it in a flying battery sent beyond the outworks to enfilade the Allies in their trenches.[106] Once back on the offensive, the French made greater use of Vauban's technique on the attack, for example at their sieges of Le Quesnoy and Douai.[107] Here too, however, we find evidence

[103] Add MSS, 61245, f. 77b, to Marlborough, Lille, 26 September 1708, point seven. Allent also concluded that the Allies used ricochet fire very rarely. *Histoire du corps impérial du génie*, p. 589.

[104] AG Article 15 Section 2, Béthune, #4bis Journal du siège, ff. 21–21v and especially 24–24v.

[105] This may be related to the fact that the Allies often chose to attack along much narrower frontages than Vauban (as well as Ath 1697 versus 1706, compare also his expansive 1691 attack of Mons to the Allies' cramped trenchworks in 1709). Their trench attacks as a result rarely allowed batteries to be sited where they could utilize ricochet fire (such as Ath's batteries G and H).

[106] AG 1M 126–3 Mémoires historiques, #13 Journal de ce qui a été fait pour l'investiture, 15 June.

[107] Anquetil, "Suite des mémoires rédigée par Anquetil," in A. Petitot and Monmerqué (eds.), *Collection des Mémoires pour servir à l'historie de France*, vol. 69 (Paris, 1828), p. 388. See also AG A[1] vol. 2384, #116, Valory, 25 September 1712; and AG Article 15 Section 2 §1 Douai folder, Précis du journal du siège de Douay fait en 1712, ff. 12–13.

that the technique's advantages were poorly understood, witness a French engineer referring to Allied howitzer fire as *à ricochet*![108] From these few examples, we must conclude that ricochet fire may have been used, but it certainly did not have the impact on siegecraft commensurate with its discussion in the historiography, at least during Louis XIV's lifetime.[109]

Much more important was weight of fire. Most Allied sieges in particular deviated from the efficient ideal of ricochet fire by replacing it with sheer volume. Although besieging trains consistently outnumbered garrisons' guns, often by a two-to-one ratio, Allied siege trains mustered much more firepower than the French mustered when they were on the attack. As Appendix F suggests, Allied sieges generally included more guns than the French sieges. A comparison of those places besieged by both sides highlights even more starkly the Allies' preference for massed firepower by largely controlling for the strength of the fortress under attack. For the 1703 attack on Huy's small forts, the Allies mustered seventy cannon and forty-six mortars. The French, planning their own attack on Huy in 1705, judged that the Allies had amassed far more guns than were actually needed to capture the weak place.[110] They contented themselves with less than half of the artillery their opponents had used and yet their capture required only one day more than the Allied capture two years earlier—possibly an indication of the greater efficiency of the French engineers. When the Allies returned later that same year to retake the town and its outlying works, they once again transported a large number of guns to the site. Confirming this divergence in views, a later French project to capture the oft-contested town again set their requirements for another siege at only forty cannon and fourteen mortars.[111] In the most striking example, the Allies collected *five* times as many cannon for their siege of Douai in 1710 as the French did in 1712, and twice the number of mortars. More

---

108 AG 1M 2365, Caligny, Relation de ce qui s'est passé au siège de Menin, f. 2v. Either Caligny was unfamiliar with Vauban's assessment of the two types of fire, or the howitzer fire was deemed as effective as Vauban's ricochet. See later in the chapter for a discussion of howitzers and other forms of indirect fire.

109 See, for example, the marquis de Quincy's interesting rejection of ricochet fire (circa 1725) in *Histoire militaire du règne de Louis le Grand*, vol. 3, p. 298.

110 Rorive, *La guerre de siège*, pp. 217, 254.

111 For the projected artillery train, see Lecestre (ed.), *Mémoires de Saint-Hilaire*, vol. 5, p. 297 #252 Saint-Hilaire to Chamillart, 5 June 1708.

generally, it would be two decades before French siege trains consistently equaled the size of what the Allies used in the War of the Spanish Succession, Naulet identifying the siege of Turin in 1706 as the maximum effort of Louis XIV's artillery arm, where they assembled just over one hundred 24-pounders.[112] Staying in the Spanish Succession, some evidence even suggests that the Allies planned their trenches around massive firepower: one planning document for a 1706 siege of Mons determined the choice of approaches based on which terrain would give the attackers enough frontage to site their crowded artillery batteries.[113] Yet again, brute force replaced finesse.

Though Vauban preached the importance of artillery in general, he decried such inefficient use, noting of Coehoorn's famous attack on Namur:

> And if they had not had 120 cannon and 60 mortars, they would have failed miserably. I am persuaded further that they burnt as much powder, fired as many bombs and ruined as many cannon [in that one siege] as we did in our four largest sieges, and they lost more men there than we have suffered in ten sieges.[114]

Confirming the philosophical nature of this difference, a few contemporaries even expressed this reliance on overwhelming firepower in explicitly anti-Vaubanian terms. French officers at Breisach rejected such efficiency, wishing instead for "a battery *à la Cohorn*, that is to say sixty pieces of cannon, not wanting those of only eight or ten pieces according to Vauban's old manner."[115] The marquis de

---

[112] Naulet, *L'artillerie française*, pp. 223–225. The French would have slightly fewer cannon available in the 1730s than in the 1700s, suggesting that it was not simply a matter of having more cannon in the later period.

[113] BL Add MSS 61339, f. 46, Mémoire qu'on pourroit avoir en vue de faire, concernant l'attaque de la ville de Mons, written during the siege of Ath in 1706. Contemporary practice called for ten to fifteen feet of solid earth between each embrasure in a battery, while the need to target via line of sight increased frontage requirements further. Note that this is different from Vauban's ricochet batteries, which also required a large front of attack, but only in order to place batteries on the periphery that could enfilade the covered way.

[114] Rochas d'Aiglun (ed.), *Vauban*, vol. 2, p. 439 Vauban to Le Peletier de Souzy, Brest, 25 September 1695. Allent wrote that "Vauban only employed the minimum amount of artillery necessary" (p. 316). This is in contrast with historians who argue that Vauban's siege was the more expensive of the two (e.g. Lynn, *Giant*, p. 78).

[115] Archives affaires étrangères (AAE) CP Autriche 83, ff. 268–296v, Torcy's agent Noblet to his patron, from Breisach, 27 August 1703. My thanks to Dr. John Rule for this reference.

Feuquières also contrasted Vauban's preferred method of capturing covered ways with the reliance on overwhelming firepower:

> Our enemies pursued a different Conduct in the sieges of Namur in 1695 and of Menin in 1706. They drew before those places a prodigious quantity of heavy artillery, which they planted in batteries, that they might be enabled at the same time to destroy the whole front of the attack under the protection of that fire. It was easy for them to advance their trenches and ruin the out-works and even the bodies of those two places in a few days, which they accordingly did, in such a manner that the places would be assaulted by all the enemy's infantry, when they should attack the covert-way.[116]

Similarly, the Duke de La Feuillade, commanding one of the French armies in Italy, wrote to the Secretary of War about his plan to besiege the fortified town and citadel of Turin with the small number of men at his disposal:

> After thinking day and night about the possibility of capturing Turin with the few troops that I have, I do not know of any option except to attack the citadel with eighty large cannon, twenty small ones in case they can be used effectively, and forty-five mortars that we have; when all this fires together in the manner of M de Coehoorn, I only see it requiring six days to ruin all the defenses.[117]

Informed of such ideas, Vauban warned the young upstart:

> I am completely against the advice that you attack à la Coehoorn; this method, which is not even a method, is only good against bicoques.... above all else do not imitate Coehoorn's method of attack: it will betray you. The sieges of Barcelona [1697] and Verrua [1704–1705] are not sieges to be emulated. These places were captured only by miracles, after much time and after suffering great losses; there is nothing else to do but to follow faithfully [*se renfermer*] the rules, which, once observed, will never fail you.[118]

In his own defense, the young La Feuillade justified his plan to his father-in-law:

---

[116] Feuquières, *Memoirs Historical and Military*, vol. 2, pp. 289–290.

[117] Vault and Pelet (eds.), *Mémoires militaires*, vol. 5, p. 188, La Feuillade to Chamillart, Veneria, 3 September 1705. Berwick had a similar plan for Nice in the same year. *Mémoires du Maréchal de Berwick écrits par lui-même*, vol. 1, p. 192. To be fair, La Feuillade had witnessed a number of sieges in his eighteen years of military service, including those of Mons, Namur and Charleroi under Vauban, as well as several earlier in the Italian theater during the War of the Spanish Succession.

[118] Vault and Pelet (eds.), *Mémoires militaires*, vol. 5, pp. 652ff., Vauban to La Feuillade, 13 September.

> It is not to be doubted that in speaking of the Coehoorn attack, M. de Vauban would not be of my advice; he forgot among the 'small' sieges that Coehoorn conducted that of Namur, which Vauban had fortified with all his *industrie*. I never claimed I would storm [the covered way] after having weakened the defenses from 300 toises [1900 feet] away, but to first damage them from this distance and then, after having quieted the enemy's fire, work without danger up to the lodgment of the covered way, which will be done after the enemy's mines have been ruined; we will then establish our batteries on the covered way in order to make a breach, after which, we will conduct the siege according to the ordinary rules and with all the precaution possible. We have thirty-five large mortars, sixty-nine 24-pounders, six 16-pounders, and a large enough number of medium-size artillery pieces. I cannot prevent myself from adding that M. *le maréchal* Vauban must have a very poor opinion of me to give me general lessons on sieges.[119]

Although the siege had to be postponed until the next year, Vauban's advice was still ignored. Critically, the town was not fully invested despite the engineer's insistence, allowing Prince Eugene of Savoy's relief army to attack a gap in the circumvallation lines and rout the besieging force. La Feuillade had promised his father-in-law the fortress by the end of August, yet it was still in Savoyard hands on 7 September when Imperial relief columns pushed into the siege camp and forced an end to the botched Bourbon siege.[120]

As we have just seen, besiegers could call upon far larger siege trains than they had in past decades, trains far larger than Vauban recommended. These brute force advocates relied increasingly on other types of weapons (rather than ricochet fire) for indirect fire capability. At the same time as batteries were targeting fortress embrasures, other batteries set to work bombarding the fortifications obscured by intervening obstacles with high-trajectory cannonballs, bombs, stones and ricochet fire. Bomb-launching mortars (from the massive 100-pounders down to the smaller 50-pounders), *pierriers* hurling

---

[119] Esnault (ed.), *Chamillart*, vol. 2, p. 33 #164, La Feuillade to Chamillart, Turin, 20 September 1705. This dialogue between the two is evidence that the historiographical confusion was mirrored among contemporaries when speaking of the attack à la Coehoorn; for Vauban it meant first and foremost (unprepared) storms of the covered way, while for La Feuillade it meant a reliance on firepower to batter the enemy into submission.

[120] Esnault (ed.), *Chamillart*, vol. 2, p. 113 #199, La Feuillade to Chamillart, Turin, 29 July 1706. His earlier estimates promised the 15th of September.

baskets full of stones, grenade-throwing hand-mortars (16-pounders and 8-pounders), and shell-lobbing howitzers all combined to rain down destruction on the garrison and its works. For days or weeks these projectiles killed and wounded men, blasted the parapets, chipped away at the masonry revetments, smashed battery platforms and carriages, and pock-marked the infantry's firing steps. National differences arose here as well, for French besieging forces used relatively few mortars. Though Vauban recognized the utility of these bomb-throwers, Saint-Rémy listed only fifty or so mortars of all sizes (the smallest 8.5-inch mortars firing a 35 pound bomb) and four to eight *pierriers* for the three sieges of Mons, Namur and Charleroi in the Nine Years' War—Ath would see only twenty-two mortars in use.[121] French attacks in the Spanish Succession war saw even fewer indirect-fire weapons in the theater most amenable to large siege trains, thirty mortars and twelve *pierriers* being the norm for their 1712 sieges.

The far greater Allied reliance on indirect-fire weapons can be seen in both their projects and their siege trains. Projected Dutch siege trains included anywhere from twelve to thirty large mortars; at their various sieges, they usually collected similar numbers of 75-pound and 50-pound mortars. For comparison, Saint-Hilaire's requirements for a siege of Huy called for only fourteen mortars, half as many as the Allies would use in their siege a few months later and a third of what the enemy had used in 1703.[122] To those who were either ignorant of or unimpressed with Vauban's ricochet fire, two other types of artillery enabled them to overcome obstructions between them and their target, though not as efficiently. In addition to the cannon and heavy mortars, a dozen or so howitzers fired explosive shells along a high-arcing trajectory—French artillery parks had no howitzers at all. Instead of *pierriers* and ricochet fire, Allied besiegers also depended on an enormous number of smaller "hand mortars" developed by Coehoorn (termed 'Coehoorn mortars'), which fired double-grenades up to 2,000 feet away. Many

[121] Saint-Rémy, *Mémoires*, vol. 2, p. 269 table. At the beginning of Louis' reign French siege trains included no more than a dozen mortars, though this number slowly increased to four dozen by the end of the Nine Years War. Naulet, *L'artillerie française*, pp. 221–223.

[122] Lecestre (ed.), *Mémoires du Saint-Hilaire*, vol. 5, p. 297 #252, to Chamillart, 5 June 1708.

Allied sieges witnessed eighty or more of these portable hand-mortars in operation. Even the *bicoques* attacked in 1702 were not spared the hail of grenade shrapnel these pieces rained down: Coehoorn amassed an astounding 250 of these grenade-launchers against Liège's citadel, Venlo had 108 targeted against it, and at the more formidable fortress of Béthune they gathered together 170 hand-mortars. With tens of thousands of rounds provided for each siege, these anti-personnel weapons made a fortress's exposed works a very dangerous place. Several French garrisons emphasized their effectiveness, claims reinforced by some of Vauban's subordinates.[123] In some cases garrisons were even forced to construct wooden shelters in an attempt to cover themselves.[124] Vauban, however, continued to insist on efficiency and therefore remained unimpressed with these Allied tools, writing in his 1704 treatise:

> The Dutch have recently utilized a number of small, short-barreled cannon that they call *obus* [howitzers], and small, portable grenade-launching mortars carried by two men, with which they produce a tremendous weight of fire. But I do not find much utility in either of them; they require too much effort and expense, and do not produce very much effect; it is better to stick with large cannon, our bombs and *pierriers*.[125]

Two years later, after witnessing the losses of the 1706 campaign, he would be forced to acknowledge their efficacy in his treatise on the defense, though he still rejected them because of their inefficiencies. Referring to the Low Countries theater, where extensive fluvial networks allowed massive siege trains to be conveyed to the siege site by water, he wrote:

> it must be acknowledged that the fury of sieges has become so great due to the quantity of cannons, bombs and stones that are used, in addition to enemy's devious mechanical devices [*diableries*] called howitzers

[123] At Lille 1708: "Journal du siège de la ville de Lille" entry of 1 September, in Deidier, *Le parfait ingénieur francois*, p. 307. For Douai 1710: AG 1M 126–3 Mémoires historiques, #13 Journal de ce qui a été fait pour l'investiture et l'attaque de la ville de Douay, 15 May. At Béthune: AG Article 15 section 2, Béthune, #4bis, Journal du siège de Béthune, f. 28v.

[124] For the continued use of these medieval hoardings, see Koninklijke Bibliotheek, Knuttel Pamflet #15958, Journal du siège de Bouchain depuis qu'elle fût investie jusques à sa prise, p. 12.

[125] Vauban, *Traité*, p. 124. Saint-Rémy, *Mémoires d'artillerie*, vol. 1, p. 257 confirms its use being limited to the Allies.

> and perdreaux which destroy the front under attack. The bombs, stones and grenades cause such chaos, killing and wounding many in little time, and shorten considerably the time needed to capture places.[126]

The disagreement once again revolved around their differing objectives—finesse for efficiency, brute force and weight of shot to force an end to siege-related delays.

Commanders also differed from Vauban in their belief that a general bombardment of the town would force an earlier surrender by pushing the townspeople to pressure the garrison into capitulating. A certain amount of damage to the town was inevitable in any siege of course, as stray shots would inevitably fly over the rampart walls and into the town itself. As a result, one military author even went so far as to warn the besieging commander not to attack a town on directly opposite sides, lest each attack's batteries overshoot their targets and accidentally strike the trenches on the other side![127] But Vauban was opposed to intentionally targeting the town, not only because of the tactic's inhumanity but also because of pragmatic concerns with efficiency: "Never fire on the buildings in towns, because this wastes time and munitions, and does nothing to contribute to their fall, while the repairs required after their capture are always costly."[128] In 1691 he reported his discussion of the issue with the *maréchal* Luxembourg:

> He spoke to me also of bombarding Brussels; but as the bombardments of Oudenarde, Luxembourg and even Liège gained the King not an inch of terrain, and that, far from this, they wasted munitions and exhausted and weakened his troops, I said nothing to him on the subject, because it seems a very poor way of reconciling ourselves to the hearts of the people in a time when this country is better inclined towards the King than ever before.[129]

This general advice would be repeated in the journal of Ath 1697:

> Monsieur de Vauban had all along taken great care to preserve the Town; a few Shells had flown over the works of the Attack, and the

[126] Vauban, *Traité de la Défense des Places*, p. 108.
[127] Marquis de Quincy, *Histoire militaire*, vol. 7, p. 138. This is repeated in P.P.A. Bardet de Villeneuve, *Traité de l'attaque des places*, p. 48.
[128] Vauban, *Traité*, p. 263 maxim 16; also p. 122.
[129] Rochas d'Aiglun (ed.), *Vauban*, vol. 2, p. 327, to Louvois, Mons, 17 July. He similarly criticized the Spanish governor defending Ath in 1697 for needlessly destroying outlying buildings shielding impervious fortifications that would never be attacked. Rochas d'Aiglun (ed.), *Vauban*, vol. 2, p. 450, Vauban to Barbesieux, Ath, 18 May.

> same chance had likewise directed a few Bullets that way; but Monsieur de Vauban had taken the Houses, Buildings, Gates, &c. under his protection, all which made people imagine it was the King's intention to keep the Place; but the true reason was, that devastations and burnings contribute nothing to the taking of a Place, only consuming Ammunition to no purpose and occasioning extraordinary charges in repairs afterwards (p. 125).

Befitting the engineering focus on efficiency, here too we are shown how a single choice could minimize several inconveniences and offer advantages as well.

Vauban was only partially successful in his efforts at Ath and elsewhere, not only because of their rejection of the cult of efficiency, but also because few others shared his compunction and tended to be much more callous about the matter. If generals and politicians were willing to sacrifice more of their own men in order to speed up a siege, they rarely agonized over the prospect of setting towns aflame if they could gain several more days of campaign time as a result.[130] Louis and Louvois had both expressed their satisfaction with the tactic on many occasions.[131] Representatives from besieged towns received only vague promises that the attackers would protect the property of individual persons or (usually religious) institutions, but ending the siege quickly was always the first priority. The burgomeisters of Bonn, hoping to avoid the widespread destruction that they witnessed under siege fourteen years earlier, could only extract from Marlborough a promise to preserve the town's churches and public edifices "as much as possible."[132] The Duke of Burgundy

[130] This demands we reconsider the frequently-expressed view that general storms were no longer popular in the later 17th century because they threatened the resources of the captured place, and that sieges were conducted primarily as a way of gaining these assets (e.g. Chandler, *The Art of Warfare in the Age of Marlborough*, p. 267). Military commanders and their political masters both considered the time saved worth both the financial damage the town would suffer and the hatred such a tactic might prompt among the inhabitants, potentially their future subjects and tax-payers. As is often the case, military expediency outweighed longer-term political and fiscal interests, especially in the minds of time-conscious generals.

[131] Wolf, *Louis XIV*, p. 449; and Louvois's correspondence in late 1683 and early 1684 regarding the city of Luxembourg, printed in Jacques Hardré, *Letters of Louvois*, (Chapel Hill, 1949).

[132] Murray (ed.), *Letters and Dispatches*, vol. 1, p. 91, Marlborough to Alègre, Bonn, 7 May 1703. In the previous war, the Confederate army under the Elector of Brandenburg bombarded this same town for more than a month with one hundred cannon and thirty mortars before finally opening trenches. *An Exact and Complete Diary of the siege of Keyserwaert and Bonne.*

assured his tutor Beauvillier that he would fire on the town of Breisach as little as the siege would allow, although he was forced to admit that destroying the cavaliers on top of the town walls would inevitably lead to much collateral damage.[133] More often, intentional targeting of civilians was embraced by impatient commanders in the hopes it would make life miserable for the inhabitants and generate an uprising. Ouwerkerk, assisted by the English fleet, coolly rejected the Oostende representatives' appeals for an end to the bombing. Only when the garrison surrendered would the bombs cease falling, he declared. True to his word, most of the town was set ablaze by the time the French garrison capitulated.[134] At Menin the bombardment set fires that quickly consumed half of the town.[135] A garrison account from within Lille reported rumors that the besiegers were venting their frustration over the slow pace of the siege by firing all their balls and bombs into the town in the hopes of forcing a revolt—the resulting ammunition shortage left them with little to target the outworks, just the situation Vauban had warned about.[136] With the end of the 1708 campaign stretching into December, the Allies sought to use the same tactic to recover Ghent, which had been surprised earlier in the year and now sheltered a large French garrison of thirty battalions and nineteen squadrons. Receiving a deputation of magistrates entreating him to "not destroy their buildings with our bombs and fire-balls," Marlborough informed them that "since they had brought this misfortune upon themselves by their own folly or negligence, they must either assist us against the garrison or expect we should use all manner of extremity to reduce them to their duty."

---

[133] Vogüé (ed.), *Le duc de Bourgogne et le duc de Beauvillier*, p. 198 #47, Breisach, 20 August 1703.

[134] Veenendaal, Jr. (ed.), *Briefwisseling Heinsius*, vol. 5, p. 349 #658, Ouwerkerk to Heinsius, Oostende, 28 June 1706. On the bombardment, vol. 5, p. 371 #694, Ouwerkerk to Heinsius, Oostende, 3 July. See also Royal Commission on Historical Manuscripts, *The Manuscripts of the Duke of Athole, K.T., and of the Earl of Home*, (London, 1891), p. 63 #155, Tullibardine to Athole, Tournai, 17 July. Marlborough's biographer Jones notes that Marlborough forced the Dutch to "ensure . . . a maximum bombardment" against their pecuniary wishes. *Marlborough*, p. 127.

[135] Vault and Pelet (eds.), *Mémoires militaires*, vol. 6, p. 535 #204, Journal du siége de Menin, 9 August 1706. Cuper described the massive damage done to the town's walls, buildings, barracks, churches and homes in Veenendaal, Sr. (ed.), *Het Dagboek van Gisbert Cuper*, p. 242 #9, letter to van den Bergh, 6 September.

[136] Henri Malo (ed.), "Journal du siège de Lille rendu aux Alliés le 23 octobre 1708," *Revue du Nord*, 8 (1922), p. 283. Recall that it was at Lille that Marlborough blamed the *engineers* for the lack of munitions.

With the townspeople unable to effect a surrender, a short siege ensued, the Allies using red-hot cannonballs heated on iron grates to bring about its speedy reduction.[137]

A garrison account of Mons in the next year noted how the Allies hoped to precipitate an uprising among the townspeople by bombarding the town.[138] In the first siege of Douai, the besiegers "played very furiously upon the toun onely, & did great execution, setting fire to the toun wch. burned feircely for a great while & for a great space."[139] One French bombardier, a veteran of thirty-five sieges, described this three-day bombardment as a "firestorm [*déluge de feu*]," the like of which he had never witnessed. When the French returned on the offensive two years later, Villars' cannoniers bragged that they were intentionally targeting the houses of well-known Allied sympathizers within the town. After it was returned to its rightful sovereign, the French intendant estimated the destruction from this siege at over 200,000 *écus*, twice the cost of the damage incurred during the Allied siege.[140] As these examples illustrate, time-pressed commanders frequently erased the distinction between combatant and non-combatant in their attempts to put a quick end to the war. Despite the persistent use of the tactic, however, there is little evidence that such massive destruction of civilian property and loss of innocent life forced many garrisons to surrender prematurely, as is evidenced by the frequency with which garrisons held out till a storm was imminent.

Even Vauban's method for breaching the fortifications was contested when a 'quicker' alternative was available. The previously-onerous task of breaking up the thick walls could be abbreviated by once again ignoring Vauban's instructions. With batteries established on the covered way (or with a height advantage from beyond it), a

[137] Murray (ed.), *Letters and Dispatches*, vol. 4, p. 362 Marlborough to Boyle, Melle, 17 December; and vol. 4, p. 381, Marlborough to Lottum, Merelbeke, 29 December.

[138] AG Article 15 Section 3 Mons folder, #9 Relation de la deffense de Mons en l'année 1709, f. 10.

[139] David Chandler (ed.), *A Journal of Marlborough's campaigns during the War of the Spanish Succession 1704–1711*, (London, 1984), p. 104. A garrison source reports that "bombs flew into the town in such large numbers that it appears to have been expressly intended to ruin the town." AG 1M 126–3 Mémoires historiques, #13 Journal de ce qui a été fait pour l'investiture et l'attaque de la ville, 31 May.

[140] On targeting the houses, see AG A^1 2382, #28, Lefebvre d'Orval to Voysin, 17 August; for the estimate, AG A^1 2383, #193, Bernières to Voysin, Valenciennes, 12 September.

regular fortress front of two corner bastions and an intervening ravelin had no further obstacle to shield the bastion faces from breaching fire—their fall was only a few days of breaching fire away, as we have seen with Ath in 1697. In a few Flanders fortresses, however, the bastions were still hidden from the counterscarp batteries opposite them by demi-lunes or counterguards. A besieging commander strictly following Vauban's theory would be forced to capture these outworks first before establishing breaching batteries on them to target the bastion salients and faces directly behind them. The curtain wall between two bastions was visible, though attempting to attack there would invite enfilading fire from both ravelins and bastions. On the other hand, Vauban's safer method might require several days of extra time and men would likely fall in the assault on the outworks. Given the choice, the Allies not surprisingly chose speed and targeted the curtain wall between the bastions with batteries sited in the salient angles of the covered way, firing through the gaps between the outworks. Vauban had explicitly rejected this technique, arguing in 1704 that it was rarely done and could not be very effective.[141] Nevertheless, with this tactic Allied gunners were able to aim breaching fire against both the ravelins and the curtain wall at the same time from batteries on the angles of the covered way. Utilizing the technique at Menin, Des Rocques refused the advice of the English gunner/engineer Blood, who had insisted on capturing an adjacent ravelin before breaching the enceinte. In the end, the Huguenot was proven correct, saving several days and undoubtedly a number of casualties as a result.[142] The Allies used the same tactic at Lille—Du Mée recounted in his journal that 68 pieces fired against the town's bastion faces between the Magdalene and Saint-André hornworks.[143] So too at Douai in 1710 do we find Des Rocques using the same strategy, overturning French expectations that all of the approach's outworks would have to be taken before the corps could be breached.[144]

---

[141] Vauban, *Traité*, p. 119.

[142] Veenendaal, Jr. (ed.), "De Mémoires van Sicco van Goslinga," 32.

[143] De May, *An exact journal of the siege of Lille*, p. 3. See French confirmation in Journal du siège de la ville de Lille, printed in Deidier, *Le parfait ingénieur François*, p. 306.

[144] For its success at Douai 1710, see Veenendaal, Jr. (ed.), *Briefwisseling Heinsius*, vol. 10, p. 430 #851, Des Rocques to Heinsius, Douai, 12 June. Louis' expectations that all the outworks had to be captured first are in AG A[1] 2215, #148, to Villars, 4 June.

Vauban feared such a technique would demand a much higher price when the besiegers crossed the ditch towards the breach—a price that many commanders were willing to pay in any case. The risk, though, was not as great as might be thought. The cumulative attritional effect of the massive firepower brought to bear against the garrison (usually undermanned in the first place) seriously degraded their ability to resist in the later stages of the siege. A garrison was likely to surrender before the last ditch had been fully traversed—by that point, there would be little reason for the attackers to grant honorable conditions as they had already expended the effort crossing the moat, while the results of any defense of the breach was a foregone conclusion. In this case, embracing a technique that Vauban warned against was quite reasonable for the impatient siege commander. Popular when fortifications allowed it, this tactic was one of the more efficient examples of how brute force supplanted Vauban's methods and sped up the pace of siegecraft.

While Vauban, Coehoorn and besieging generals all relied on artillery to overcome the enemy's defenses, they disagreed on how the arm was to be used. Vauban preferred ricochet fire and trench cavaliers in order to maximize the attack's effectiveness at the same time as he saved powder, ammunition and lives. Many Allied and French commanders, on the other hand, used such tools from time to time, but did not do so consistently. More fundamentally, they disregarded the goal of ever-improving efficiency and focused their massive batteries on the general destruction of the place—the outworks, the town walls, even the buildings within the town were targeted. Both systems could capture towns, although the non-Vauban system generally required far greater resources to sustain such brute force tactics over the length of a long, attritional war.

3. *Resource Mobilization*

As Vauban and other contemporaries attested, brute force sieges were a much more expensive option in both armaments and lives. With little concern for calculating the minimal amount of force required, hundreds of siege cannon were hauled on site, their deafening fire often placating the besieger's psychological needs as much as advancing their material cause. The Spanish Succession's larger sieges could consume 100,000 rounds of ammunition or more, ten times as much as was available to besiegers a century earlier and four times as much as what Vauban consumed at Ath in 1697. In addition to the costs of mustering, transporting and firing these pieces, the greater use of such weapons dictated by brute force tactics also led to significant attrition among the guns themselves. Even the less taxing sieges, such as those undertaken by the French in 1705, could put up to half of their battering pieces out of action, especially if the gunners did not give their pieces time to cool down between shots.[145] From the beginning of the war the Allies too found themselves complaining about damaged pieces: at the conclusion of each of their sieges a third of their breaching bombards (numbering sixty pieces or more) could require repair before being sent back into service.[146] Some of these losses were due to defective metal and substandard casting, but brute force techniques accelerated this damage with their greater reliance on a rapid rate of fire—the more pieces used, the more shots fired per gun and the less time allowed for cooling off the barrels between each round, the more stress a greater number of cannon would experience and the more likelihood of structural failure. Impatient generals encouraged such stresses in the hopes that a greater volume of fire would accelerate a fortress's fall. Despite such attrition, the Dutch managed to keep their army provided with the hundreds of cannon necessary for vigorous attacks in siege after siege. The rapidity of the 1706 sieges took a particularly heavy toll on the artillery, but here too, the Dutch would eventually replace their losses.[147] Their vigorous use of firepower at Oostende, Menin, Dendermonde and Ath had also consumed all of the munitions in the Dutch magazines, though autumn rains forced an end

[145] AG A[1] 1835, #256 Villeroi to Chamillart, 11 June; #266 Bernières to Chamillart, Huy, 12 June. For a later example when the French returned to the offensive, see AG Article 15 Section 2 §1 Douai folder, #19, Siége de Bouchain 1712, f. 22.

[146] For Kaisersweert, Veenendaal, Jr. (ed.), *Briefwisseling Heinsius*, vol. 1, p. 275 #477, Wassenaar-Obdam to Heinsius, 9 June 1702; for Venlo, vol. 1, p. 419 #840, Wassenaar-Obdam to Heinsius, 8 September 1702. Ath: Veenendaal, Sr. (ed.), *Het Dagboek van Gisbert Cuper*, p. 160. Tournai: Veenendaal, Jr. (ed.), *Briefwisseling Heinsius*, vol. 9, p. 79 #158, Vegelin van Claerbergen to Heinsius, 21 July 1709. Mons: vol. 9, p. 354 #749, Vegelin van Claerbergen to Heinsius, 13 October 1709. Douai 1710: ARA RvS 688–1, Vegelin van Claerbergen to the Raad van State, 30 June 1710.

[147] After eight days of firing, Dutch administrators reported that twenty-four of their heavy cannon were already out of service. Veenendaal, Jr. (ed.), *Briefwisseling Heinsius*, vol. 5, p. 590 #1135, Goslinga to Heinsius, Ath, 7 October.

to the campaign season in any case.[148] Nevertheless, their energetic pace, validated by the fruits of the 1706 campaign, allowed Marlborough to hope that "if we can agree upon carrying on the warr this next yeare with vigor, I no ways doubt but we should have such peace, as that we might end the rest of our days in quietness."[149] There would, in fact, be no Flanders sieges in 1707 due to a strategic focus on the Italian theater, but in the next year Dutch guns would again thunder forth in ever-greater numbers. Serving as the 'arsenal of Europe,' the United Provinces' warehouses would provide the necessary firepower. The Dutch riverine transportation network also made it relatively easy to ship cannon to the front, while the polder-riddled country boasted one of the continent's largest arms markets, and further maintained a large reserve of naval guns from the fleet that it could call upon in an emergency. With such resources the Dutch were particularly well-adapted to maintain such large siege trains in the Flanders theater.[150] This type of war was sustainable, but expensive nonetheless. In a materiel sense, Clausewitz's metaphor applies quite well to the cult of vigor: "Iron will-power can overcome this friction; it pulverizes every obstacle, but of course it wears down the machine as well."[151]

Vigorous sieges also consumed more human lives than an efficient siege, a critical failing in the minds of engineers and other, more efficiency-minded, contemporaries. To the long-established charge that slow, systematic attacks by the sap masked indecisiveness and cowardice, Vauban countered that blindly rushing the trenches forward and precipitously storming well-fortified covered ways illustrated the opposite extreme—a wasteful disregard for the King's troops and money that more often than not lengthened a siege rather than shorten it. He approved of accelerated attacks (*brusqué*) when

---

[148] Concerns over the supply of munitions surfaced in August. Veenendaal, Jr. (ed.), *Briefwisseling Heinsius*, vol. 5, p. 451 #859, Albemarle to Heinsius, Helchin, 6 August; Royal Commission on Historical Manuscripts, *Report on the Manuscripts of the Earl of Mar and Kellie*, p. 276 Dalrymple to the Earl of Mar, 9 September.

[149] Snyder (ed.), *Marlborough-Godolphin Correspondence*, vol. 2, p. 669 #676, Marlborough to his wife Sarah, Grandmetz, 16 September.

[150] H. Vogel, "Arms production and exports in the Dutch Republic, 1600–1650," in M. van der Hoeven (ed.), *Exercise of arms: warfare in the Netherlands, 1568–1648* (Leiden, 1997), pp. 197–210. While the Dutch could focus their efforts on Flanders, the French had to distribute their pieces amongst garrisons and armies in four separate theaters.

[151] Clausewitz, *On War*, p. 119 (Book 1 Chapter 7).

conditions warranted, but he condemned the techniques commanders chose almost reflexively, for they planned and implemented them poorly and wasted far too many lives for the purchase of a few additional yards of terrain:

> What I find ridiculous is that when these officers leave the trenches, they brag with a satisfied and content air about how they lost 100 or 150 men during their watch, among these perhaps eight or ten officers and several brave engineers who could have yet served elsewhere. Is this something to celebrate? And the Prince, is he in debt to those who accomplish with the loss of one hundred men what could be done as easily with a little *industrie* and the loss of only ten men?. . . . There is nothing so common among us as this brutality that depopulates our troops of veterans, so that a war of ten years exhausts the entire kingdom.[152]

He further warned his engineering peers that they would find many who would reject their proposals to save men's lives due to tradition and ignorance. He was even more scathing, however, of those "who have such a bizarre and depraved mind that they consider their achievements victorious only to the extent that they are reddened by the blood of their own men."[153] Vauban was not the only one to notice this callousness towards loss of life. Fontenelle's eulogy in honor of Vauban noted that the *maréchal* had frequently been compelled to resist the impatience of generals who wanted to sacrifice lives in order to achieve a more rapid victory and a glory worthy of their greatness.[154] One account of Turin noted that the gunner Vallière's request for a week to prepare a demi-lune for assault did not suit La Feuillade; his impatience to attack it carried the day, as well as carrying many men to the next life unnecessarily.[155] Other engineers' opposition to brute force, much like their opinions in general, rarely found expression in the public realm, but in accounts destined for their peers we find their occasional justification of efficiency over vigor. One French engineer recalling the casualties suffered by his peers in their 1712 siege of Douai contrasted the two opposing viewpoints bluntly: "the precipitation with which the engi-

[152] Vauban, *Mémoire pour servir*, p. 17.
[153] Vauban, *Mémoire pour servir*, pp. 20–21.
[154] Virol, *Les Oisivetés*, Annexe 2, p. 480.
[155] Grandchamp (Gatien de Courtilz de Sandras), *Mémoires de la dernière guerre d'Italie: avec des remarques critiques & militaires. Par Monsieur D****, (Cologne, 1728), pp. 106–107.

neers were pushed cost many lives without any additional progress."[156] The chevalier de Quincy also lamented the French officers' penchant for stupidly wasting the lives of their own men in poorly-planned actions.[157] These late echoes of Vauban show how little had changed in the minds of officers by the end of Louis XIV's reign, and we even find confirmation in the generals' own statements. The tactical callousness of the vigorous ideal is encapsulated most memorably by one of its most energetic proponents, Villars. Despite his claim that "I always had for principle to conserve the troops, especially the officers," he had earlier indicated the disposability of individual lives *when time was short*:

> I returned immediately to my infantry, which was in battle order: but as soon as I joined them, I saw the enemy army dashing towards the Escaut River in several columns. The marquis Albergotti proposed to make fascines to cross Denain's retrenchments. "Do you think," I responded, showing him the enemy, "that these gentlemen will give us the time? Our fascines will be the bodies of our first men to fall into the ditch." There was not an instant, not a minute to lose.[158]

While the resulting victory at Denain has been applauded by many as decisive, such a statement would have undoubtedly horrified the engineers and their fellow travelers.

Those who had to pay for the generals' obsession with time certainly recognized the resulting costs, as disputes among the upper echelons over discourses of efficiency and vigor were literally matters of life and death for them. The consequence of the invariable recourse to vigorous measures became evident late in the war—a serious decline in Allied troop morale that led to widespread desertion by 1711. In a field battle, soldiers could reassure themselves that however intense the danger, it would be over by the end of the day, and they could even hope that victory might end the war. At major sieges on the other hand, the outcome of an assault on the covered way was rarely the end of the siege, but only the prelude to possibly more dangerous fighting in the ditches and outworks. Particularly discouraging must have been their knowledge that generals wished to throw them into an assault on the covered way while the

[156] Article 15 Section 2 §1 Douai folder, Précis du journal du siege de Douay fait en 1712 . . ., f. 4; see similar comments on f. 7.
[157] For example, Lecestre (ed.), *Mémoires du chevalier de Quincy*, vol. 3, p. 159.
[158] Anquetil, *Suite des mémoires rédigée par Anquetil*, p. 383; p. 375.

engineers wished to proceed more cautiously. It was even more deflating to see how little the assaults often gained them. Only in the weakest of fortresses would the garrison give up the covered way entirely after a storm; normally the most that could be hoped for was a number of lodgments at the salient angles of the covered way. Defenders would have to be evicted from the rest of the counterscarp by yet more digging and firing, grenade-throwing and hand-to-hand combat. And once this town was taken, two more fortresses stood waiting their turn. Blackadder's lament captured the weariness a war of sieges could quickly produce: "on whatever side France has the thickest and strongest nests of garrisons, Providence turns our arms that way."[159]

The major sieges conducted from 1708 onward slowly drained the fight from the men. One English lieutenant expressed a fatalistic acknowledgment of the high costs of Lille's capture (12,000 casualties in front of both the town and citadel): "our generals if posable are designed to have it, cost what it will."[160] While Lille was the Allies' single most bloody siege of the theater, the next year's sieges of Tournai and Mons added 5,200 additional casualties to the total. Vegelin van Claerbergen detected an interesting shift in the attitudes of the rank-and-file at the attack on the former. The States' chief engineer Des Rocques had broken his leg at the outset, leaving the conduct of the siege up to the impetuous Du Mée. This was "a significant setback. Not a single director is to be found here; everyone, officer and soldier alike, is concerned about [Des Rocques'] health, I would never have believed he was as popular as he seems now."[161] Another observer was more forthright in writing: "Thus the conduct of the siege will undoubtedly fall on Du Mée, who has lost some of the troops' confidence for having sacrificed them excessively in the attack on the town."[162] The citadel's extensive system of coun-

[159] Crichton (ed.), *The Life and Diary of Lieut.-Col. J. Blackadder*, p. 413, 31 July 1711 O.S.

[160] Royal Commission on Historical Manuscripts, *Report on the Manuscripts of the Earl of Mar and Kellie*, p. 464, Lt. William Nodding to the Earl of Mar, Lille, 20 September 1708. He then added that the engineers were generally believed to have been "very much mistaken."

[161] Veenendaal, Jr. (ed.), *Briefwisseling Heinsius*, vol. 9, p. 27 #56, Vegelin van Claerbergen to Heinsius, Tournai, 7 July 1709; also p. 37 #79, Goslinga to Heinsius, Tournai, 11 July.

[162] Veenendaal, Jr. (ed.), *Briefwisseling Heinsius*, vol. 9, p. 23 #49, Caesar Caze

termines forced a more measured pace, and this was welcomed by the troops who were increasingly fearful of siege duty, one English lieutenant-colonel speaking approvingly of "the reason we have to not sacrifice men's lives for a few days more or less."[163] The last siege of the season, Mons, saw similar concerns. Lt.-Col. Blackadder readily embraced the engineering caution that raised such disgust among the general officers: "We hardly wish [the siege] to be over so soon, for fear we be employed at another siege."[164]

Things only got worse the next year, for the campaign of 1710 saw a dramatic increase in the numbers wounded and killed in the trenches. A thankful Blackadder initially reported the relief his unit felt when the roster for trench duty at Douai stopped one regiment short of theirs. His regiment was soon enough called to serve in the trenches, taking the place of another that was cut to pieces in a garrison sally, but the God-fearing officer did have one piece of good news to report to his wife late in the siege: "now most of our labour is sapping, which is a sure and a slow work; and therefore [we] do not grudge the siege lasting eight or ten days longer. The saving of men will recompense the loss of time, though we will still be losing men every day at working."[165] Douai's defenders inflicted 8,000 casualties on the enemy, the second most costly siege in the theater. When the three later sieges are included, the 1710 campaign cost the Allies 8,000 at Douai, 3,300 casualties at Béthune, 900 at the short siege of the weak place of Saint-Venant, and another 6,500 at Aire—18,700 wounded and killed in all, coming close to what the Allies suffered in the bloodiest battle of Louis' many wars. As the campaign of interminable sieges dragged on, several thousand fled

d'Harmonville (a Huguenot in Dutch service) to Heinsius, Lille, 7 July 1709. Jacob Hop, the Dutch *tresorier-generaal* for the Allied Condominium of the Spanish Netherlands reported along similar lines on p. 24 #52, Lille, 7 July. Vegelin van Claerbergen hoped that the injured director-general would still be able to sketch out his attacks on paper so they could be implemented in his absence: p. 27 #56, to Heinsius, Tournai, 7 July.

[163] *MSS Portland*, vol. 4, p. 497, James Cranstoun to Cunningham, Tournai, 5 August O.S.

[164] Crichton (ed.), *The Life and Diary of Lieut.-Col. J. Blackadder*, p. 366, to his wife, 29 September 1709 O.S.

[165] On missing trench duty, Crichton (ed.), *The Life and Diary of Lieut.-Col. J. Blackadder*, p. 379, to his wife, Douai, 24 April 1710 O.S. He also wrote of "a new instance of the Lord's goodness to us in disappointing our expectations of going upon this siege" (p. 378). On the desirability of sapping, p. 387, letter to his wife before Douai, 16 May O.S.

the ranks while many others succumbed to the diseases breeding in wet, cold trenches. Douai provides the only estimate of desertions during a siege that I have discovered, some 2,000 troops in addition to the 8,000 casualties, or a quarter of their casualties.[166] The later siege of Aire saw fewer combat casualties than Douai, but, hampered by the swampy terrain and torrential downpours, it may have been even worse for the troops' health and morale overall:

> Aire is going to take up the rest of the campaign, because it is one of the best places that the French have, and I expect that we will not have it for the entire month of November, which will dissipate our army, especially through desertion and disease. We have never seen so many desertions as we have among our troops, whereas the enemy army does not suffer as we do.[167]

A mid-October review of Dutch regiments confirms the toll taken by both desertion and widespread illness: battalions which were at full strength now averaged only 400 men fit for duty, or two-thirds of their size at the beginning of the campaign.[168] Combat, desertion and disease all took their toll.

With this long sequence of bloodletting fresh in their minds, the Allied army continued to hemorrhage throughout the winter, its troops deserting in droves at the mere prospect of another siege. The Rhenish lieutenant-general of Dutch cavalry and quartermaster-general of the States' army Daniël Wolf Dopff complained in late April on the eve of the 1711 campaign:

> The desertion is terrible in the infantry.... The worst is that desertion is occurring not only among the recruits but among the veterans; even among the Danes and Saxons [i.e. well-paid veteran mercenaries], in general no nation's troops are exempt. Here is the fruit of the siege of Aire and I can tell Your Excellency in confidence that the foot soldiers are so disheartened that if we were to open our campaign with a siege, some countries' troop contributions would be at risk and the officers would have difficulty preventing the ensuing deser-

[166] ARA familie Surendonck, #146, Jacob Surendonck, Eenige consideratien over de verdene operatien van de tegenwoordige campagne, 5 July. If this rate held for the other sieges in this campaign, the number of desertions would equal 4,600 men, putting the year's tally of dead, wounded and deserted at over 23,000 men, or a quarter of the initial field force.

[167] Veenendaal, Jr. (ed.), *Briefwisseling Heinsius*, vol. 11, p. 107 #192, Albemarle to Heinsius, Tournai, 31 August.

[168] Veenendaal, Jr. (ed.), *Briefwisseling Heinsius*, vol. 11, p. 264 #487 Vegelin van Claerbergen to Heinsius, Lambres 16 October.

tion.... We must give the troops a different idea about the upcoming campaign to make them forget the last one.[169]

Daniel Defoe's anti-war pamphlet of the same year, relying on reports from the *Post Boy* paper, claimed that 10,000 Allied troops deserted upon hearing of the prospect of a siege of Ypres or Saint-Omer.[170] Three months into the campaign the Allies turned to capture the relatively-weak town of Bouchain, prompting the lower ranks to perform an informal risk-benefit analysis: "Some of us are wishing to be on this siege [of Bouchain], as being easier than it would be at the end of the campaign, at a stronger town and worse weather." And while the rank-and-file may have been happy that this siege was easier than expected, they did not share Marlborough's impatience for its trenches to be repaired so that further towns might be attacked. "We are at an utter uncertainty still about our future operations. Some are wagering we shall make another siege, others that we shall not; and every body wishes the last may gain. I see none so public-spirited in the army as to wish for another siege."[171] After another winter of repose, two more sieges of moderate difficulty would be attempted by the Allies in 1712, but the English withdrawal from the fight indicated that they had clearly shot their bolt. Allied forces had reached the last impediment in Vauban's *pré carré* (Landrecies) just in time: it is questionable how many more major sieges their troops could have sustained had the war continued much longer. Villars' victory against the entrenched camp of Denain put an end to any such debate.

By emphasizing the cult of vigor's relative insensitivity to casualties we do not mean to suggest that those in command were totally

---

[169] Veenendaal, Jr. (ed.), *Briefwisseling Heinsius*, vol. 11, p. 763 # 1334, Dopff to Heinsius, Aire, 29 April 1711.

[170] *Reasons why this nation ought to put a speedy end to this expensive war*..., (London, 1711).

[171] On preferring the easier siege: Crichton (ed.), *The Life and Diary of Lieut.-Col. J. Blackadder*, p. 417, 10 August 1711 O.S. On the lack of public spirit: p. 425, Bouchain, 8 September. Almost two centuries later an English officer studying the Anglo-Boer War would identify this basic fact of combat motivation. Colonel G.F.R. Henderson explained the results of using British close-order tactics against long-ranged, deadly-accurate repeating rifles in South Africa: "When the preponderant mass suffers enormous losses; when they feel, as they will feel, that other and less costly means of achieving the same end might have been adopted, what will become of their morale?" Quoted in Michael Howard, "Men against Fire: The Doctrine of the Offensive in 1914," in P. Paret (ed.), *Makers of Modern Strategy from Machiavelli to the Nuclear Age*, (Princeton, 1986), p. 516.

oblivious to the sufferings of their men, witness the attention paid to medical services and hospitals (e.g. Louis' famous institution *Les Invalides*), as well as the work to ensure an adequate food supply and allow plunder when necessities were lacking. Rather, the vigorous mindset steeled them to accept the fact that a successful war effort would require sacrificing many of their men on occasion, and that, at the moment of tactical decision, this resolution to act had to be made without hesitation and without regret. Undoubtedly the social divide between the men and their leaders made this easier (though nobles died in large numbers as well), and this may also partially explain why Vauban and other engineers, closer to the troops in social status, empathized more readily with them.[172] This psychological acceptance of high(er) casualties in order to gain immediate tactical objectives also fits well with Lynn's concept of a battle (i.e. combat) culture of forbearance, though the generals themselves did not embrace the corollary of stoic patience; this, apparently, was only expected of their men under fire.[173] This went far beyond the confines of siegecraft, for officers impatient with sieges were equally irritated with inaction on the campaign. The generals who pressured engineers to accelerate sieges were also the same as those who sought battles, for the belief in decisive battle also required a corresponding acceptance of possibly high casualties.[174] Ideally the victor would suffer only a few thousand casualties while the enemy's army was crushed, but battle was a dangerous venture, abhorred by the risk-averse. While the length of a battle was far more certain than the duration of a siege, the casualties suffered in it were as unpredictable as the lengths of many sieges. For example, the Allies suffered only 3,600 casualties at Ramillies, but five times as many at the bloody battle of Malplaquet. Even this carnage on the field of battle—twice as many Allied losses as the enemy, or some 20,000 soldiers wounded and killed—did not disturb Marlborough enough to waken him from his reverie that this 'victory' would force the French to give them

[172] The social history of the French military has received much study in the past several decades. See particularly the many works by André Corvisier and Jean-Pierre Bois' *Les anciens soldats dans la société française au XVIIIe siècle*, (Paris, 1990).

[173] Lynn, *Giant*, pp. 513–515. Chandler argues the opposite (with little evidence), that Marlborough was "equally careful with his men's lives—a trait of which they thoroughly approved." *Marlborough as Military Commander*, p. 315; see also pp. 328–329.

[174] For example, Jeremy Black, *European Warfare, 1660–1815*, (New Haven, CT, 1994), p. 112.

"what peace we please."[175] One contemporary made the connection between a siege and one of the bloodiest battles of the Old Regime explicit in his description of the aftermath of the Allied assaults on French field fortifications at Malplaquet:

> I can liken this battle to nothing so much as an attack of a counter-scarp from right to left: and I am sure you would have thought so, if you had seen the field as I did ye day after. In many places they lye as thick as ever you saw a flock of sheep.[176]

For generals who accepted and even embraced the risk of battle's losses as an opportunity, capturing a covered way at the cost of two thousand further casualties was seen as a bargain. It allowed them the consolation that, even if they could not avoid the siege altogether, they could at least buy several additional days with the lives of their men, and that these days might be put to better use later on. Generals such as these interpreted success solely in terms of the rapidity of victory—casualties were largely irrelevant when time was short. Buttressed by the (unrealistic) hope that a decisive victory was just around the corner, time-conscious commanders pushed the siege machine faster and faster.

Commanders, perhaps for the first time in modern European history, could consistently win with such a mindset because their brute force approach was facilitated by the European fiscal-military state's new ability to replace those incapacitated in their vigorous assaults. Brute force tactics had been common enough in siegecraft whenever time or resources were short, or when technical expertise was lacking. What was new to the late 17th century, however, was not only that Vauban's efficient siege offered a viable alternative approach, but also that centralizing states could now afford to reject such efficiency and sustain a successful vigorous siege in attack after attack, with little need to slacken its pace due to a shortage of funds. Yet the growing ability of the fiscal-military state to mobilize resources for its war-waging was not immediately obvious to observers. At the

---

[175] Snyder (ed.), *Marlborough-Godolphin Correspondence*, vol. 3, p. 1360 #1392, Marlborough to Godolphin, Blaregnies, 11 September 1709. Contrast Marlborough's celebration with Villars' perspective: "if God graces us with the loss of another such battle, Your Majesty can be assured that your enemies will be destroyed." Vault and Pelet (eds.), *Mémoires militaires*, vol. 9, p. 377, to Louis, Le Quesnoy, 14 September.

[176] H.H.E. Cra'aster, "Letters of the first Lord Orkney during Marlborough's campaigns," *English Historical Review*, (1904), p. 320 from Bellian, 16 September 1709.

very beginning of the Spanish Succession war the semi-official *Mercure galant* predicted the Allied effort would soon falter due to its profligate ways. Regarding the rapid collapse of Liège's forts and citadel, it reaffirmed the old adage that money and credit were the sinews of war by noting that the works had been:

> besieged by an army four times as large as would have been necessary for a similar siege. . . . the enemy also used an extraordinarily large siege train for such a small place. This shows how formidable the French troops are, and how much the enemy feared engaging them in close combat; when cannon and bombs capture places, there is little glory for the besiegers, and the method is so costly that a State cannot take many fortresses in this fashion without ruining itself in very little time![177]

The rest of the war would deflate the Gallic monthly's early optimism: it was correct about the costs of such brute force sieges, but it was wrong about how long fiscal-military mercantile states could (and would) sustain the outlays necessary to continue this type of fight. The demands for manpower did not necessarily come through an increase in the sizes of the armies needed to besiege major fortresses. Though there has been no resolution to a recent debate over whether sieges of *trace italienne* fortresses increased early modern European army sizes, provisional statistics gathered by Lynn suggest that the strength of besieging forces did not increase significantly over the course of the sixteenth and seventeenth centuries, with the largest siege armies consistently numbering 30,000–40,000 men even in the War of the Spanish Succession.[178] Instead it appears that increasing army growth over this same period was siphoned into armies for multiple theaters and, within each theater, into garrisons and ever-expanding observation forces. The real strains brute force sieges placed on manpower, then, came from the need to replace the heavy losses suffered by those regiments engaged in siege service and from the indecisiveness that stretched wars out into decade-long contests. Both sides fought through their exhaustion in the hopes of forcing their opponent to "see reason." The Allied mercantile

---

[177] *Le Mercure Galant*, octobre 1702, p. 430.

[178] For this debate between Parker and Lynn, see Lynn, "The trace italienne and the Growth of Armies" and his evidence for overall French establishment growth in "Recalculating French Army Growth During the Grand Siècle, 1610–1715."

approach sought new lines of credit through an ever-increasing public debt. In contrast, the French relied on ever-more creative fiscal contortions, applied most successfully from 1708 onward when Chamillart was replaced as *contrôleur général des finances* by a nephew of Colbert, Nicolas Desmaretz.[179] The efficacy (if not efficiency) of the Allied juggernaut is evident when we summarize the last half-dozen years of the war: Vauban's *pré carré* was pierced in a matter of a few years, the major fortresses of Lille, Tournai and Mons, Douai, Béthune and Aire all captured in marathon campaigns of brute force sieges. Even more impressive is France's ability to survive. Combating most of Western Europe and struck by widespread famines in 1693–1694 and 1708–1710, Louis managed to hang on throughout the long war and still return to the offensive in 1712, quickly retaking four fortresses from Allied forces that were no longer able to call on Marlborough or his English regiments. We should also note that despite decades of war over the course of Louis' long reign, field army sizes in the Spanish Succession—which were slightly lower for the French than in the Nine Years' War and probably larger than in the previous war for the Allies—peaked in the last years of this final war. These efforts must also be placed within the context of the many wars the Grand Alliance fought against the much-dreaded 'Lewis the Baboon.' The Allies fought (and largely won) the War of the Spanish Succession in the field after overcoming significant obstacles in the Nine Years' War: the Dutch had seen an explosion of their public debt (facilitated by the Bank of Amsterdam), while England had been forced to recoin its currency and establish its own Bank of England along the Dutch model. Despite these fiscal crises, after only a few years of peace they returned again to the warpath against the 'hegemon of Europe,' and they even managed to surpass their previous efforts by accumulating more debt. With

[179] Unlike the 'demand' side of siegecraft covered here, much research has been done on the 'supply' side of early modern military administration and finance more generally. The most important authors (many of whom have been cited already) include the pioneer Geoffrey Parker on the Spanish during the Dutch Revolt, and James Tracy on Charles V; on the English earlier in the century James Scott Wheeler and Mark Charles Fissel, and D.W. Jones and John Brewer for Louis XIV's age. For the French, see particularly the overviews by John Lynn and Richard Bonney, David Parrott on the Richelieu period, Guy Rowlands for Louis XIV's reign, and Gary McCollim on Desmaretz particularly. The Dutch side has been examined by J. Aalbers and Olaf van Nimwegen among others.

such financial efforts, two different versions of the early-eighteenth century fiscal-military state had proven their ability to sustain the cult of vigor.

Thanks to a combination of systematic trenchworks, a heavy reliance on firepower, and the ability to mobilize the resources needed to sustain such attacks, sieges conducted both according to and in opposition to Vauban's precepts were more often successful from the late 17th century onward than in earlier eras. Time-conscious generals, rarely successful in their attempts to avoid a siege altogether, were usually forced to acknowledge the necessity of sieges by attacking fortresses *dans les formes*. This did not mean they had to accept Vauban's conception of what the siege should look like. Vauban's method may have been more efficient, but efficiency required prerequisites that few armies of the day could or would meet, particularly a large number of well-trained engineers who were given unquestioned authority over the conduct of the siege. Yet ignoring Vauban's tactics did not doom a siege to failure, even those targeting the strongest fortresses of Europe. It did, however, increase the costs of capture, measured primarily in terms of the besieger's casualties and their siege artillery. Many generals were willing to pay this additional price if they believed it would expedite the siege. Vauban could rail against the inefficiencies and argue that his method saved lives *and* time, but such counterfactuals were impossible to prove to impatient generals. As long as these commanders could point to examples where breaking his rules led to success (as measured by their criteria), and as long as they received the necessary guns, munitions, manpower and funds, the attack *à la Coehoorn* survived and even prospered. The *Raad van State* field deputy's discussion of Aire neatly summarizes the true nature of Spanish Succession sieges as well as identifying the real reasons for their successes:

> The siege continues as always like Douay and consequently badly. I hardly know how much we will have advanced in two or three nights and even less when we will capture the counterscarp. The harmony between the Prince of Anhalt [the commander of the attack] and our directors is very small and they squabble amongst themselves. Nevertheless, the troops, the ammunition and the States' purse will suffice.[180]

[180] Veenendaal, Jr. (ed.), *Briefwisseling Heinsius*, vol. 11, p. 271 #507, Vegelin van Claerbergen to Heinsius, Lambres, 20 October.

CHAPTER NINE

THE VIGOROUS FUTURE

For centuries fortifications presented an unavoidable obstacle to mobility and decision. This strength of the medieval tactical defensive was only briefly interrupted by an interlude of crumbling castle walls as gunpowder was put to use in the sieges of the early- to mid-fifteenth century. The gradual development and deployment of the *trace italienne* from 1470 onward once again returned warfare to stasis by the middle of the sixteenth. Yet the pendulum would swing back once more as anonymous engineers and soldiers worked throughout the next century to capture these works more efficiently than before. Vauban was the culmination of this collective effort, combining old techniques and new improvements into a rationalized system that epitomized the efficient balancing of delays, lives and costs. His attack offered, for the first time, a successful alternative to brute force. The reality of sieges, however, was quite different from the rhetoric of scientific, clockwork sieges. Even Vauban had difficulty conducting the attacks according to his wishes, and those engineers that came after him had even less authority. Furthermore, the ranks of the French, Dutch, English, Spanish, Austrian, and German engineering corps were constantly depleted thanks to the dangerous duty, low pay and even lower standing, making application of the efficient ideal a difficult task. In contrast to the rhetoric of siege history, the engineers did not, in the end, dictate the conduct of the siege attack. Without Vauban's personal presence, his ultra-efficient ideal was rejected by many of the generals who oversaw the sieges. Accelerating the opening of the trenches, ignoring Vauban's recommendations for artillery use, storming the covered way rather than rely on the sap, bombarding the town indiscriminately—all these tactical decisions, as much as their frequent complaints, illustrate the general officers' utter impatience with the engineers. In opposition to the Vaubanian ideal, many generals supplemented efficient tactics with brute force, tactics which suited their wide-ranging belief in the importance of vigorous action. The resulting amalgam of techniques was enough to force most towns to submit, yet the results were unsatisfactory for both sides. The engineers were compelled to unnecessarily risk lives through the use of impatient tactics. Generals who had wished to avoid sieges in the first place were just as frustrated by the delays imposed by the overly-cerebral technicians and their time-consuming sieges.

Many of the most successful (and most trusted) field commanders quickly tired of the slow pace of operations and looked for other alternatives. Their most commonly stated preference was for a field battle that might decide the matter within the timeframe of a single campaign, though this was rarely a possibility. If fortified positions could not be bypassed, perhaps a storm, surprise or bombardment could achieve the same results as a siege, but with fewer delays. More inventive plans to overcome Vauban's *pré carré* were also discussed. Among the most well-known to English historians is Marlborough's 1708 alternative to a siege of Lille—a project to break free of their supply lines by somehow skirting around the *pré carré* while being supplied by the Royal Navy. The Dutch Lieutenant-General Hompesch also focused on the logistical part of the equation when faced with the same prospect, witness his musing that hand mills might replace bread convoys and allow the army to march past the enemy's belt of fortifications straight into their heartland.[1] These remained only fanciful hopes, however, and the attritional war of sieges continued for another four years.

The struggle between efficiency and vigor would continue well beyond the War of the Spanish Succession, and the victory of vigor would inform more modern views of the 'proper' conduct of war. Not every general was a vigorous one, but an energetic outlook has been a requirement for ever-lasting fame. Great captains like Gustavus Adolphus, Condé and Turenne, but particularly the Spanish Succession heroes Marlborough and Eugene, left their mark on the rest of the century. In France the aged yet still vigorous Villars would continue to influence the French army while serving on young Louis XV's Regency Council, and he would even command an army in the field as late as 1733. More broadly, biographies and campaign histories of the twin Allied captains provided models for battle-seekers the

[1] Veenendaal, Jr. (ed.), *Briefwisseling Heinsius*, vol. 7, p. 484 #1053, Hompesch to Heinsius, Amougie, 27 August 1708. An earlier Allied attempt in 1705 to strategically bypass the *pré carré* by marching through the Moselle ended in stalemate and starvation.

world over.[2] Moving further afield to the most reckless commander of them all, the Swedish King Charles XII's love of cavalry charges with cold steel would earn him his own cult following, despite the fact that such head-strong behavior lost Sweden its great power status.[3] Maurice de Saxe, France's victor in the War of the Austrian Succession, first experienced combat in the armies of Marlborough and Eugene, and was placed by his father (newly-elected) Augustus II of Poland under the particular charge of Schulenburg; no doubt this environment's disparaging view of the engineers had an impact on the impetuous youth. Twenty years later, and well before French engineers began quantifying the cost-effectiveness of various fortification schemes, thinkers like Maurice were already questioning the decision to fortify cities at all, critiquing engineers like Vauban and Coehoorn for designing fortifications that were so easily captured, and at the same time declaring a clear preference for battles over sieges.[4]

---

[2] On Marlborough and Eugene: *England's triumph; or the glorious campaign in the year 1704 . . . More particularly of the forcing the lines of the enemy at Donawert, and the glorious victory obtained by His Grace the Duke of Marlborough, in conjunction with the confederate forces, over the French and Bavarians, at the famous battle of Hochstet and Blenheim 1704*, (London, 1704); Francis Hare, *The conduct of the Duke of Marlborough during the present war*, (London, 1712); Arthur Maynwaring, *La vie du Prince et Duc de Marlborough*, (Amsterdam, 1714); Philo-Strategos, *Eugene's annals: being an exact and compleat history of the life and glorious actions of Prince Eugene of Savoy*, (London, 1715); Philo-Strategos, *Churchill's annals: being a compleat view of the glorious actions of the great Duke of Marlborough, both in the field and cabinet*, (London, 1722); *The life of the late victorious and illustrious prince, John Duke of Marlborough, and Pr. of Mindelheim*, (London, 1723); Jean Dumont, *Histoire militaire du prince Eugène de Savoye, du prince et duc de Marlborough, et du prince de Nassau-Frise, ou l'on trouve un détail des principales actions de la dernière guerre, & des batailles & sièges commandez par ces trois généraux*, (La Haye, 1729)—translated into both Dutch and English; Millner's *A Compendious Journal* (1733); Thomas Lediard, *Life of John, Duke of Marlborough: Prince of the Roman Empire*, (London, 1736); John Bancks, *The history of John, Duke of Marlborough*, (London, 1741); Eléazar Mauvillon, *Histoire du prince François Eugene de Savoye*, (Vienna, 1741); Richard Kane, *Campaigns of King William and Queen Anne; from 1689 to 1712*, (Dublin, 1748); and W.H. Dilworth, *The life and heroic actions of John Churchill, Duke of Marlborough . . . Published for the Improvement and entertainment of the British Youth of both Sexes*, (London, 1758).

[3] At least six works (largely laudatory) on Charles would be published in the 18th century. Most famously, Voltaire wrote in 1731 a biography of the "Lion of the North" that both recognized his greatness and at the same time bemoaned the losses of life his wars engendered.

[4] See his *Mex Rêveries: ouvrage posthume de Maurice Comte de Saxe, Duc de Courlande et de Sémigalle, Maréchal Général des Armées de Sa Majesté Trés-Chrétienne*, (Amsterdam, 1757) vol. 2, pp. 1–7 of fortifying cities, p. 40 on Vauban and Coehoorn, and p. 16 on how easily a battle could overturn the results of several sieges. He also spent many pages providing his own solution to the problem of the offensive imbalance in siegecraft by designing new field fortifications.

Commanders of various nationalities competed with one another to declare their particular ethnic group the peerless inheritors of the decisive mindset.[5] The most eminent warriors of the rest of the eighteenth century, like their compatriots of previous eras, would also demand a vigorous approach to the waging of war, with decisive field victories taking pride of place. Practitioners of field operations elaborated their opposition to positional warfare in the decades following the War of the Spanish Succession and sought out vigorous alternatives. The strategic offensive may have been preferable, but vigorous tactics were just as relevant to those forced onto the strategic defensive. Further to the east, the century's most famous commander, the Prussian King Frederick the Great, followed in his forceful forerunners' footsteps. Surrounded by enemies, his preference for "lively and short" wars was enlisted in the service of a defensive strategy of interior lines, defeating one enemy in battle before rushing off to fight another. He embodied as well the vigorous vice of denying his own engineers the resources needed to perform their functions effectively and then criticizing them for their resultant failings when he was confronted with fortified works. Sieges conducted by the Prussians in Frederick's wars consequently relied even more on brute force than those of the War of the Spanish Succession—most in fact depended on the non-siege tactics of bombardment, storm and blockade rather than competent trenchwork.[6]

These tendencies towards vigor were manifested in more than just the thoughts and actions of a few great captains. The various military branches, too, continued to separate over the following decades according to their experiences established during Louis XIV's reign. Military engineers would carry on their quest for the perfect siege and the perfect fortress, pursuing ever more elaborate and mathematical systems, all the while paying lip-service to the master Vauban. The French *Génie* would lead the way for Europe's military engi-

---

[5] While many Frenchmen (e.g. Villars and Maurice de Saxe) considered an attacking posture the natural condition of the French, a recent work argues that this was a particularly German trait. Robert Citino, *The German Way of War: From the Thirty Years War to the Third Reich*, (Manhattan, KS, 2005). The English could of course claim Oliver Cromwell and Marlborough for their case. For a more general discussion of 18th century perceptions of 'national characters,' see Duffy, *The Military Experience in the Age of Reason*, pp. 18–32.

[6] Duffy, *The Fortress in the Age of Vauban and Frederick the Great*, pp. 122–128; and 134–139.

neers professionally as well—French siege journals of the mid-eighteenth century highlighted this increased sense of *esprit de corps* by prominently listing all those brother engineers present at each siege, a detail unmentioned in accounts from an earlier, less-institutionalized era. These technicians, focused on slow evolutionary improvement, did manage to win occasional victories over the next century, most notably their defense against the radical 'perpendicular' fortifications of a cavalry officer named Marc-René, marquis de Montalembert. But it was a losing battle in the long run for these defenders of the conservative status quo. Even within France, where engineering had the greatest momentum and an increasingly noble composition, military engineers became further isolated from military operations and from the military itself over the course of the 18th century, as Janis Langins has recently concluded. By the French Revolution, military engineers had become masters of technical details, but their war of *industrie* could not compete with the grandiose visions of maneuver and battle.[7]

Just as siegecraft experts would continue to concentrate on their increasingly-arcane subject matter, so too did specialists of the artillery continue on the path they had already started down. The attention paid to siegecraft in Vallière's 1717 manuscript treatise would fade as the topic of field artillery rose in importance throughout the century. Engineer-Artillery conflicts would resurface as the gunners distanced themselves even further from their technical brethren—not only institutionally (as evidenced by the failed mid-century union), but functionally as well. While fortification engineers like Cormontaigne and Fourcroy de Ramecourt worked to institutionalize and extend Vauban's best practices, artillery engineers like Bernard Forest de Bélidor and Jean-Baptiste de Gribeauval extended the early 18th century reforms of Vallière, improving particularly the cannon and their carriages. The result by late century was an artillery arm more mobile and more effective than ever before, and this would make gunners and their guns particularly pertinent to military thinkers of the 18th century, who were themselves expressing a growing interest in field operations at the expense of siegecraft.[8]

---

[7] Langins, *Conserving the Enlightenment*, p. 118.

[8] Lynn has discussed how an earlier French attempt to make lighter cannon failed due to the predominant siege-centric view of warfare in "Forging the Western army in seventeenth-century France," in M. Knox (ed.), *The Dynamics of Military Revolution, 1300–2050*, (New York, 2001), pp. 40–44.

Those who survived Louis XIV's later wars reinvigorated the attack by calling in their treatises for more lively operations. A younger generation articulated how the human heart was the soul of war by, among other things, contrasting the intense emotional impact of shock tactics (*l'arme blanche*) on the enemy's psyche with the rational, even mechanical methods of siegecraft. The modern rhetoric of siege history has largely adopted this particular interpretation of vigor, drawing a strict dichotomy between battle and siege. In France, the most important early Military Enlightenment theorist, Jean-Charles chevalier de Folard enlisted the Ancients in his attempts to return decision to the battlefield. The ensuing debate that raged between proponents of column tactics like Folard and supporters of linear tactics that emphasized firepower shared one important belief, that war needed to be taken out of the hands of the engineers (or at least to disengage from the engineering cult of efficiency) by returning attention to the battlefield—Jean-Pierre Bois refers to this debate as an effort to resolve the period's exasperating *blocage tactique*.[9] The Military Enlightenment would also include its share of those enamored with the engineering ideal (e.g. Adam Heinrich Dietrich von Bülow), though we should be careful not to assume that all members of the Military Enlightenment shared the engineering desire for efficiency, nor should we presume that they likewise rejected the cult of vigor. One of the quintessential Enlightened military thinkers, the Welshman Henry Lloyd, served as an engineer, yet as young Du Mée's battle against his senior Des Rocques illustrated, it was not always easy to inculcate engineering values in those engaged as engineers, especially as the corps found itself increasingly besieged over the course of the century. Lloyd was no believer in applying Vaubanian efficiency to warfare more broadly. Instead, we find a military theorist who was deeply concerned with lost time, who believed in the importance of offensive operations, and who even criticized Vauban's influence on field operations.[10] The end of the War of the Spanish

[9] Bois has written a number of articles on this concept, e.g. "Approche historiographique de la tactique à l'epoque moderne," *Revue historique des armées*, (1997): 23–30. This tactical debate has been covered extensively over the decades; the classic account is Robert Quimby, *The Background of Napoleonic Warfare: The Theory of Military Tactics in Eighteenth Century France*, (New York, 1957). On Folard, see Jean Chagniot, *Le chevalier de Folard, la stratégie de l'incertitude*, (Paris-Monaco, 1997); and Bois, *Maurice de Saxe* (pp. 184–187 on Folard's views).

[10] Patrick Speelman, *Henry Lloyd and the Military Enlightenment of Eighteenth-Century*

Succession, as we have seen, did not initiate critiques of positional warfare. What *was* new by the mid- to late-eighteenth century was the explosion of theoretical publications promoting an offensive mindedness and a focus on field operations (witness as well the explosion of literature on partisan warfare, *petite guerre*). The complaints of siegecraft found in the letters of impatient Spanish Succession generals (and likely in earlier wars as well) would be echoed and reinforced in later treatises reflecting upon the Sun King's wars. These complaints would later be transformed from grumbling and theoretical rumination into constructive action by the time of the French Revolution.

The competition between efficiency and vigor extended to defending fortresses as well as attacking them. For many, a garrison's fate depended most importantly on the vigor with which they defended themselves, rather than the money spent on their works or the number of men and the supplies they were given.[11] As the power of the siege attack grew, many looked to lengthen their resistance by demanding a more vigorous defense. Both Louis and Vendôme believed that a royal order to repulse at least one assault on the breach before surrendering would in and of itself encourage stronger defenses. Almost ninety years later the revolutionary Committee of Public Safety would also demand adherence to a strict 'no surrender' policy, but it too was disappointed with the results. Sorties were also considered a good measure of a garrison's vigor, forget the harsh reality of trench parallels. Discussing how to defend a fortress, Folard warned his readers not to shrink from a project because it might be considered foolhardy or reckless (*téméraire*), for there were in fact extremely few things in war that were impossible—elsewhere he practically enthused over the massive sallies of the Ancients. Maurice de Saxe also argued for the psychology of vigor, arguing that in the defense of a place, "the more vigor you display, the more the enemy will be discouraged."[12] Not surprisingly, this particularly vigorous

Europe, (Westport, CT, 2002), pp. 43–44, 51–52, esp. 101–104, 109–110, and 124–125. Lloyd had attended the Spanish engineering academy and served as engineer on a number of campaigns.

[11] For example, the English-language literature on the 1706 campaign attributes the collapse of the Spanish Netherlands fortresses to the demoralizing shock felt by the various Bourbon garrisons after the defeat at Ramillies, and ignores the many other factors that were even more important.

[12] De Saxe, *Mes Rêveries: ouvrage posthume de Maurice Comte de Saxe*, (Amsterdam, 1757), vol. 2, p. 8.

interpretation of a garrison's proper behavior differed greatly from Vauban's own solution to the quandary of offensive domination. Instead of stoking the passions of the men, patience was required. He recommended his garrison commanders manage the *vigueur* of their troops by not risking them unnecessarily, and in 1706 he expressed his concern that the commander of Menin was making inefficient use of his troops by fatiguing them with constant sorties.[13] Struggling to rein in impulsive officers inside the fortress as well as in the trenches outside it, in the last years of his life Vauban contrasted these attempts at instilling vigor with the real solution, *science*: "Many governors believe that it is sufficient to have exposed their life in every possible situation in order to do their duty. They have mistakenly believed this based on the many examples of places which were captured due to negligence even though they had been defended with much valor and *éclat*. They would not need to fear the blame they rightly deserve for surrendering early if they only deigned to combine their valor with the science that they have neglected to learn."[14] Here too, the conflict between efficiency and vigor continued.

As the eighteenth century progressed, conditions started to catch up with the mindset of the more aggressive commanders, leading to a dramatic shift towards battle-centered field operations. Scholars have already discussed the improvements in roads, the dismantling of fortifications (particularly in the Austrian Netherlands from 1781), the increased mobility of artillery, the development of agile marching formations (e.g. the division), the development of permanent staff organizations, and the increased agricultural productivity that, all combined, changed the logistical and operational nature of warfare by the end of the 18th century. These increasingly-amenable conditions allowed the expression of such vigorous desires to be fully applied in the field. So too, increasingly powerful bureaucracies and nascent military-industrial complexes further strengthened the ability of fiscal-military states to raise, equip and supply armies. Many of these technical details of Napoleonic-style warfare had to be perfected in the decades after the Spanish Succession war, but the pre-

[13] Rochas d'Aiglun (ed.), *Vauban*, vol. 2, p. 505 in a letter to Blainville on how to defend Kaisersweert; vol. 2, p. 584 to Chamillart.

[14] Vauban, *Traité de la Défense des Places*, p. 145. Here too, Vauban would adopt familiar themes such as the importance of application and experience.

requisite of vigorous commanders constantly pushing their troops forward and being frustrated when they could not had long been present.

By the time of the French Revolutionary wars, the vigorous ideal finally had the chance to be implemented on its own terms. Motivated in part by wartime shortages and raw recruits, French Revolutionaries would embrace once again the use of cold steel weapons like bayonets and even pikes, while Revolutionary fervor (and Terror) overcame the conservatism of Old Regime operations. The firebrand revolutionary Georges Danton would call for *l'audace, encore l'audace, et toujours l'audace* against counter-revolutionary invaders, and the Revolution's most famous military engineer would be Lazare Carnot, known more for his mathematical and organizational skills than for his direct personal contributions to military victories. Far to the east Russia too had embraced the offensive-minded cult of the bayonet, following the dictum of its famous Generalissimo Aleksandr Vasil'evich Suvorov: "Money is dear; human life is still dearer; but time is dearest of all."[15] Napoleon would of course represent the pinnacle of this offensive attitude, seeking and forcing decisive field battles on his enemy at every opportunity. His advice to others: "The first consideration with a general who offers battle should be the glory and honour of his arms. The safety and preservation of his men is only second."[16] The philosophical elaboration of this reaction against the Vaubanian alternative was stated most graphically in Clausewitz's *On War*:

> We are not interested in generals who win victories without bloodshed. The fact that slaughter is a horrifying spectacle must make us take war more seriously and not provide an excuse for gradually blunting our swords in the name of humanity. Sooner or later someone will come along with a sharp sword and hack off our arms.[17]

---

[15] Bruce Menning, "Train Hard, Fight Easy: The Legacy of A.V. Suvorov and his 'Art of Victory'," *Air University Review*, 38 (1986): 79–88.

[16] Quoted in David Chandler (ed.), *The Military Maxims of Napoleon*, (New York, 1988), pp. 126–127. On Napoleon's preference for the "shortest and sharpest methods available" (and Frederick the Great's as well), see David Chandler, *The Campaigns of Napoleon: The Mind and Methods of History's Greatest Soldier*, (New York, 1966), pp. 141–142. As with Villars a hundred years earlier, Napoleon then goes on to argue that, over the long run, casualties will be fewer overall because a decisive field victory would end the war more quickly.

[17] Clausewitz, *On War*, p. 260, Book 1, chapter 11. See a Napoleonic era example of this applied to siegecraft in John Brooks Edwards, "'We must take it by storming yet!' A letter from the siege of St. Sebastian by Ensign Wright Edwards,"

As much as his message, Clausewitz's particularly violent image illustrates how inured cultists of vigor had become to the necessary prerequisite of rapid victory, a ready acceptance of bloodshed.

But the disagreement—dare we sent disgust?—with Vauban went much further than his humanitarian impulse, for it went back to the fundamental principles that the cult of vigor was based upon (or perhaps grew from?): war is about spirit, about heart, about action. This too, was hardly a new idea for those with a vigorous mindset: we could easily trace Villars' claim that hiding behind entrenchments (or fortress walls) was enfeebling and demoralizing back to Machiavelli, to knights of the medieval world who would abandon their castle to fight on the more honorable field of battle, and even to the Classical Greek hoplites. The engineer's siegecraft, in the rhetoric of siege history and in the experience of many Spanish Succession veterans, was not about vigor or drive; it was instead mechanical, materialistic, and practically inert. Folard argued in his later reflective work that although Vauban had achieved impressive results with the attack, it was still merely a "pure mechanism" that could be learned in only one or two sieges, and it was an activity in which *l'esprit* rarely mattered. "A vigorous defense," on the other hand, required "more talent, more art, more genius, more intelligence and more valor than the attack." Any attack on such a defense, therefore, would be useless.[18] We find Clausewitz taking much the same view, both of the nature of siegecraft and its relevance to real war. Even though he recognized that methods of war varied according to their political and technological contexts, in his discussions specifically focused on sieges he nonetheless viewed them as matters of purely material objects (e.g. trenchworks and batteries) rather than the "actual conduct of war" dictated by moral forces. Siege warfare was therefore "intellectually uncreative," its success a matter of "mechanically" flattening the defenders with artillery and therefore governed by geometry rather than moral effort. As a result, siege theory was not much more than a "refined mechanical art" that assumed (along

Journal of the Society for Army Historical Research, 80 (2002): 281–286. The traditional complaint of engineering incompetence is raised here again, though it is more plausible in this case as the engineers were English, a corps known for its lack of technical skills and experience.

[18] Folard, *Histoire de Polybe*, vol. 3, p. 59. Garrison sorties are of course the perfect example of this.

with Enlightened tactical theorists) that human beings were automata, "pieces of clockwork set off by a mere word of command."[19] Thus the rhetoric of siege history acquired philosophical justification, and the diametrical opposition between siegecraft and field battles became entrenched.

Future centuries would come to be dominated even more by the cult of vigor, an approach seemingly validated by Napoleon's battlefield victories and the philosophical justifications provided by Clausewitz's *On War*. Many of the most renowned commanders of the nineteenth and twentieth centuries, Robert E. Lee and George Patton foremost among them, would also gain their celebrity from their offensive-minded leadership—Patton frequently cited *l'audace, l'audace, toujours l'audace* (attributed to Frederick the Great) as his mantra. Less successful—than Patton at least—were the offensive-minded strategists and commanders who fought World War I: Ferdinand Foch's resurrection of the cult of the bayonet and the German General Staff's strategy of annihilation (*Vernichtungsstrategie*) come most immediately to mind. Both the justifications for, and the results of, wave after wave of assaults across no-man's-land were not dissimilar to the repeated covered way storms from an earlier age, though Vauban at least offered his century a practical alternative. Certainly, the immense human losses suffered by repeated over-the-top assaults on the Western Front would simply have been unsustainable without a corresponding acceptance of casualties as the necessary cost of long-term victory, even after the hopes of immediate success dimmed. As it turned out, modern militaries drawing upon nationalistic fervor and modern bureaucratic enforcement also found morale as difficult to sustain over time as the Allies did in the late stages of the War of the Spanish Succession. As befits a vigorous mentality, the eventual tactical solution to the strength of modern trench defenses was

[19] *On War*: on the material products of siegecraft and its limited intellectual effort see p. 133; the mechanical use of artillery, p. 393; for sieges governed by geometry, p. 214; for automata and clockwork, pp. 133–134. The conflation of siegecraft with Enlightened battlefield theorists is justified by the fact that eighteenth century writers sought to replicate Vauban's 'predictable' successes in the trenches on the battlefield. Gat, *The Origins of Military Thought*, p. 35. For more on Clausewitzian friction and siegecraft, see Jamel Ostwald, "Like Clockwork? Clausewitzian Friction and the Scientific Siege in the Age of Vauban," in S. Alton (ed.), *Instrumental in War: Science, Research, and Instruments between Knowledge and the World*, (Leiden, 2005), especially pp. 93–95.

found in the vigor of tank-led assaults—Churchillian outflanking operations against the weak underbelly of Gallipoli were as unsuccessful as early modern projects for English descents along the French coast. The cult of the offensive in the form of blitzkrieg and panzer maneuver would largely be vindicated, tactically and operationally at least, in the next 20th century war. With much less maneuver an earlier Churchill managed to lead a confederate army through most of Vauban's *pré carré* in a matter of a few years. The total warfare of the Second World War also reprised the indiscriminate bombardment of civilian populations as well as a more general Allied reliance on overwhelming firepower. As in the 18th century, however, each historical period would also see competing approaches supported by their own proponents, for example Hans Delbrück and his strategy of attrition (which denied the strict dichotomy between battle and maneuver) around the turn of the twentieth century. Depending on the context, such methods could even defeat a battle-centric strategy. By 1975, the United States had been thrown into confusion and even disbelief as an enemy that had never won a major battlefield victory still managed to defeat one of the world's two superpowers and absorb its South Vietnamese ally. And, even more recently, the debate about the suitability of a vigorous art of war would publicly resurface yet again over the current war in Iraq, a campaign in which Iraqi insurgents quickly recovered from the "shock and awe" of conventional military maneuvers that had been so successful in capturing the enemy's capital in early 2003. Two years into the ensuing low-intensity conflict, the U.S. Army found itself publicly criticized for a dangerous infatuation with what is now termed 'kinetic' warfare, a "pre-disposition to offensive operations, and a sense that duty required all issues to be confronted head-on."[20] While so much of warfare has changed in the past three hundred years, this aggressive approach and its reliance on fire-and-maneuver tactics would have been familiar to the cultists of vigor in Louis XIV's

[20] This criticism (that has received widespread American media coverage) came from Brigadier Nigel Aylwin-Foster, "Changing the Army for Counterinsurgency Operations" *Military Review*, Nov.-Dec. 2005, quote on p. 2. This British officer noted as well that non-Americans tend to consider the U.S. Army "too 'kinetic'. This is shorthand for saying U.S. Army personnel were too inclined to consider offensive operations and destruction of the insurgent as the key to a given situation" (p. 4).

age. Many of the political, social, economic and cultural conditions have changed radically since the 1700s, yet an offensive mentality remains one of the fundamental features of the dominant Western military tradition.

APPENDICES

APPENDIX A

ALLIED CAMPAIGN LENGTHS IN FLANDERS

Table A.1 Allied Campaign Lengths in Flanders

Year	Begin Campaign	To Winter Quarters	Campaign Length (days)
1702	May 15	November 3	172
1703	April 20	October 31	194
1704	April 23	November 23	214
1705	May 15	October 30	168
1706	May 16	November 8	176
1707	May 22	October 30	161
1708	May 26	January 4	223
1709	June 23	October 28	127
1710	April 20	November 15	209
1711	April 30	October 26	179
1712	May 26	October 22	149
Mean	–	–	179
Median	–	–	176

Source: Dates taken from George Murray (ed.), *The Letters and Dispatches of John Churchill, First Duke of Marlborough, from 1702–1712*. 5 vols. (London, 1845) and J.W. Wijn, *Het Staatsche Leger: Het Tijdperk van de Spaanse Successieoorlog, 1702–1715*, vol. 8, 3 parts (The Hague: Martinus Nijhoff, 1956–1964).

APPENDIX B

SIEGE DATASET METHODOLOGY

The methodology behind the siege data presented in Appendix C requires explanation. Many scholars have judged duration an especially salient characteristic of sieges, using it as a useful proxy variable for the balance between attack and defense. Geoffrey Parker's interest in the revolutionary nature of the *trace italienne* led him to search for a convenient way to measure this balance over time; the time needed to capture a place is one of the most self-evident, readily-available and easily-quantified of all siege variables. Accordingly, lengthy attacks required more men as desertion and sickness whittled away at blockading forces, while assaulting troops were slaughtered in the many attempts to storm a breach. The longer the siege lasted, the more likely the chance of relief, which necessitated in turn extensive lines of circumvallation and the troops to man them. Clifford Rogers similarly used siege length as a measure of the offensive-defensive balance in the 15th century—informing us of the dramatic decline in durations following the introduction of effective gunpowder artillery.[1] A number of other scholars have also emphasized siege lengths in their own works, further reinforcing our conviction that this measure reflects the fundamental essence of siege warfare.[2]

Although historians recognize the importance of siege lengths, few have offered the data behind their conclusions for public scrutiny.[3] Looking at the period's prototypical war of positions, the Dutch Revolt, we find that two important military historians of the war, Geoffrey Parker and Jonathan Israel, fail to provide the underlying data they used to develop their conclusions. Instead, we find generalizations either based on a handful of cited examples or on an

[1] See his "The Military Revolutions of the Hundred Years' War," pp. 64–67.

[2] For a general discussion of long sieges, see M.S. Anderson, *War and Society in Europe of the Old Regime 1618–1789*, (New York, 1988), pp. 40–42.

[3] Lynn's "The *trace italienne*" is an important exception to this that will be discussed below. Parker criticized Lynn's data from this article on army sizes, but did not challenge the accuracy of length data, conceding that sources usually agreed on siege lengths. Parker, "In Defense," p. 351.

unspecified number of cases spread over hundreds of pages of text.[4] Parker provided readers with a general sense when he argued that "Normally, the capture of a stronghold defended by the *trace italienne* required months, if not years. . . . Numerous examples of each technique fill the annals of early modern warfare; but the one thing they all had in common was longevity."[5] Confusingly, Jonathan Israel concludes quite the opposite. Nothing epitomizes the unsettled state of research, and the need for explicit quantification, better than their views on the attack of Breda in 1624–1625: Parker argues that its nine-month defense was "relatively short by the standards of the Low Countries' War," while Israel stresses the "exceptional duration" of the very same defense![6] Part of a larger argument over the nature of the war's siegecraft, these two scholars were clearly not considering the same data, or the same time frame, or using the same definitions. Claims about how many fortresses were attacked over the course of an eighty-year contest, what methods were used to attack them, how long they defended themselves—all are either illustrated by a few anecdotes or remain unsupported altogether.[7] The historiography's failure to present all the evidence and state their assumptions makes it impossible for an impartial reader to decide without redoing much of their research.

Turning to the late seventeenth century and the Age of Vauban, here too siege data is surprisingly scarce. Even recent biographies of Vauban fail to discuss how exactly he changed the siege attack, much less provide systematic data on the sieges themselves. Sketchily-drawn stereotypes of pre-Vaubanian chaotic trenches abound, while Vauban's own judgment on prior French siegecraft (presented to the reader in a sentence or two) is taken to encapsulate the entirety of Renaissance

---

[4] In Parker's *Military Revolution* synthesis, one page out of 154 is dedicated to anecdotal evidence supporting his proposition that 16th century sieges were measured in months rather than weeks, and another two paragraphs discuss the sizes of the armies needed to capture these places (pp. 13–14).

[5] *The Military Revolution*, p. 13.

[6] *The Military Revolution*, p. 13; Israel, *The Dutch Republic*, p. 107.

[7] Israel limits himself to the second half of the Dutch Revolt; Parker focused more closely on the earlier phases. Parker mentions 208 towns having permanent garrisons in the Spanish Netherlands, thus the potential for a *very* large number of sieges. In the most recent edition of his *Army of Flanders* (p. 8), Parker gives us a sense of the large number of cases involved by displaying a table of 95 towns captured by the Spanish between 1578 and 1590; sixty of these were captured through military means.

siegecraft.[8] The details of the period's siegecraft have remained hidden, in small part, because access to Vauban's personal papers has been extremely limited, although wider access has been granted in the past decade. More important, though, is the almost exclusive reliance on biographical studies which guarantees a continuing narrow view of Vauban. Sieges conducted by other French engineers do not fit easily into a chronological portrayal of his life, while non-French sieges (as well as those before and after his tenure) are almost completely outside such a framework. In contrast with the more cosmopolitan discussions of the Military Revolution, Vauban studies tend to be parochial, rarely venturing beyond mid- to late-seventeenth century France and they rarely explore how others viewed him. The biographical format of these studies further limits our vision of tactical change by demanding a holistic accounting of the myriad aspects of their amazingly prolific, multi-faceted subject—Vauban the architect, the strategist, the Court client, the religious pluralist, the proto-Enlightened reformer. All these leave little space for a discussion of his tactics and their context in the wider field of military history. Vauban's contribution to the siege attack appears almost self-evident then, taking the form of a ritual retelling of his tactical innovations as they unfolded across the decades, buttressed by the illustrative siege accounts of his various innovations.

A final body of literature with potential relevance focuses on the great commander Marlborough, but it too is somewhat contradictory. An early student of Marlborough explained that the reduction of fortresses "was always slow and usually a costly process."[9] In 1976 the premier authority on the Duke, David Chandler, wrote that "so effective were [Vauban's] methods that the average time taken to bring sieges to satisfactory conclusions was *reduced* considerably in his day, although some were very lengthy."[10] All the same, earlier he had chosen to emphasize their length, arguing that many lasted 40 to 60 days, and still required "lengthy and elaborate procedures."[11]

---

[8] The best attempts to place Vauban in a larger context come from Lazard's and Blomfield's biographies of Vauban, and Child's brief comments in *The Nine Years War*, p. 93.

[9] C.T. Atkinson, "Marlborough's Sieges," *Journal of the Society for Army Historical Research*, 13 (1935), p. 196.

[10] Chandler, *The Art of Warfare in the Age of Marlborough*, p. 276, emphasis in the original.

[11] Chandler, *Marlborough as Military Commander*, p. 81.

This earlier conclusion is, in fact, the prevalent view among English scholars: that sieges tended to be quite long in spite of Vauban's efforts. Thus John Childs described the late 17th century "siege in form" of "considerable duration."[12]

Despite the general lack of interest in quantifying the period's sieges, quantitative information on sieges has been compiled by two modern scholars: David Chandler's list of more than fifty sieges between 1688 and 1745, and John Lynn's list of 135 French sieges conducted from the 15th through 18th centuries.[13] A closer look at the existing datasets will illustrate the issues more clearly. Unlike Lynn's far-ranging, Gallo-centric dataset, Chandler is interested more narrowly in the sieges of Louis XIV's reign, especially those conducted during the Duke of Marlborough's tenure. Both of these datasets (Lynn's especially) have been used by a number of other historians, and they therefore deserve a closer look. Testing these datasets by independently collecting data on sieges in the War of the Spanish Succession, we uncover a serious measurement error in both sets of data.[14] As Figure B.1 illustrates, there are substantial omissions which amount to a hidden bias in these existing datasets: of the 115 distinct sieges in the War of the Spanish Succession I have identified thus far, Lynn lists 58 of them (half of the actual number), while Chandler includes only 24 (a fifth of the total).[15]

---

[12] Childs, *Warfare in the Seventeenth Century*, p. 148. Though not a part of the Marlborough literature, a very broad survey of siegecraft throughout history also suggests a stagnation of the attack c. 1700. Bruce Allen Watson, *Sieges: A Comparative Study*, (Westport, CT, 1993), p. 4.

[13] These works are based largely on 19th century statistical studies. The most important is Bodart, *Militär-historisches kriegs-lexicon (1618–1905)*, (Vienna, 1907–1908), whose findings have been largely ignored by most modern historians. Another 19th century example, L.-H.-C. Vauvilliers's *Recherches historiques sur le rôle et l'influence de la Fortification* (Paris, 1845), argues that fortifications were essentially a waste of time, although his conclusions are clearly focused on the contemporary debate over whether or not to fortify Paris. A very brief but even more ambitious example of the 19th century macro-statistical approach is Sicard, "Recherches historiques sur les guerres, sièges et traités de paix depuis l'origine du nom français," *Le Spectateur militaire* 7 (1829), pp. 619–649.

[14] An exception to this is the largely-accurate table of Spanish Succession Low Countries sieges in Olaf van Nimwegen, *De subsistentie van het leger*, pp. 100–101. However, as van Nimwegen's focus was on logistics and not siegecraft, he too failed to include a few of the smaller sieges in the Low Countries; nor did he analyze this data in any systematic fashion. The non-Flanders theaters were also beyond the scope of his book.

[15] These 115 events are all strictly-defined sieges, as distinguished from blockades,

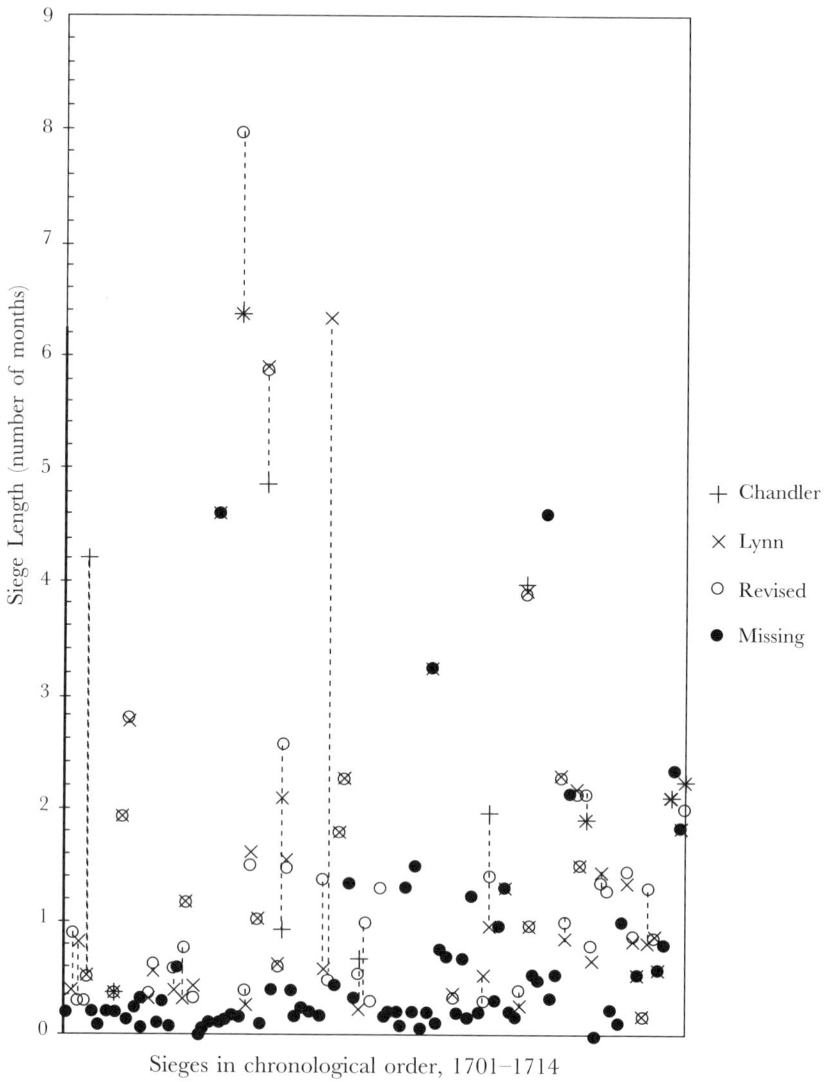

Figure B.1 A Comparison of Low Countries Siege Lengths, by Source

A large number of sieges have been missed in previous surveys, threatening to completely undermine the usefulness of existing datasets.

If their data were randomly drawn from the entire population of early modern sieges, their conclusions would remain valid. However, almost all of the missing Spanish Succession sieges are of short duration. As the histogram in Figure B.2 illustrates, both authors missed a particularly large number of sieges lasting six weeks or less. Not only were there two- to four-times as many sieges conducted as previously claimed, but the median length of this comprehensive list of sieges is half of the previous, incomplete lists. In fact, half of the entire war's sieges lasted two weeks or less.[16] The shorter sieges in Iberia and Italy, seen as peripheral theaters by many English and French scholars (and thus underrepresented in the literature), are particularly absent: Lynn's database lists only eleven of forty-four sieges in Iberia and nine of the twenty-five conducted in Italy, while all twelve sieges in France are accounted for, as are thirteen of the twenty sieges in the Spanish Netherlands and thirteen of the nineteen sieges in Germany. The result? The extreme examples of the longest sieges have been privileged by historical scholars, while the shorter sieges are relegated to obscurity.[17] Such a skewed sample is extremely misleading in understanding siege warfare as a whole, particularly given the centrality of siege lengths. When we limit our sights to the Low Countries theater, the general pattern of brief Spanish Succession sieges continues to hold true—only a quarter held out for even a month, while very few met, much less exceeded Chandler's 40–60 day range.[18]

---

bombardments, surprisals and storms. Chandler, Lynn and van Nimwegen used place-level data, i.e. if both a town and its citadel were besieged consecutively, they measured the start date from when the trenches were opened against the town, and the end of the siege when the citadel surrendered. I have aggregated my data up to the place-level to make valid comparisons.

[16] The median (.5 months as compared to Lynn's uncalculated .9 and Chandler's 1.3) is the best measure of central tendency for this data, since the distribution of the sieges is highly skewed toward the longer sieges. The mean is one month, compared to Lynn's 1.4 and Chandler's 1.8. Luh repeats Chandler's data, claiming an average siege length of 47 days. *Ancien Regime Warfare and the Military Revolution*, p. 108.

[17] Lynn has more recently stated that "throughout the wars of Louis XIV, most towns and fortresses fell in much less time than one would expect, and only a minority of cases required the full process of a lengthy formal siege." *Wars of Louis XIV*, p. 114.

[18] Note that these lengths are cumulative, e.g. the siege of Lille's town required

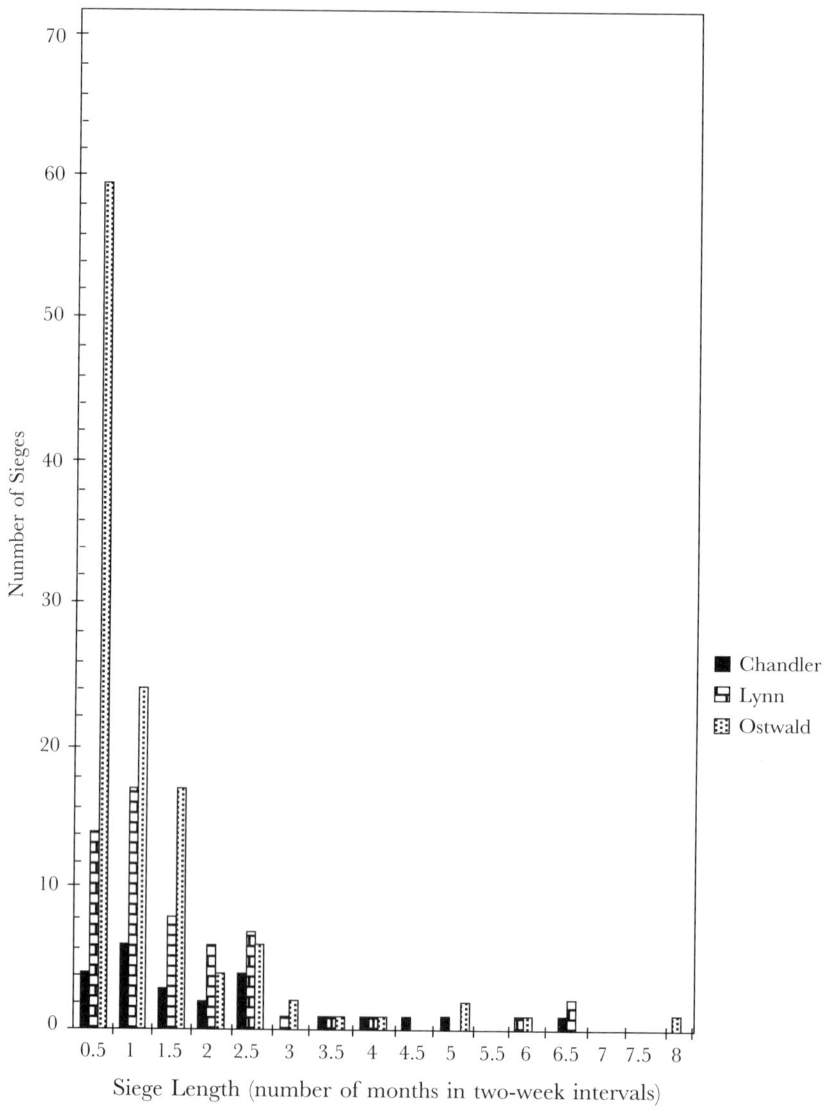

Figure B.2 A Frequency Comparison of Siege Lengths, by Source

There are several possible explanations for this weakness in the historiography. A suspicious reader might infer that a few historians have focused attention on the longer sieges in order to emphasize the plodding, indecisive nature of many early modern campaigns. Supporters of a Marlburian decisive-battle strategy constantly criticize the reliance on sieges and contrast it with Marlborough's "will to battle."[19] More generally, historians emphasize long delays as an inherent element of siege warfare—sieges become *by definition* an incredibly time-consuming practice, thus emphasizing the right extreme of the distribution, i.e. the longest sieges.[20] The rhetoric used to describe sieges often emphasizes their extreme length: "a single siege could sometimes consume an entire campaigning season."[21] Such generalizations only beg the question of whether such long sieges were common or rare—how often is "sometimes?" Chandler's efforts to emphasize the extreme lengths of sieges are particularly zealous: "Sieges were as a rule extremely expensive in terms of both material and time. Their length varied enormously. . . . Many, however, fit into a 40–60 day bracket."[22] Yet when we analyze his own data, only one-quarter of his listed sieges lasted 43 days or longer.[23] Such a relatively small percentage does not seem to justify characterizing sieges *in general* as extremely long. Figure B.3 displays the corrected

two months to capture and its citadel (in many ways an independent operation that required separate attacks) another two months to capture, yet Lille is represented here as a four-month siege rather than two two-month sieges.

[19] See Ostwald, "The 'Decisive' Battle of Ramillies," pp. 649–656 for a discussion of this literature.

[20] On the importance of considering the entire distribution of a phenomenon, see Stephen Jay Gould's accessible *Full House: The Spread of Excellence from Plato to Darwin*, (New York, 1996).

[21] Archer Jones, *The Art of War in the Western World* (New York, 1987), p. 278. John Childs repeats this (they "often consumed an entire campaigning season") in "The Military Revolution I: The Transition to Modern Warfare," in Charles Townshend (ed.), *The Oxford Illustrated History of Modern War* (Oxford, 1997). For that matter, a battle could end a war, but this happened about as often as a single siege consumed an entire campaign season.

[22] *Marlborough as Military Commander*, p. 86. Elsewhere he wrote that "a great many sieges" were 40–60 days in length (*The Art of Warfare in the Age of Marlborough*, p. 245).

[23] Seventy-one percent of his sieges did not even reach forty days. *Marlborough as Military Commander*, Appendix B: "Marlborough's Major Sieges." If we remove the incorrect Lüttich entry (see below), the mean decreases 12%, to only twenty-eight days.

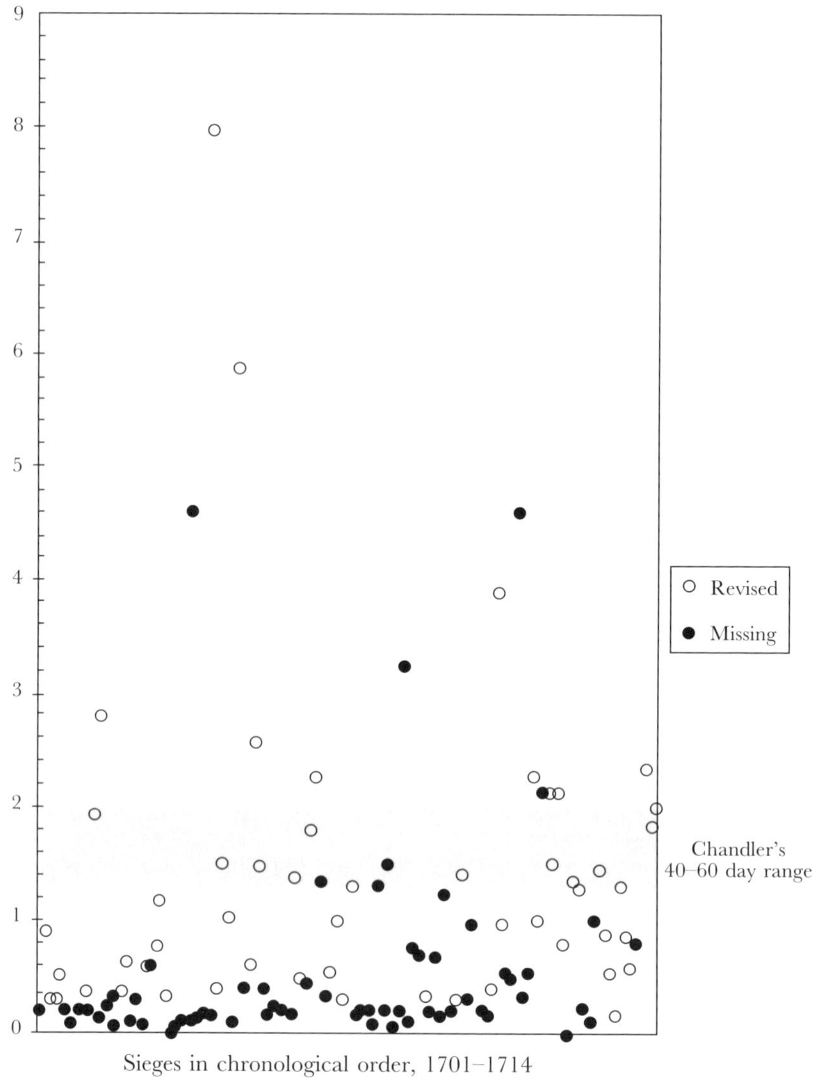

Figure B.3 Revised Spanish Succession Siege Lengths

dataset, allowing us to see that few of this war's sieges met or exceeded Chandler's forty to sixty day range, while a far greater number capitulated before a month had passed.

The danger of focusing on one extreme is found in the quantitative works mentioned above, but they are just as common in histories that do not explicitly rely on numerical evidence. Even the least battle-centric of historians misleadingly emphasize the longevity of sieges by confining their textual generalizations *a priori* to the larger sieges. Returning to the Dutch Revolt, even if Parker and Israel had provided their data, it is likely that the issue still would not have been resolved, for both proffered caveats that raise the question of how representative their samples are. Both confine their generalizations to "main" or "great" actions or sieges, implying that there were minor ones as well, but without informing the reader as to how these were handled.[24] Without a presentation of the underlying data and an estimate of its representativeness, curious readers must return anew to often-illegible papers tucked away in distant archives in order to uncover their full variety.[25] Lynn also discusses "major" sieges, but does not inform us of his criteria or even why this distinction is necessary.[26] Chandler also specifies that his appendix only includes Marlborough's "major" sieges, yet his data includes four sieges less than ten days long—are there even shorter sieges that were not included?[27] Even in his epic survey of early modern siegecraft

[24] Israel's sample probably included far more short actions, as he argues for shorter sieges. Or perhaps he used a more strict definition of 'siege' to exclude blockades, thus eliminating from consideration some long blockades that Parker included. Without the underlying data, it is difficult to tell—another example of the need to disclose the methodology and data.

[25] Common qualifications to historians' accounts of major sieges, such as 'many sieges were not, of course, as large as Lille 1708,' suggest only that the author has not deemed siege warfare worthy of much study. What proportion of sieges were like Lille? In what way were they similar to Lille? What impact did these similarities and differences have on the period's warfare? These questions go to the very heart of early modern warfare, but we remain ignorant of the answers to even these basic questions.

[26] For example, in *Giant*, pp. 575–576, Lynn goes back and forth in successive paragraphs. First he stresses the length of "major sieges" (averaging four to eight weeks) that "could consume an entire campaign season." Then he discusses the fact that "many fortresses and towns fell quickly." Then he concludes by returning to the "costly, time-consuming and generally indecisive character of siege warfare." Chandler does similarly in *The Art of Warfare in the Age of Marlborough*, pp. 244–246. Without strict criteria, it is impossible to judge the validity of such generalizations.

[27] Chandler, *Marlborough as Military Commander*, Appendix B.

(which does not include a dataset), Duffy also focuses on large sieges as case studies rather than exploring the full variety of sieges. When ascertaining the duration of sieges, one cannot simply ignore the shorter sieges without justification, particularly if one is claiming to summarize the essence of siegecraft. Not only should each dataset be presented to the reader in full, but scholarship should also include an estimate of the relationship between the sample and the larger population. How many of these minor sieges were there? How long did they last? Were they besieged? Blockaded? Stormed? Why are they so different from longer sieges that they deserve separate treatment? These answers are critically important, particularly if we make assumptions about sieges based off of their length—do short sieges always involve small armies and few casualties?[28] Thus the definition of "major" siege threatens to undermine best efforts. Measuring change across time is impossible without a clear idea of the data's relation to reality and a clear idea of how exactly the data might be biased. Cumulative and comparative scholarship seeking long-term patterns cannot advance very far in an environment that fails to address such issues.

Other explanations of this bias against short sieges are more benign and more likely to be true, yet they are just as damaging. The publication of Chandler's and Lynn's underlying data facilitates further scholarship by making mistakes easy to find and correct.[29] Since their data (and sources in Lynn's case) are available for public scrutiny, we can pinpoint the flaws: over-reliance on a single mathematical measurement, a few traceable typographic errors, inconsistent start and end dates, as well as a much more damaging weakness in the sources being relied upon. The reliance on the statistical mean as *the* measure of central tendency combined with a small sample size can lead to minor mistakes significantly impacting the results. As the

---

[28] In the case of Lynn's "trace italienne" conclusions, the impact of missing cases revolves around the definition of "major" siege. If a dataset includes all of the lengthy sieges, it does not *necessarily* follow that it also includes all of the sieges that included large besieging armies, as besiegers could conceivably increase their army sizes to decrease the siege's duration—in such a case Lynn's dataset would not accurately reflect trends in besieging army sizes over time. Only a closer examination of the data and an explanation for the cause of bias can clarify the issue.

[29] Lynn deserves our thanks for pushing the debate onto more solid ground, and his contributions (both this siege database and his more detailed work on French army sizes) have received widespread praise, including Parker's commendation of Lynn's article on army growth. "In Defense," 362 footnote 59.

dashed vertical bars in Figure B.1 indicate, only ten percent of Chandler's and Lynn's siege lengths differ significantly from mine. Most striking of all these errors is the length Chandler assigned to the siege of Lüttich (the German name for Liège) in 1702, an enormous 127 days instead of the actual nine days.[30] This mistake significantly skews the average siege length upward in a small sample already biased towards the larger sieges.[31] In fact, someone as knowledgeable about the Spanish Succession as Chandler is should have been struck by the fact that the strongest of Louis' Flanders fortresses (Lille) appears to have taken a week less time to capture than the feeble fortifications of Liège, and that both Liège and Roermond were apparently besieged at the same time, a very rare occurrence. With little analysis of the data, however, the resulting error is not surprising. Worst of all, and sadly indicative of military historians' occasionally cavalier attitude towards numbers, other scholars have perpetuated this error in their own work.[32] Lynn catches the particularly egregious error of Liège, although a few other dates are marred by typographical errors.[33]

Errors also come from the inconsistencies of measuring the beginning and end of sieges. Durations are obviously calculated from a start and an end date, but there were several possible dates for each of these stages, which can cause confusion and incompatibility. Most simply, a siege could begin when a besieging army invested the town or when it opened the trenches; the siege could end either when the capitulation was signed, when the garrison evacuated the fortress, or when the main besieging force left the town. All of these dates were

---

[30] *The Art of Warfare in the Age of Marlborough*, p. 308 (both in the original edition and in the 1994 reprint by Sarpedon). This error is also found in *Marlborough as Military Commander*, Appendix B. The 1995 reprint of *Marlborough as Military Commander* reproduced this error, but the 2000 Penguin edition attempted to remedy it: a corrected entry for Liège 1702 was added, but the erroneous Lüttich entry was retained!

[31] Jürgen Luh's perpetuation of this error adds an entire week to his average siege length, a 15% increase. This large a deviation comes from a dataset that includes some of the largest sieges in the Spanish Succession and almost none of the shortest. The median of his original flawed data is only 34 days compared to a mean of 47. *Ancien Regime Warfare and the Military Revolution*, p. 108.

[32] Frank Tallett (*War and Society in Early Modern Europe*, p. 51) even cites Chandler's inflated Liège length to showcase how long sieges could last.

[33] For example, he adds an extra 18 days to the siege of Stevensweert (capitulating 2 October instead of 20 October), and at the siege of Badajoz, he probably transposed the start date as 10 April (4/10) instead of 4 October (10/4). "The *trace italienne*," 327.

identified as discrete events by contemporaries, although the besieger's arrival, the investment, opening the trenches and the capitulation stages are most frequently mentioned. Each of these stages is a valid measure depending on the purpose, but when comparing siege lengths we must consistently use the same criteria. The 1667 siege of Lille in the War of Devolution illustrates the problem. Lynn's 1991 dataset identifies the siege lasting from 8 July to 28 August, while his recent work on Louis' wars corrects the error: the town was actually invested 28 August, the trenches were opened 18 September and it finally capitulated on 27 September.[34] The siege in fact lasted only nine days of open trenches (18–27 September), but held out for a full month, three times as long, when measured from investment to capitulation. And Lille is not exceptional: in Lynn's 1991 dataset, the War of the Spanish Succession Low Countries sieges include seven where the start date is measured from the opening of the trenches, whereas the rest of the sieges are measured from the date of investment. When we start to compare the dates given by other scholars, we soon discover a multitude of (often-contradictory) dates being offered. With numerous errors, we must return to the original sources for much of our data in order to assure ourselves of the reliability of the data collection process, or at least to appreciate the nature of any resulting bias.

The larger and more fundamental explanation for the under-representation of short sieges derives from the way they are reflected in the historical record. The major sieges were of immense interest to contemporaries, and thus received widespread comment. As is painfully obvious to anyone who attempts to construct a systematic siege dataset, witnesses were much less interested in the smaller, 'inconsequential' attacks that did not require either side's full attention or resources for very long.[35] Far fewer sources provide far less information on them as a result. Noteworthy also is the fact that

---

[34] Lynn, *Giant of the Grande Siècle*, p. 575; *The Wars of Louis XIV*, p. 108. Duffy corroborates this correction (*The Fortress in the Age of Vauban*, p. 7), although he too condemns to obscurity the short sieges of 1667. In general, information on sieges given in the context of a specific campaign are more likely to be correct than data collected in a decontextualized dataset.

[35] Whether one or both of these are true will help determine the bias in a dataset that under-represents short sieges, i.e. did sources miss these cases because they involved few resources or because they were short? Once again, an empirical question whose answer might even vary by period or place.

both secondary and published primary sources for the Spanish Succession are biased heavily towards the Flanders and German theaters, theaters which had a smaller proportion of short sieges. The cumulative effect of all these flaws—primary sources skewed towards the larger sieges, both primary and secondary sources interested only in those theaters in which their national army saw service, typographical errors, confusing multiple measures for start and end dates—is data of uncertain accuracy. At a minimum, we need to test the reliability of existing datasets, but when there is a question of sample bias, we have no choice but to return to the original sources, preferably placing this data in the context of the campaigns in order to ferret out mistakes. Only with a detailed examination of the campaigns through primary sources can we accurately measure trends in siege warfare.[36]

Responding to Mahinder Kingra's effort to document the small sizes of garrisons during the Dutch Revolt, Parker replied that: "This may be true, but his figures do not prove it."[37] This comment applies, unfortunately, to the entire discussion of siegecraft in early modern Europe. No one has presented the solid figures necessary to outline adequately the trends of siegecraft over time—neither the participants in the *trace italienne* debate, neither Vauban's or Marlborough's biographers, nor the secondary campaign narratives of Louis XIV's wars. Without this essential information, scholars have been forced to choose whatever examples were foremost in their minds when hypothesizing about trends in early modern siegecraft. Readers are inevitably unsure of whether the conclusions are really supported by *all* the underlying data. With a clearer view of the pitfalls of such datasets, we can now turn to the next Appendix for the raw data on the sieges in the War of the Spanish Succession.

---

[36] Parker criticizes Lynn's use of encyclopedic sources; only a systematic, empirical comparison of these disputed sources with the actual values can resolve this issue.

[37] "In Defense," 364 note 72.

APPENDIX C

SIEGE LENGTHS DATASET

Having discussed the methodological issues dealing with the collection of siege data (Appendix B), we can now turn to the data itself. This database is incomplete, but it does provide enough information to examine siegecraft during the War of the Spanish Succession. There is an inherent margin of error for each of these dates, as different sources might vary in their dating of night-time events (is 2 AM the night of the 23rd or the 24th?). Thus, there is a small margin of error for these lengths, up to two days per siege. Nevertheless, the representativeness of this dataset far exceeds any existing collection. The data for the Low Countries includes almost every siege conducted in the war (with the exception perhaps of an isolated fort or two that quickly surrendered), but the Iberian and German theaters in particular may very well have a dozen or so very short sieges that have not been included here. Many of these sieges lasted only a day or two, and the secondary literature does not always give enough specificity to determine whether they were strict sieges or storms or whether they surrendered upon being summoned. They have therefore been excluded from this dataset. If a large number of short sieges are indeed missing, they would further bolster the argument that the 'average' siege was more likely to be a brief affair than the epic struggles so prevalent in the literature.

Key

NOTE: All dates are New Style, displayed in Month/Day format.
I: Date of Investment
OT: Date of open trenches
C: Date of Capitulation
I-C: Length (in days) measured from Investment to Capitulation
OT-C: Length (in days) measured from Open Trenches to Capitulation
#mo: 'Standardized' length of siege. Calculated as (I-C)÷30, or (OT-C)÷30 if I-C data is unavailable. Rounded to the tenth of a month,

they are therefore grouped together in 3-day intervals. In a few non-Flanders sieges, secondary sources provide little information except for the overall length of the siege.
Thtr: Theater (SN = Spanish Netherlands, Fr = France, Sp = Spain, It = Italy, Ge = Germany)
sideB: Side of besiegers (Fr = French and allies, All = Allies)

Table C.1 Siege Lengths, War of the Spanish Succession

Place	Year	I	OT	C	I-C	OT-C	#mo	Thtr	sideB
Kaisersweert	1702	4/18		6/15	58		1.9	Ge	All
Saint-Donas	1702	5/11		5/17	6		0.2	SN	All
Castiglione	1702	5/27	5/28	6/1	5	4	0.2	It	Fr
Landau	1702	6/18		9/10	84		2.8	Ge	All
Borgoforte	1702	8/22	8/24	8/28	6	4	0.2	It	Fr
Venlo	1702	8/28	9/10	9/23	26	13	0.9	SN	All
Guastalla	1702	8/29	8/31	9/9	11	9	0.4	It	Fr
Stevensweert	1702	9/23	9/24	10/2	9	8	0.3	SN	All
Roermond	1702	9/26	10/2	10/6	10	4	0.3	SN	All
Liège	1702	10/13	10/20	10/29	16	9	0.5	SN	All
Rheinberg	1702	10/21	10/22	10/30	9	8	0.3	SN	All
Hulst	1702	10/27		11/2	6		0.2	SN	Fr
Trarbach	1702	10/30	10/30	11/6	7	7	0.2	Ge	Fr
Andernach	1702	11/15		11/17	2		0.1	Ge	All
Governolo	1702	12/17	12/17	12/22	5	5	0.2	It	Fr
Neubourg	1703	1/31		2/2	2		0.1	Ge	Fr
Kehl	1703	2/20	2/25	3/9	18	13	0.6	Ge	Fr
Bonn	1703	4/27	5/3	5/15	18	12	0.6	Ge	All
Nago	1703	7/31		8/4	4		0.1	It	Fr
Arco	1703	8/8	8/10	8/17	9	7	0.3	It	Fr
Breisach	1703	8/15	8/23	9/7	23	15	0.8	Ge	Fr
Huy	1703	8/15	8/17	8/26	11	9	0.4	SN	All
Limburg	1703	9/8	9/23	9/27	19	4	0.6	SN	All
Landau	1703	10/13	10/17	11/17	35	31	1.2	Ge	Fr
Augsburg	1703	12/3	12/7	12/13	10	6	0.3	Ge	Fr
Castello de Vide	1704				4		0.1	Sp	Fr
Barcelona	1704	5/30		6/1	2		0.1	Sp	All
Susa	1704	5/31		6/12	12		0.4	It	Fr
Portalegre	1704	6/2		6/8	6		0.2	Sp	Fr
Vercelli	1704	6/5	6/14	7/20	45	36	1.5	It	Fr
Rain	1704			7/16	12		0.4	Ge	All
Villingen	1704	7/16		7/22	6		0.2	Ge	Fr
Isabella	1704	8/1		8/3	2		0.1	SN	All
Gibraltar	1704	8/1		8/4	3		0.1	Sp	All
Ulm	1704	8/23		9/10	18		0.6	Ge	All
Gibraltar	1704	8/24	10/21	4/30	249	191	8.3	Sp	Fr
Ivree	1704	8/30	9/2	9/30	31	28	1.0	It	Fr
Landau	1704	9/9	9/13	11/25	77	73	2.6	Ge	All
Verrua	1704	10/14	11/7	4/8	176	152	5.9	It	Fr

Table C.1 *(cont.)*

Place	Year	I	OT	C	I-C	OT-C	#mo	Thtr	sideB
Trarbach	1704	11/4		12/18	44		1.5	Ge	All
Valencia de Alcantara	1705	5/2	5/3	5/8	6	5	0.2	Sp	All
Albuquerque	1705	5/16		5/21	5		0.2	Sp	All
Huy	1705		5/30	6/11		12	0.4	SN	Fr
Liège	1705						1.0	SN	Fr
Huy	1705	7/6		7/11	5		0.2	SN	All
Chivasso	1705			7/29	40		1.3	It	Fr
Mirandola	1705	7/16		9/8	54		1.8	It	Fr
Barcelona	1705	8/23		10/6	44		1.5	Sp	All
Zoutleeuw	1705	8/29	8/31	9/5	7	5	0.2	SN	All
Hagenau	1705	9/27	9/28	10/5	8	7	0.3	Ge	All
Badajoz	1705	10/3	10/4	10/17	14	13	0.5	Sp	All
Zandvliet	1705	10/23	10/26	10/29	6	3	0.2	SN	All
Diest	1705	10/25		10/25	1		0.1	SN	Fr
Nice	1705	10/31	11/17	1/7	68	51	2.3	It	Fr
San Mateo	1705	12/27		1/9	13		0.4	Sp	Fr
Barcelona	1706	4/3		5/12	39		1.3	Sp	Fr
Alcantara	1706	4/9		4/14	5		0.2	Sp	All
Hagenau	1706	5/1		5/11	10		0.3	Ge	Fr
Ciudad Rodrigo	1706				7		0.2	Sp	All
Turin	1706	6/2	6/3	9/7	97	96	3.2	It	Fr
Oostende	1706	6/20	6/28	7/6	16	8	0.5	SN	All
Menin	1706	7/23	8/4	8/22	30	18	1.0	SN	All
Alicante	1706			9/4	15		0.5	Sp	All
Dendermonde	1706	8/27	8/29	9/5	9	7	0.3	SN	All
Ath	1706	9/16	9/20	10/2	16	12	0.5	SN	All
Pavia	1706	9/29		10/2	3		0.1	It	All
Cuenca	1706	10/3		10/9	6		0.2	Sp	Fr
Pizzigetone	1706	10/4	10/17	10/26	22		0.7	It	All
Elche	1706				3		0.1	Sp	Fr
Cartagena	1706	11/11		11/17	6		0.2	Sp	Fr
Casale	1706	11/15	11/24	12/6	21	12	0.7	It	All
Milan	1707	2/12		3/12	29		1.0	It	All
Villena	1707	4/18			6		0.2	Sp	All
Toulon	1707	7/14		8/22	39		1.3	It	All
Ciudad Rodrigo	1707	9/18		10/4	16		0.5	Sp	Fr
Susa	1707			10/30		6	0.2	It	All
Lérida	1707		10/2	11/10		39	1.3	Sp	Fr
Morella	1707		12/2	12/11		9	0.3	Sp	Fr
Tortosa	1708	6/4	6/21	7/11	37	20	1.2	Sp	Fr
Exilles	1708	8/2		8/12	10		0.3	Ge	All
Lille	1708	8/13	8/22	12/8	117	108	3.9	Fr	All
Fenestrelles	1708	8/15	8/17	8/31	16	14	0.5	Ge	All
San Felipe	1708	9/16		9/29	13		0.4	Sp	All

Table C.1 (*cont.*)

Place	Year	I	OT	C	I-C	OT-C	#mo	Thtr	sideB
Leffinghe	1708	10/17		10/24	7		0.2	SN	Fr
Denia	1708		11/7	11/17		10	0.3	Sp	Fr
Saint Ghislain	1708	11/28		12/1	3		0.1	SN	Fr
Brussels	1708	11/22	11/24	11/27	5	3	0.2	SN	Fr
Alicante	1708	11/30		4/18	139		4.6	Sp	Fr
Ghent	1708	12/18	12/24	12/30	12	6	0.4	SN	All
Tournai	1709	6/27	7/7	9/3	68	58	2.3	SN	All
Mons	1709	9/20	9/25	10/20	30	25	1.0	SN	All
Douai	1710	4/23	5/4	6/26	64	53	2.1	Fr	All
Béthune	1710	7/15	7/23	8/29	45	37	1.5	Fr	All
Aire	1710	9/6	9/12	11/9	64	58	2.1	Fr	All
Saint Venant	1710	9/6	9/16	9/30	24	14	0.8	Fr	All
Gerona	1710	12/12		1/24	43		1.4	Sp	Fr
Aren fort	1711	7/10		7/30	20		0.7	Sp	Fr
Bouchain	1711	8/6		9/13	38		1.3	Fr	All
Venasque	1711	9/1	9/11	9/16	15	5	0.5	Sp	Fr
Castel-Leon	1711	9/25	9/28	10/9	14	11	0.5	Sp	Fr
Cardona	1711	11/12	11/14	12/22	40	38	1.3	Sp	Fr
Le Quesnoy	1712	6/8	6/19	7/4	26	15	0.9	Fr	All
Landrecies	1712	7/17		8/2	16		0.5	Fr	All
Marchiennes	1712	7/25		7/30	5		0.2	Fr	Fr
Douai	1712	7/31	8/14	9/8	39	25	1.3	Fr	Fr
Le Quesnoy	1712	9/8	9/18	10/4	26	16	0.9	Fr	Fr
Bouchain	1712	10/1	10/9	10/19	18	10	0.6	Fr	Fr
Gerona	1712	11/1		1/3	63		2.1	Sp	All
Landau	1713	6/11	6/24	8/20	70	57	2.3	Ge	Fr
Freiburg	1713	9/22	9/30	11/16	55	47	1.8	Ge	Fr
Barcelona	1714	7/7	7/12	9/13	68	63	2.3	Sp	Fr

Table Sources and Notes

I have not cited each primary source consulted for each date for every siege (an average of 5 sources × 3 dates per siege × 108 sieges = 1,620 citations!). For the reasons explained in Chapter 4, I have focused my efforts on the Low Countries sieges. Therefore I rely heavily on secondary sources for those sieges in the non-Flanders theaters and cite them accordingly. For Flanders, rather than citing sources for each siege, I have included the most important collections of published correspondence. It is important here to compare the various sources' account for each of the stages; hence one should look at all of the relevant sources wherever possible. Even the simplest of measures—a date—can be surprisingly complicated, especially for sieges where the stages themselves were short, i.e. most sieges that are missing from existing secondary accounts. Sources that give the wrong date may also provide clues that help you determine the correct date—the more sources you add, the greater the certainty. With the example of a unique source that refers to a date only as a day of the week (e.g. "last Tuesday we captured the covered way") with an unknown calendar, you begin to see the necessity of comparing all possible sources to create a robust dataset. A perfect dataset would employ a similar methodology for every theater's sieges.

That being said, the individual documents can easily be looked up using a combination of the two lists. Use the information in the Table C.1 to determine the month, date and year of the relevant documents, then determine which volume of correspondence to look in by consulting the date ranges listed below.

Low Countries Siege Sources

- Murray (ed.), *Letters and Dispatches*
  Vol. 1: 17 April 1702–28 March 1705
  Vol. 2: 31 March 1705–16 July 1706
  Vol. 3: 16 July 1706–5 May 1708
  Vol. 4: 5 May 1708–22 April 1710
  Vol. 5: 23 April 1710–18 March 1712
- Snyder (ed.), *Marlborough-Godolphin Correspondence*
  Vol. 1: 4 March 1701–25 June 1706
  Vol. 2: 17 June (O.S.) 1706–3 January 1709
  Vol. 3: 24 December 1708 (O.S.)–10 November 1711
- Van 't Hoff (ed.), *Marlborough-Heinsius Correspondence* (1700–1712 in one volume)
- Veenendaal, Jr. (ed.), *Briefwisseling Heinsius*
  Vol. 1: 1702
  Vol. 2: 1703
  Vol. 3: 1704
  Vol. 4: 1705
  Vol. 5: 1706
  Vol. 6: 1707
  Vol. 7: 1 January 1708–30 September 1708
  Vol. 8: 1 October 1708–30 June 1709
  Vol. 9: 1 July 1709–31 December 1709
  Vol. 10: 1 January 1710–31 July 1710
  Vol. 11: 1 August 1710–30 April 1711
  Vol. 12: 1 May 1711–31 January 1712
  Vol. 13: 1 February 1712–31 August 1712
  Vol. 14: 1 September 1712–13 April 1713

The standard collections of Vault and Pelet (eds.), *Mémoires militaires*, Wijn, *Het Staatsche Leger*, and the *Spanischer Successions-krieg: Feldzüge des Prinz Eugen von Savoyen* series also provide good summaries of the events in the Low Countries theater. Other primary sources (archival and published) are cited throughout the text.

Non-Flanders Siege Length Sources

Castiglione 1702 (Castiglion-delle-Stiviere): Lecestre (ed.), *Mémoires du chevalier de Quincy*, 1:204–205; Vault and Pelet (eds.), *Mémoires militaires*, 2:213–214; Marquis de Quincy, *Histoire militaire du règne de Louis le Grand*, 3:647–649.

Landau 1702: Marquis de Quincy, *Histoire militaire du règne de Louis le Grand*, 3:579–592; Vault and Pelet (eds.), *Mémoires militaires*, 2:332ff.; Veenendaal, Jr. (ed.), *Briefwisseling Heinsius*, 1:554ff.; "Journal du blocus et du siège de la ville et du Fort de Landau" in *Mercure Galant* after octobre 1702.

Borgoforte 1702: Vault and Pelet (eds.), *Mémoires militaires* 2:255ff.
Guastalla 1702: Vault and Pelet (eds.), *Mémoires militaires* 2:258ff.; Lecestre (ed.), *Mémoires du chevalier de Quincy*, 1:244–245; Marquis de Quincy, *Histoire militaire du règne de Louis le Grand*, 3:684ff.
Trarbach 1702: Vault and Pelet (eds.), *Mémoires militaires*, 2:124; Marquis de Quincy, *Histoire militaire du règne de Louis le Grand*, 3:607ff.
Andernach 1702: Vault and Pelet (eds.), *Mémoires militaires*, 2:125.
Governolo 1702: Vault and Pelet (eds.), *Mémoires militaires*, 2:274–279.
Neubourg 1703: John Garland, "Irish Officers in the Bavarian Service During the War of the Spanish Succession," *The Irish Sword*, 14 (1981), 240–255; Cocula (ed.), *Mémoires de Monsieur de La Colonie*, 134; Vault and Pelet (eds.), *Mémoires militaires*, 3:510.
Kehl 1703: Vogüé (ed.), *Mémoires du Maréchal de Villars publié d'après le manuscrit original*, 2:49ff.; Vault and Pelet (eds.), *Mémoires militaires*, 3:515–524.
Bonn 1703: Vault and Pelet (eds.), *Mémoires militaires*, 2:17–35; Snyder (ed.), *Marlborough-Godolphin Correspondence*, 1:163ff; Murray (ed.), *Letters and Dispatches*, 1:75ff.
Nago 1703: Vault and Pelet (eds.), *Mémoires militaires*, 3:240–244.
Arco 1703: Vault and Pelet (eds.), *Mémoires militaires*, 3:247–252.
Breisach 1703: DuMoulin, *Campagne de Monsieur le maréchal de Villars en Allemagne l'an MDCCCIII. Contenant les lettres de ce maréchal & celles de plusieurs autres officiers-généraux au roi & à Mr. de Chamillart, ministre au Département de la guerre, avec les réponses du roi, & de ce ministre. Recueil . . . formé sur les originaux qui se trouvent en dépôt au Bureau de la guerre de la Cour de France*, 2 vols. (Amsterdam, 1762), 2:142ff.; Rochas d'Aiglun, *Vauban*, 2:527ff.
Landau 1703: Vault and Pelet (eds.), *Mémoires militaires*, 4:472ff.
Augsburg 1703: Cocula (ed.), *Mémoires de Monsieur de La Colonie*, 134; DuMoulin, *Campagne de Monsieur le maréchal de Villars en Allemagne l'an MDCCCIII*, 2:205–206, 425–427. Vault and Pelet (eds.), *Mémoires militaires*, 3:713ff.
Castello de Vide 1704: Sir Charles Petrie, *The Marshal Duke of Berwick: The Picture of an Age*, (London, 1953), 171; Berwick, *Mémoires du Maréchal de Berwick*, 1:159–160; J.A.C. Hugill, *No Peace without Spain*, (Oxford, 1991), 72.
Barcelona 1704: *Daily Courant*, 19 June O.S. #679; Petrie, *The Marshal Duke of Berwick*, 113ff.; Hugill, *No Peace without Spain*, 80ff.
Susa 1704: Vault and Pelet (eds.), *Mémoires militaires*, 4:120, 126, 130–136; DuMoulin, *Campagne de Monsieur le maréchal de Tallard en Allemagne l'an MDCCIV contenant les lettres de ce maréchal & celles de plusieurs autres officiers-généraux au roi & à Mr. de Chamillart . . . avec les réponses . . . Recueil formé sur les originaux qui se trouvent en dépôt au Bureau de la guerre de la Cour de France*, (Amsterdam, 1763) 1:300.
Portalegre 1704: Hugill, *No Peace Without Spain*, 69; *Daily Courant* 13 June–7 July.
Vercelli 1704: Lecestre (ed.), *Mémoires du chevalier de Quincy*, 2:12–29; Vault and Pelet (eds.), *Mémoires militaires*, 4:223; DuMoulin, *Campagne de Monsieur le maréchal de Tallard en Allemagne l'an MDCCIV*, 2:122, 133.
Rain 1704: Cocula (ed.), *Mémoires de Monsieur de La Colonie*, 198–203; Snyder (ed.), *Marlborough-Godolphin Correspondence*, 1:338ff.
Villingen 1704: DuMoulin, *Campagne de Monsieur le maréchal de Tallard en Allemagne l'an MDCCIV*, 2:82–87; Vault and Pelet (eds.), *Mémoires militaires*, 4:521ff.
Gibraltar 1704: Hugill, *No Peace Without Spain*, 87–99; David Francis, *The First Peninsular War, 1701–1713*, (London, 1975), 109ff.
Ulm 1704: Murray (ed.), *Letters and Dispatches*, 1:471ff., DuMoulin, *Campagne de Monsieur le maréchal de Tallard en Allemagne l'an MDCCIV*, 2:106–108; Vault and Pelet (eds.), *Mémoires militaires*, 4:603ff.
Gibraltar 1704: Hugill, *No Peace Without Spain*, 126–146; Francis, *The First Peninsular War, 1701–1713*, 128–149.

Ivree 1704: Vault and Pelet (eds.), *Mémoires militaires*, 4:183, 262–270; Lecestre (ed.), Mémoires du chevalier de Quincy, 2:33–51; DuMoulin, *Campagne de Monsieur le Marechal de Marsin en Allemagne l'an MDCCIV*, 2:140; Veenendaal, Jr. (ed.), *Briefwisseling Heinsius*, 3:322.

Landau 1704: Vault and Pelet (eds.), *Mémoires militaires*, 4:490; DuMoulin, *Campagne de Monsieur le Marechal de Marsin en Allemagne l'an MDCCIV*, 2:240ff.

Verrua 1704: Lecestre (ed.), *Mémoires du chevalier de Quincy*, 2:54–87; Vault and Pelet (eds.), *Mémoires militaires*, 4:278ff.

Trarbach 1704: Vault and Pelet (eds.), *Mémoires militaires*, 4:656–659.

Valencia de Alcantara 1705: Hugill, *No Peace Without Spain*, 152–153; Veenendaal, Jr. (ed.), *Briefwisseling Heinsius*, 4:148, 150; Francis, *The First Peninsular War, 1701–1713*, 160–161.

Albuquerque 1705: Hugill, *No Peace Without Spain*, 153; Veenendaal, Jr. (ed.), *Briefwisseling Heinsius*, 4:165; Francis, *The First Peninsular War, 1701–1713*, 161.

Chivasso 1705: Vault and Pelet (eds.), *Mémoires militaires* 5:153ff.; Lecestre (ed.), *Mémoires du chevalier de Quincy*, 2:189.

Mirandola 1705: Vault and Pelet (eds.), *Mémoires militaires*, 5:263–264, 269ff.

Barcelona 1705: Hugill, *No Peace Without Spain*, 171ff.; Veenendaal, Jr. (ed.), *Briefwisseling Heinsius*, 4:375; Francis, *The First Peninsular War, 1701–1713*, 171–193.

Hagenau 1705: Vault and Pelet (eds.), *Mémoires militaires*, 5:801–806.

Badajoz 1705: Portland, *Manuscripts of the Duke of Portland*, 4:263; Veenendaal, Jr. (ed.), *Briefwisseling Heinsius*, 4:364; Francis, *The First Peninsular War, 1701–1713*, 167ff.

Nice 1705: Vault and Pelet (eds.), *Mémoires militaires*, 5:116–123, 224ff.

San Mateo 1705: Hugill, *No Peace Without Spain*, 195–197; Francis, *The First Peninsular War, 1701–1713*, 201ff.; H.T. Dickinson, "The Earl of Peterborough's Campaign in Valencia 1706," *The Journal of the Society for Army Historical Research*, 45 (181) 1967, 38ff.

Barcelona 1706: Wijn, *Het Staatsche Leger*, 8 2:762ff.; Hugill, *No Peace Without Spain*, 202; Veenendaal, Jr. (ed.), *Briefwisseling Heinsius*, 5:230ff.

Alcántara 1706: Petrie, *The Marshal Duke of Berwick*, 191; Hugill, *No Peace Without Spain*, 213ff.; Francis, *The First Peninsular War, 1701–1713*, 224; *Daily Courant* 25 April.

Hagenau 1706: Vault and Pelet (eds.), *Mémoires militaires*, 6:419ff.

Ciudad Rodrigo 1706: Petrie, *The Marshal Duke of Berwick*, 193; Hugill, *No Peace Without Spain*, 216; *Daily Courant* 8 June 1706; Francis, *The First Peninsular War, 1701–1713*, 225.

Turin 1706: Lecestre (ed.), *Mémoires du chevalier de Quincy*, 2:190ff.; Vault and Pelet (eds.), *Mémoires militaires*, 6:173ff.

Carpi 1706: Veenendaal, Jr. (ed.), *Briefwisseling Heinsius*, 5:450, 453.

Reggio 1706: Veenendaal, Jr. (ed.), *Briefwisseling Heinsius*, 5:465, 481.

Alicante 1706: Francis, *The First Peninsular War, 1701–1713*, 233, 240; Hugill, *No Peace Without Spain*, 230; Veenendaal, Jr. (ed.), *Briefwisseling Heinsius*, 5:595; *Daily Courant* 2 September, 10 October.

Pavia 1706: Veenendaal, Jr. (ed.), *Briefwisseling Heinsius*, 5:630; Vault and Pelet (eds.), *Mémoires militaires*, 6:322.

Cuenca 1706: Veenendaal, Jr. (ed.), *Briefwisseling Heinsius*, 5:662, 691; Hugill, *No Peace Without Spain*, 240; *Daily Courant* 23 October.

Pizzigetone 1706: Veenendaal, Jr. (ed.), *Briefwisseling Heinsius*, 5:577, 636, 638, 630–631, 682.

Elche 1706: Hugill, *No Peace Without Spain*, 241.

Cartagena 1706: Petrie, *The Marshal Duke of Berwick*, 202; Hugill, *No Peace Without Spain*, 241–242; *Daily Courant* 5 December.
Casale 1706: Vault and Pelet (eds.), *Mémoires militaires*, 6:356ff.
Milan 1707: Veenendaal, Jr. (ed.), *Briefwisseling Heinsius*, 6:70, 86, 98, 114, 130.
Villena 1707: Hugill, *No Peace Without Spain*, 249–250; J.W. Wijn, "Une bataille classique du XVIIIe siecle: Almansa1707 25 avril 1707," *Revue internationale d'histoire militaire*, 22 (1960), 92–104.
Toulon 1707: Lecestre (ed.), *Mémoires du chevalier de Quincy*, 2:258ff.; Veenendaal, Jr. (ed.), *Briefwisseling Heinsius*, 6:476; Y.J. Saint-Martin, "Le pseudo-siège de Toulon en 1707," *Provence historique*, 176 (1994), 199–209.
Ciudad Rodrigo 1707: Hugill, *No Peace Without Spain*, 270; Veenendaal, Jr. (ed.), *Briefwisseling Heinsius*, 6:496, 577; Francis, *The First Peninsular War, 1701–1713*, 249.
Susa 1707: Lecestre (ed.), *Mémoires du chevalier de Quincy*, 2:283–284; Veenendaal, Jr. (ed.), *Briefwisseling Heinsius*, 6:577, 589, 601, 615.
Lérida 1707: Hugill, *No Peace Without Spain*, 265–268; Francis, *The First Peninsular War, 1701–1713*, 249.
Morella 1707: Hugill, *No Peace Without Spain*, 269; Francis, *The First Peninsular War, 1701–1713*, 249.
Tortosa 1708: Hugill, *No Peace Without Spain*, 274–276; Veenendaal, Jr. (ed.), *Briefwisseling Heinsius*, 6:275–276, 284, 379, 474; Francis, *The First Peninsular War, 1701–1713*, 266.
Exilles 1708: Albert de Rochas d'Aiglun, "La campagne de la Succession d'Espagne dans les Alpes (1707–13)," *Bulletin de la Société d'études des Hautes-Alpes*, 7 (1888), 104.
Fenestrelles 1708: Albert de Rochas d'Aiglun, "La campagne de la Succession d'Espagne dans les Alpes (1707–13)," 107; Lecestre (ed.), *Mémoires du chevalier de Quincy*, 2:313ff.
San Felipe 1708: Hugill, *No Peace Without Spain*, 276–278; Veenendaal, Jr. (ed.), *Briefwisseling Heinsius*, 6:512; Francis, *The First Peninsular War, 1701–1713*, 267.
Denia 1708: Hugill, *No Peace Without Spain*, 283; Francis, *The First Peninsular War, 1701–1713*, 272.
Alicante 1708: Hugill, *No Peace Without Spain*, 283–294; Francis, *The First Peninsular War, 1701–1713*, 272; Antoine-Marie Augoyat, "Siege du chateau d'Alicante et surprise de Tortose pendant la guerre de la Succession d'Espagne," *Le Spectateur militaire*, 43(254) 1847, 129–145.
Gerona 1710: Hugill, *No Peace Without Spain, 336*; Francis, *The First Peninsular War, 1701–1713*, 320.
Aren fort 1711: Hugill, *No Peace Without Spain*, 342; Delorme, "Le journal d'un capitaine de mineurs (1711)," *Carnet de la Sabretache*, 4/6 (1896), 426–431.
Venasque 1711: Hugill, *No Peace Without Spain*, 342ff.
Castel-Leon 1711: Augoyat, *Aperçu historique sur les fortifications, les ingénieurs et sur le corps du Génie en France*, (Paris: J. Dumaine, 1860), 1:383–384.
Cardona 1711: Hugill, *No Peace Without Spain*, 345ff.; Veenendaal, Jr. (ed.), *Briefwisseling Heinsius*, 12:517, 579, 629.
Gerona 1712: Hugill, *No Peace Without Spain*, 349ff.; Veenendaal, Jr. (ed.), *Briefwisseling Heinsius*, 14:145.
Landau 1713: Vault and Pelet (eds.), *Mémoires militaires*, 13:259ff.
Freiburg 1713: Vault and Pelet (eds.), *Mémoires militaires*, 13:350ff.
Barcelona 1714: Petrie, The Marshal Duke of Berwick, 254; Hugill, No Peace Without Spain, 387.

APPENDIX D

NON-SIEGE POSITIONAL TACTICS

The skirmish between Parker and Israel over the length of Dutch Revolt sieges mentioned in Appendix B highlights another fundamental issue that needs to be addressed: what exactly constitutes a siege? Their disagreement is indicative of the widespread confusion over its precise definition—at several points Israel in fact argues against Parker's characterization of the Dutch Revolt as a war dominated by blockades, although Israel himself slightly confuses the issue by continuing to refer to them as sieges. Later he comes closest to explicitly distinguishing between the two: "Almost alone among the great sieges of the 1621–48 war, Breda was simply an exercise in starving the defenders out. Hence its exceptional duration. . . . There was scarcely any fighting, little bombardment, and few casualties."[1] Length, then, is necessarily conflated with the method used to capture a town. The resolution to historiographical confusion is to avoid the *generic* use of the term "siege" and explicitly state what type of tactic was used according to concrete criteria.

Historians have begun to emphasize the distinction contemporaries made at the time. Duffy has led the way, clearly delineating the "means of reducing a fortresses short of a formal siege," including storm, surprise, bombardment and starvation.[2] Progress among historians not focused solely on siegecraft can be seen best in the evolution of Parker's works, where his original emphasis on blockades branches out into the wider variety of tactics used to attack a place as he has elaborated upon various aspects of the argument.[3] In his early *The Army of Flanders*, he does not clearly discriminate between

---

[1] *The Dutch Republic and the Hispanic World, 1601–1660*, p. 107. See also p. 101 and his accounts of the sieges on pp. 318–321.

[2] Duffy, *Fire and Stone*, pp. 94–101.

[3] For recent acknowledgments of the different tactics mentioned in different periods, see: Chandler, *The Art of Warfare in the Age of Marlborough*, p. 245; Pepper and Adams, *Firearms and Fortifications*, p. 169; Bruce Allen Watson, *Sieges: A Comparative Study* (Westport, CT, 1993), pp. 138–141; and Geoffrey Parker's entry "Sieges" in G. Parker and R. Cowley (eds.), *The Reader's Companion to Military History*, (Houghton Mifflin, 1996).

siege and blockade, using "siege" to refer generically to any attack on towns, but it is clear his emphasis is on blockades.[4] His showcase example of Amiens 1597 purports to illustrate the "standard procedure" of "blockade and attrition," yet the contemporary engraving of the combat shows several batteries of cannon firing at the fortifications themselves, something not usually associated with a blockade.[5] In a later work he distinguishes more clearly between different methods used to capture a town—including surprise, storm, treachery, starvation, assault, and mines—and he suggests through several examples that the last three methods all shared longevity. His clearer focus is now indicated by the replacement of Amiens 1597 with Breda 1624–1625, which was a 'pure' blockade in the sense that "Not a shot seems to have been fired against the bastions and hornworks of Breda itself: the city surrendered . . . through simple starvation."[6] In the recent (2004) revision of his *The Army of Flanders*, he includes new data on the method of capture for 95 towns between 1578 and 1590 (pp. 7–8), explicitly distinguishing the various positional tactics. Further refinement of this taxonomy of tactics is a starting point for measuring the conditions of positional warfare throughout the ages.

Developing a viable definition of what is, and is not, a siege requires a very broad view, but by the War of the Spanish Succession the distinctions are rather clear. The most common tactics included siege, blockade, bombardment, storm, and stratagem or surprise. More precisely, a siege required that several specific criteria be met. The goal was usually, with a few exceptions, to capture the fortification under attack.[7] Other ways of attacking a town, bombardment particularly, were often used purely for intimidation or retaliation, or to achieve some other objective (such as levying contributions, or destroying a forage magazine) without having to go to the expense

---

[4] Page 9, and plate 4.

[5] It is possible that the engraving was meant to illustrate several different types of attack (siege, blockade . . .) in a single asynchronous image.

[6] *The Military Revolution*, p. 13. Whether blockades and sieges (starvation, assault and mine) were both inherently lengthy is a matter for empirical research to confirm or disprove, although Parker does not directly contest Lynn's contention that siege lengths decreased throughout the 17th century.

[7] Occasionally a siege might be undertaken solely with the intention of diverting the enemy's attention from elsewhere, as the French did at Liège in 1705. In the medieval period, sieges might also be conducted in order to extract tribute from a town. Richard Rogers, *Latin Siege Warfare in the Twelfth Century* (Oxford, 1997), p. 9.

of actually capturing the town. The defenders had to start with a desire or intention to resist the would-be attacker, whether garrison, militia, or both. Otherwise, urban authorities would simply admit representatives of the approaching army into the town, as might happen after a successful field battle such as Ramillies in 1706. This seemingly self-evident requirement has significant repercussions, however, since a number of factors could influence the decision to defend a town or not, including:

- the overall operational context
- the state of the fortifications
- the availability of garrison supplies
- the steadfastness of the garrison commander
- the presence of a relief army (i.e. the likelihood of external help)
- the morale of the garrison
- the inclination of the townspeople.

Any one of these factors might make a town indefensible. Sieges therefore, by definition, resulted only when the defenders thought that the town could resist for at least a short time. In other words, only the comparatively-stronger towns were besieged, while weaker ones would submit without putting up a fight.[8] A siege's first prerequisite then was a willingness to defend the works and the belief that it was reasonable to do so.

More practically, I will confine my use of the term 'siege' to an attack on fortifications that were attacked with artillery or underground mines for a day or more.[9] This definition of a siege separates it from four other related positional tactics that could also be used to capture a town. In a storm, a fortified position was attacked by infantry assault (often by escalade) without artillery preparation

[8] Having said this, it is important to further distinguish between the minimum threshold for defending a town (the garrison considered the fortifications strong enough to delay the besiegers) and the conditions which would allow a garrison to force the besiegers to abandon the siege. Just because a town decided to withstand a siege does not mean it could be expected to hold out very long.

[9] Trenches were almost universally present in sieges, but it is not a primary criteria, as the surrounding terrain might offer a besieger's troops natural cover, or the hardness of the ground might make digging impossible. In most cases, some type of material was needed to shield the attacking troops from the garrison's fire. Sixteenth century engravings depict soldiers shielded by walls of gabions built from the ground up instead of trenches dug into the earth, but this might only be an artistic convention.

or trenches. Storms were usually conducted against weakly-defended towns (either in terms of their fortifications or by an under-manned garrison) and more frequently against smaller forts, where the small size of the works allowed little supporting fire by the garrison. A blockade surrounded a fortified position with fortified lines or small forts and cut it off from outside support in order to starve the defenders into submission. Blockades tended to be used in one of five cases: against towns too strong to be taken by siege/assault; when the attacker could afford to wait out the garrison; when the attacker was afraid of the casualties they might sustain in a siege or assault; when enough resources could not be dedicated to a formal siege to assure its success; or when several towns were to be captured at the same time.[10] Bombardments fired field pieces and mortars indiscriminately into a town in order to terrorize the inhabitants and garrison into surrendering (contrasted with the siege's targeting of the town's fortifications) and were usually used against towns which were expected to need only a token show of force before they would surrender. They might also be used by attackers that lacked proper siegecraft expertise or equipment (e.g. heavy siege artillery). They were also frequently conducted by naval vessels against coastal fortresses since they posed little risk to the attackers.[11] Finally, the tactic of stratagem or surprise could include sneaking or bluffing one's way into a town, with or without help from inside the town. The following table summarizes the essential characteristics of each positional tactic.

Table D.1 Positional Tactics Used to Capture a Town

Tactic	Method of Attack	Target of Attack	Time Frame
"Formal" Siege	Artillery, mines	Walls and fortifications	Long (>1 day)
Storm (escalade, assault)	Infantry assault	Overrun walls	Brief (< 1 day)
Blockade	Isolate from outside contact	Garrison's supplies	Longer
Bombardment	Field artillery, including mortars and howitzers	Buildings or population (Morale)	Brief (several days)
Stratagem (treachery, surprise, ruse)	Stealth, Negotiation	Town's security measures	Brief (< 1 day)

[10] For a good discussion of the use of blockades in an earlier war, see Satterfield, *Princes, Posts and Partisans*.

[11] Jean Peter, *Les Artilleurs de la Marine sous Louis XIV* (Paris, 1995).

With this typology, a siege is distinguished from a storm by the use of siege artillery and/or mines, from a blockade by actively attacking the garrison and its fortifications, and from a bombardment by targeting the town's walls and fortifications rather than (or in addition to) the buildings and morale within the town. The goal of a stratagem was to avoid fighting for the town altogether.

These categories are distinct but not mutually exclusive: the method the attacker intended to use to capture the place is the fundamental criteria. For example, a siege's beginning could mimic a blockade if there was a lengthy period between when the besiegers cut the town off from the outside (the investment) and when they started to attack the town (opening of the trenches). Similarly, a siege might conclude with an assault (i.e. storm) of a breach made in the main wall, and might well include firing shells into the town itself (a bombardment) in order to encourage the townspeople to pressure the besieged garrison into surrendering. Non-siege tactics could also be used together, such as a blockade-bombardment. These different tactics existed along a continuum rather than as hermetically discrete categories, but any attack that used heavy artillery and/or mines against the *fortifications* is classified as a siege, regardless of whatever other tactics were also used.[12]

At the other extreme are scholars who define 'siege' so strictly that they practically classify it out of existence. Several historians have argued, for example, that the Allied attack on Toulon in 1707 was not a proper siege at all, but rather a "pseudo-siege," as the town was not fully invested.[13] This view stems from an unnecessarily strict conception of what constituted a 'siege' in the 17th and 18th centuries, one tied to a strict interpretation of Vauban's theory instead

---

[12] The classification of these earlier attacks as "sieges" may change when we attempt to resolve the issue of siege versus non-siege tactics and their applicability to pre-Vaubanian warfare. It is conceivable, for example, that the relatively distinct categories of positional tactics (siege, blockade, bombardment, storm, surprise) may have only slowly solidified over the course of the 17th century. Thus many of the longer "sieges" of the earlier 17th century might in fact be better categorized as blockades rather than sieges, or a hybrid of the two. Similarly, Bousmard argued that sieges and bombardments were synonymous c. 1790.

[13] Y.J. Saint-Martin, "Le Pseudo-Siège de Toulon en 1707," *Provence historique* 176 (1994), pp. 199–209. Fernand Braudel made a less categorical although similar statement in *The Identity of France, 1: History and Environment*, (New York, 1993), p. 361. In *Marlborough as Military Commander*, p. 199, Chandler described Toulon as less of a regular siege than a partial blockade.

of the messier reality—the stereotype of "scientific" Vaubanian sieges encourages scholars to confuse besieging with besieging well. To take Saint-Martin's criteria of incomplete investment as an example, several other besiegers in the War of the Spanish Succession also failed to completely cut off the town, for example at Kaisersweert in 1702, Verrua in 1704–5, Turin in 1706, and at several coastal fortresses such as Gibraltar in 1705. Yet no contemporary would argue that these were not sieges, only that they may have been poorly-conducted ones. Whereas contemporaries made distinctions between sieges, blockades, bombardments, storms and surprisals, they made no distinction between a *real* siege and a *pseudo*-siege. There was oftentimes a wide gap between theory and practice, and we should not use an idealized conception of a siege as the standard when constructing a siege database.

The distinction between the various tactics used to capture a town, and the elevation of sieges to a distinct plane in particular, is justified for a number of reasons. First, contemporaries themselves made this distinction. When a formal siege with heavy artillery and trenches was required, contemporaries usually spoke of besieging *dans les formes*, their language suggesting a more systematic and regulated attack than a simple storm, blockade or bombardment.[14] Contemporary manuals also discussed each tactic separately.[15] The distinction between these tactics is further reinforced by the many instances where an attacker progressed from one tactic to another in an attempt to

[14] For examples, see: Veenendaal, Jr. (ed.), *Briefwisseling Heinsius*, vol. 1, p. 123 #188, 23 April 1702; vol. 1, p. 416 #835, 6 September 1702; vol. 1, p. 455 #932, 2 October 1702; vol. 4, p. 125 #346, 12 April 1705; *Mémoires militaires*, vol. 3, p. 173, 22 April 1703; AG A[1] 2084 #136, 28 November 1708; AG A[1] 2084 #158, 29 November 1708. An English newspaper translated *dans les formes* as "in the usual form." *Postman*, #1875, From the camp before Douai, 22 April 1710. In modern French it has acquired the connotation of "as is usually done."

[15] See, for example, Feuquières, *Memoirs Historical and Military*, vol. 1, pp. 311–318 on surprisals; vol. 2, pp. 201–206 on blockades, and a much larger section on formal sieges. The abbé Deidier makes a similar distinction between attacking places by surprise, by force (bombardment, storm, siege *par forme*), or by famine (blockade). *Le parfait ingénieur francois*, nouvelle éd. (Paris, 1757), pp. 178ff. Papillon offers a similar categorization of "six different wayes, how strong holds may be reduced." Papillon, *Practical Abstract of the Art*, p. 99. Montecuccoli also discussed several techniques, categorizing them according the effort and delays they entailed. *Mémoires de Montecuculi*, pp. 135–136. The chevalier de Folard included an Article IV in vol. 3 of his Polybius commentaries entitled: "That both ancient and modern historians confuse assaults with escalades."

capture a town; each tactic was used independently of the others.[16] Vauban, to give an example, discusses blockades being converted into a "*siége réglé*."[17] Correspondents frequently noted attackers switching from one unsuccessful tactic to another in hopes of accelerating a town's capture, indicating a different beast altogether.[18] In the Low Countries, Rheinberg (1702) was transformed from a siege into a blockade and Gelders (1703) from a bombardment into a blockade. The French commander of Dendermonde in 1706 reported that the Allied "continuation of the bombardment makes me think they do not yet want to besiege us."[19] After the bombardment failed, the town was blockaded before a formal siege eventually forced its surrender.

The most important reason for making this distinction, highlighted by Monck's observation quoted in the Introduction, is that a siege required a degree of commitment and resources that the other tactics did not.[20] Unlike a siege, many of the other positional operations could be conducted by a small number of troops, or even by irregular

[16] This appears to have been particularly common in the medieval period, when many besiegers lacked the siege artillery necessary to breach thick stone walls: Bernard Bachrach (following Jim Bradbury) describes the "six S's" of subverting, scaring, sapping, starving, storming and shelling in "Medieval Siege Warfare: A Reconnaissance," 125. Period surveys, such as R.L.C. Jones, "Fortifications and Sieges in Western Europe c. 800–1450," in M. Keen (ed.), *Medieval Warfare: A History*, (Oxford, 1999), p. 183 and Michael Prestwich, *Armies and Warfare in the Middle Ages: The English Experience*, (New Haven, 1996), p. 296 also mention a similar variety of tactics. Several of Bradbury's siege accounts indicate that many medieval attackers were forced to alternate between storm and blockade due to their inability to breach the walls.

[17] Vauban, *Traité*, pp. 225–226.

[18] For examples of contemporaries distinguishing between these tactics, see: Veenendaal, Jr. (ed.), *Briefwisseling Heinsius*, vol. 4, p. 31 #68, 17 January 1705; vol. 6, p. 615 #1220, 28 October 1707; Vault and Pelet (eds.), *Mémoires militaires*, vol. 5, pp. 124 138, and 197; AG A[1] 2152, #64, 14 August 1709; AG A[1] 2384 #147, 28 September 1712; AG A[4] Carton 8 2, Alègre, 18 September 1712; and S.F.H. Johnston, "Letters of Samuel Noyes, Chaplain of the Royal Scots, 1703–4" in *The Journal of the Society for Army Historical Research*, 37 (1959), p. 152.

[19] AG A[1] 1938 #66, 2 July 1706. The trenches were only opened at Dendermonde on August 29, after the garrison had withstood an initial bombardment and then a month-long blockade.

[20] Papillon recognized this: "The sixth and most certain and method-like way [of how strongholds may be reduced], is by an intrenched camp used in Caesar's dayes, and at this time in Flanders by the French and Hollanders [c. 1640s]; because if all the five former wayes having been attempted and faile, this will assuredly carrie away the garrison, either by storme or famine." Papillon, *Practical Abstract of the Art*, p. 99. His failure to clearly distinguish between a siege/storm and blockade may suggest that the distinction only evolved later in the 17th century—possibly explaining Parker's early conflation of the two tactics.

partisan forces. Although blockades were usually reserved for towns whose "advantageous situation" precluded a siege,[21] most could be maintained with few men (several battalions and squadrons) once several small forts were built to fence in the place, allowing the main field force to operate elsewhere.[22] After several Flanders towns were taken during the war, there was talk of simply leaving a small blockading force to isolate their citadels rather than going to the effort of besieging them. Such was the speculation, for example, during the siege of Tournai in 1709, while the French commander at Rheinberg noted that "the enemies expect to capture us by famine since they were unable to take us by force."[23] A blockade also required few troops, an advantage Marlborough highlighted in his 1703 complaint over the Prince of Baden's plans for Bonn: "[its position] will make it necessary to have almost as many troupes for the bloquad as might serve for the siege."[24] Bombardments promised the benefit of requiring only the more mobile field pieces (e.g. 8-pounders) and mortars to lob bombs and red-hot shot into a town, rather than the much heavier breaching cannon needed for a proper siege.[25] The French bombardment of Liège in 1691, for example, required only twenty-four cannon and twelve mortars and two days of fire.[26] The English envoy Richard Hill explicitly contrasted the uniquely resource-intensive nature of sieges with bombardment in a report from Savoy:

> Ever since the Duke de La Feuillade has been very busy, as if he were resolved to besiege, or to bombard us. We cannot think that he is

[21] As assumed by the Spanish military writer Santa Cruz de Marcenado in *Réflexions militaires*, vol. 10, p. 159. In the 1630s the engineer De Ville had already noted that "long sieges" (his term for blockade) were reserved for those places "so well fortified that there was no hope of capturing them by force." *De la charge des Gouverneurs des Places*, pp. 442–443.

[22] This is in contrast to earlier in the 17th century, when logistical limitations often forced the main field army to encamp in a blockading position.

[23] AG A[1] 2151 #236, 21 July 1709 and AG A[1] 2151 #276, 26 July 1709. A similar choice was needed at Lérida in Spain in 1707: Veenendaal, Jr. (ed.), *Briefwisseling Heinsius*, vol. 6, p. 615 #1220, La Sarraz to Heinsius, The Hague, 28 October 1707. On Rheinberg, see *Mémoires militaires*, vol. 2, pp. 620–621, response to Détail du siège de Rhinberg.

[24] van 't Hoff (ed.), *Marlborough-Heinsius Correspondence*, p. 61 #101, Marlborough to Heinsius, Köln, 21 April 1703.

[25] As Goslinga wrote of one enemy party of 20 battalions and as many squadrons: "As they have no heavy artillery, it will only be by treachery or *un coup de main* that they claim to be able to succeed." Veenendaal, Jr. (ed.), *Briefwisseling Heinsius*, vol. 12, p. 244 #407, Goslinga to Heinsius, Lillers, 30 July 1711.

[26] Naulet, *L'artillerie française*, pp. 245–246.

> strong enough to execute the first of those designs; but his Royal Highness [the Duke of Savoy] is not strong enough to prevent the latter. The enemies have employed the last ten days in fortifying their camp, and filling it with artillery, and ammunition, which they bring from Chivas [Chivasso]. They have enough for a bombardment. They have not enough, by any means, for a siege.[27]

Another method, assault, might result in more attacking casualties than either a bombardment or a blockade, but if successful it would be over within a day or so, thus avoiding the days, weeks or even months that a formal siege or blockade might demand.[28] The tactic of surprise was the best of all, offering the chance of gaining a town with minimal losses of both time and casualties, while requiring only a small number of troops that could be quickly assembled by concentrating detachments from several garrisons together. The Dutch governor of Bergen-op-Zoom, the count de Noyelles, explained to Heinsius how the constraints he operated under limited his choice of tactics: "If I can attempt something with the few troops that have been left to me, I will, that is, if I see an opportunity to surprise Zandvliet and capture it by *coup de main*, because I do not have the troops necessary to attack it *dans les formes*."[29] In fact, the success of these three non-siege tactics usually depended on there being little opposition from an enemy relief force, i.e. they had to take place away from the main concentration of field forces.

A siege, in contrast, was not to be undertaken lightly since it required significant additional planning and preparations. Among other things it demanded competent engineers to direct the attacks, thousands of men for the besieging army and many more for an observation force, the resources and time to gather and transport the numerous artillery and tons of supplies to the site, time to approach and breach the walls, while along with all this effort in

---

[27] W. Blackley (ed.), *The diplomatic correspondance of the Right Hon. Richard Hill . . . envoy extraordinary from the court of St. James to the duke of Savoy in the reign of Queen Anne: from July 1703 to May 1706*, (London, 1845), p. 621, Hill to Hedges, Turin, 16 September 1705. Two years previously, Louis had decided to blockade the citadel of Nice since siege artillery was lacking. Vault and Pelet (eds.), *Mémoires militaires*, vol. 5, p. 124.

[28] Villeroi preferred Zandvliet be captured by storm after a few hours of breaching fire, rather than by siege, which would give the enemy time to come and attempt relief. AG A[1] 1838, #331, Villeroi to Chamillart, 26 October 1705.

[29] Veenendaal, Jr. (ed.), *Briefwisseling Heinsius*, vol. 1, p. 123 #188, Noyelles en Fallais to Heinsius, Roosdaal, 23 April 1702.

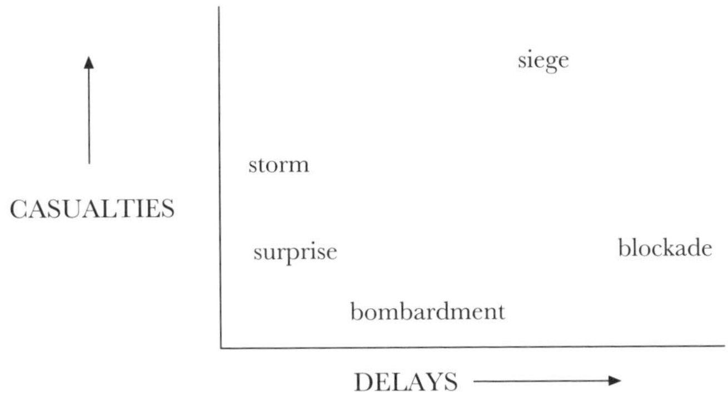

Figure D.1 Theoretical Results of Positional Tactics

time and money a significant amount of blood would likely be spilled as well. Given these requirements, attempting to convert a blockade, bombardment or surprise into a formal siege required additional time, resources and planning, as the French learned to their loss at Brussels in September 1708.[30] Less-proficient armies, unable to muster the necessary resources and expertise, tended to replace formal sieges with bombardments, blockades, or storms.[31] Figure D.1 illustrates the theoretical demands of each tactic.

Sieges multiplied many-fold the artillery and munition requirements of a bombardment, the stamina and logistical infrastructure of a blockade, as well as the manpower of a storm. For the offensive-minded commander, a siege was a unique challenge to be avoided whenever possible.

[30] The Elector of Bavaria led 14 battalions and 18 squadrons to surprise the town with the assistance of the inhabitants. The promised fifth column never materialized, and the French attempted a brief siege that was lifted upon the approach of Marlborough's relief army. The Spanish minister Jan van Brouchoven, graaf van Bergeyck commented that a proper siege of Brussels required far more men and artillery than they had at the site (AG A[1] 2083 #96, 20 September 1708). For his part, Voysin questioned the Elector's decision to conduct a siege *dans les formes* when the garrison had as many troops as the attackers (AG A[1] 2084 #136, 28 November; and AG A[1] 2084, #136, 28 November). See also Burgundy's comments in Baudrillart (ed.), *Lettres du duc de Bourgogne au roi d'Espagne Philippe V*, vol. 1, p. 335 #103, Burgundy to Chamillart, Saussois, 21 September.

[31] For an earlier example from Britain, see James Burke, "The New Model Army and the problems of siege warfare, 1648–51," *Irish Historical Studies* 27(105) (May 1990), pp. 7–29.

A preponderance of sieges (and, to a lesser extent, blockades) then might indicate a far more static tempo of warfare than one of bombardments, surprisals and storms. The broader offensive-defensive balance across the early modern period remains a mystery without a comparison of the type of tactics: perhaps siege tactics *per se* did not improve much over time, because one period's siegecraft was actually dominated by non-siege techniques (e.g. Parker's long blockades) while in a later period sieges became dominant (Israel's shorter sieges), perhaps due to improved logistics. Considering the costs of positional warfare to early modern society, it would be critical to identify each of the tactics independently: later 17th century sieges might be shorter but still more costly than the longer blockades of previous eras (measured in absolute terms rather than relative to the State's ability to pay, another issue worth investigating).

The following table lists various non-siege attempts in the Low Countries theater.

Table D.2 Non-Siege Tactics in the Flanders Theater

Place	Year	Side	Tactic	Capture
Huy	1702	Fr	Surprise	No
Namur	1702	All	Surprise	No
Fort Isabelle	1702	All	Bombard	Yes
Middelburg château	1702	Fr	Surprise	Yes
Rheinberg	1702	All	Blockade	Yes
Tongeren	1703	Fr	Bombard	Yes
Gelders (Guelders)	1703	All	Bombard	No
Gelders (Guelders)	1703	All	Blockade	Yes
Jülich	1703	All	Storm	No
Namur	1704	All	Bombard	No
Bruges	1704	All	Bombard	No
Liège	1705	Fr	Surprise	No
Dendermonde	1706	All	Bombard	No
Dendermonde	1706	All	Blockade	No
Fort Rouge	1708	Fr	Storm	Yes
Saint-Ghislain	1708	Fr	Surprise	Yes
Ath	1708	Fr	Surprise	No
Bruges	1708	Fr	Surprise	Yes
Ghent	1708	Fr	Surprise	Yes
Fort Plassendaal	1708	Fr	Storm	Yes
Saint-Amand	1709	All	Surprise	Yes
Marchiennes	1709	All	Surprise	No
Leuven	1710	Fr	Storm	No

Table D.2 (*cont.*)

Place	Year	Side	Tactic	Capture
Menin	1710	Fr	Surprise	No
Liège	1710	Fr	Surprise	No
Ypres	1710	All	Surprise	No
Fort Scarpe	1710	Fr	Surprise	No
Menin	1711	Fr	Surprise	No
Douai	1711	Fr	Surprise	No
Arras	1712	All	Bombard	No
Le-Cateau-Cambrésis	1712	All	Surprise	No
Fort Kenoque	1712	All	Surprise	Yes
Marchiennes	1712	Fr	Storm (twice)	No

Table Sources and Notes

Huy 1702: Marquis de Quincy, *Histoire militaire du règne de Louis le Grand,* 3:540; Veenendaal, Jr. (ed.), *Briefwisseling Heinsius,* 1:#310, #340.
Namur 1702: Marquis de Quincy, *Histoire militaire du règne de Louis le Grand,* 3:539; Vault and Pelet (eds.), *Mémoires militaires,* 2:36.
Fort Isabelle 1702: *Daily Courant,* 6 June 1702.
Middelburg 1702: Vault and Pelet (eds.), *Mémoires militaires,* 2:51.
Rheinberg 1702: Vault and Pelet (eds.), *Mémoires militaires,* 2:615ff., Détail du siège de Rhinberg.
Tongeren 1703: *Europische Mercurius,* May 1703, 287. Villeroi's army arrived in early May and opened fire with several 12–pounders, whereupon the Allied garrison immediately surrendered.
Gelders 1703 bombardment: *Europische Mercurius,* January 1703, 95.
Gelders 1703 blockade: Wijn, *Het Staatsche Leger,* 383ff.
Jülich 1703: *Europische Mercurius,* September 1703, 226.
Namur 1704: *Europische Mercurius,* October 1704, 208.
Bruges 1704: *Europische Mercurius,* July 1704, 78.
Liège 1705: *Europische Mercurius,* May 1705, 312.
Dendermonde 1706 bombardment: BL Add MSS 61163, f. 223, Meredith to Marlborough 24 June.
Dendermonde 1706 blockade: AG A[1] 1937, #149, from Dendermonde 14 June.
Fort Rouge 1708: Vault and Pelet (eds.), *Mémoires militaires,* 8:43.
Saint-Ghislain 1708: Vault and Pelet (eds.), *Mémoires militaires,* 8:147.
Ath 1708: Vault and Pelet (eds.), *Mémoires militaires,* 8:119; *Europische, Mercurius* October 1708, 285.
Bruges 1708: Vault and Pelet (eds.), *Mémoires militaires,* 8:27.
Ghent 1708: Vault and Pelet (eds.), *Mémoires militaires,* 8:381–385.
Plassendaal 1708: Vault and Pelet (eds.), *Mémoires militaires,* 8:27.
Saint-Amand 1709: Murray (ed.), *Letters and Dispatches,* 4:520.
Marchiennes 1709: Lecestre (ed.), *Mémoires du chevalier de Quincy,* 2:341.
Saint-Ghislain 1709: Murray (ed.), *Letters and Dispatches,* 4:593.
Saint-Ghislain 1709: Vault and Pelet (eds.), *Mémoires militaires,* 9:87.
Leuven 1710: Lecestre (ed.), *Mémoires du chevalier de Quincy,* 3:35.
Menin 1710: AG Article 15 Section 3 Menin folder, Caligny letter, 22 August.

Liège 1710: *Europische mercurius*, July 1710, 128.

Ypres 1710: Veenendaal, Jr. (ed.), *Briefwisseling Heinsius*, 10:422 #832, Nicolaes Pesters to Heinsius, Brebières 9 June; Murray (ed.), *Letters and Dispatches*, 5:49, Marlborough to Boyle, Douai 12 June. The partisan suggesting the surprise project to the Allies may have been a French double-agent. Soulié (ed.), *Journal du marquis de Dangeau publié en entier pour la première fois . . . avec les additions inédits du Duc de Saint-Simon*, 19 vols. (Paris, 1857), 13:183, 13 June.

Fort Scarpe 1710: Veenendaal, Jr. (ed.), *Briefwisseling Heinsius*, 11:190 #353, Nicolaes Pesters to Heinsius, Aire 27 September.

Menin 1711: BL Add MSS 61315, ff. 19–20, Lynden to Marlborough, Menin 20 April.

Douai 1711: Lecestre (ed.), *Mémoires du chevalier de Quincy*, 3:100.

Arras 1712: Lecestre (ed.), *Mémoires du chevalier de Quincy*, 3:107. The bombardment was successful in its objective, which was to burn the fodder stores in the town.

Le-Cateau-Cambrésis 1712: Vault and Pelet (eds.), *Mémoires militaires*, 12:21.

Fort Kenoque 1712: Lecestre (ed.), *Mémoires du chevalier de Quincy*, 3:197.

Marchiennes 1712: Marchiennes was stormed unsuccessfully twice on the same day. It was then besieged *dans les formes*. *Daily Courant*, #3364 24 July O.S. from Tournai 27 July.

APPENDIX E

LENGTH ESTIMATES, SELECTED SIEGES

The following table lists specific estimates of siege lengths by siege, sorted chronologically from earliest to latest estimate within each siege. A few cautions are required at the outset, however, as there are several problematic aspects that caution us to consider this data *very* cautiously. Most importantly, we cannot assume that all the estimates for a particular siege come from a single source; rarely, for that matter, are the engineers specifically credited as the source of an estimate (those that explicitly state this are identified in *italics*). As the wording of many of the estimates indicates, authors usually expressed estimates only in terms of "hoping" for a capture by a certain date. Brigadier Isaac Cronström illustrated the difficulties with such information when he noted to Heinsius that Aire might not be captured by the time they had been promised it (20 October); as is often the case, those doing the promising remained unidentified, including their branch of service. He also pointed out that several engineers disputed this estimate, further complicating the matter.[1] Finally, this data does not necessarily invalidate less specific complaints of underestimation, such as non-quantified grievances, though such vague complaints are much less reliable without solid confirmation.

The validity of this data also depends on the unknown issue of how the engineers came to their estimates and, as we have just discussed, how they were reported by non-engineers. Did engineers assume that the commanders would work to accelerate the trenches and take this into account in their estimate? Did they calibrate their estimates in order to appease impatient commanders? Since the engineers did not dwell on estimating siege lengths (neither in their correspondence nor their manuals nor their siege journals), the answers to such questions are unknown, making it difficult to draw solid conclusions from this data.

---

[1] Veenendaal, Jr. (ed.), *Briefwisseling Heinsius*, vol. 11, p. 197 #361, from Saint-Venant, 29 September 1710.

Most of the columns are self-explanatory, which the exception of the following:
Siege: Name of town, type of fortification (T for town, F for fort, C for citadel), and the year of the siege.

Wording of Estimate: The phrasing (translated into English when necessary) of the estimate as found in the primary source. The vagueness of their language indicates how imprecise these estimates were.

Error: The number of days by which the estimate was off. A negative number indicates that the duration of the siege was underestimated, whereas a positive number shows that the duration of the defense was overestimated.

Error as % of length: The error expressed as a percentage of the overall length of the siege. This gives a standardized indication of how far off the estimate was, and will naturally be much higher for the shorter sieges.

+/–: The direction of the error expressed as a simple underestimation (–) or overestimation (+).

Table E.1 Estimates of Siege Durations

Siege	Author	Wording of Estimate	Error	Error as % of length	+/–
Venlo T 1702	H. Erskine	its thought	–8	–67%	–
Venlo T 1702	Marlborough	I hope	5	42%	+
Venlo T 1702	Obdam	we hope	1	4%	+
Venlo T 1702	*Marlborough*	*Coehoorn hopes to have*	*1*	*8%*	+
Limburg T 1703	Marlborough	It will be	9	225%	+
Limburg T 1703	Marlborough	I am afraid will cost	7	175%	+
Limburg T 1703	Marlborough	I hope	1	13%	+
Limburg T 1703	Marlborough	I hope	4	100%	+
Limburg T 1703	Marlborough	we may bee	3	75%	+
Limburg T 1703	*Trognée*	*won't hold out for*	*3*	*63%*	+
Huy C 1703	Marlborough	I am afraid it will be	4	44%	+
Huy 1703	H. Hamilton	will keep us about	8	89%	+
Huy C 1703	Hop	hope	–1	–6%	–
Huy C 1703	Welby	I believe	–2	–17%	–
Huy F Picard 1705	*Valory*	*hope*	*–2*	*–40%*	–
Huy C 1705	Villeroi	should be ready to assault	–4	–33%	–
Huy C 1705	*Valory*	*will force capit*	*–2*	*–13%*	–
Menin T 1706	Marlborough	I should be happy if	9	50%	+
Menin T 1706	Marlborough	thay promis	2	11%	+
Menin T 1706	Marlborough	I am afraid it must last	3	17%	+
Menin T 1706	Marlborough	we shall not be masters this	4	22%	+
Menin T 1706	Geldermalsen	I hope	3	14%	+

Table E.1 *(cont.)*

Siege	Author	Wording of Estimate	Error	Error as % of length	+/–
Dendermonde T 1706	Marlborough	if we have no rain... may	–5	–64%	–
Dendermonde T 1706	Marlborough	we hope to have	0	0%	0
Dendermonde T 1706	Ouwerkerk	I believe	6	79%	+
Ath T 1706	Marlborough	I hope we may be	–2	–17%	–
Ath T 1706	Marlborough	I hope we may be	2	17%	+
Ath T 1706	Marlborough	we hope to be masters of	–1	–8%	–
Ath T 1706	Marlborough	we hope to have it by	4	33%	+
Ath T 1706	Geldermalsen	no more than	3	25%	+
Lille T 1708	Rivière	Allies claim	–37	–61%	–
Lille T 1708	*Stair*	*our engineers tell us*	*–3*	*–5%*	–
Lille C 1708	*Des Rocques*	*reason to believe will finish*	*3*	*6%*	+
Tournai T 1709	Hop	reports say no more than	–12	–55%	–
Tournai T 1709	*Marlborough*	*engineers give us hopes*	*–5*	*–20%*	–
Tournai T 1709	Vegelin	one hopes	–5	–20%	–
Tournai T 1709	Goslinga	I hope	4	18%	+
Tournai C 1709	Revett	we fear it will take up	–3	–9%	–
Tournai C 1709	Marlborough	I hope	–4	–12%	–
Tournai C 1709	Cranstoun	not probable before	15	44%	+
Tournai C 1709	Albemarle	I have reason to believe	2	6%	+
Tournai C 1709	French source	Allied officers not hope before	17	50%	+
Tournai C 1709	Laws	hope at the end of	–3	–9%	–
Douai T 1710		confidently reported here it cannot defend itself above	–41	–77%	–
Douai T 1710	Bernières	Cadogan said won't last above	–33	–62%	–
Douai T 1710	Cadogan	hope before end of month	–26	–49%	–
Douai T 1710	Cardonnel	Marlborough so sanguine to hope	–26	–49%	–
Douai T 1710	Camp letter	we hope by	–20	–38%	–
Douai T 1710	*Marlborough*	*our engineers*	*–18*	*–34%*	–
Douai T 1710	*Marlborough*	*our engineers*	*–16*	*–30%*	–
Douai T 1710	Hooft	it is hoped that	–25	–47%	–
Douai T 1710	Godolphin	we apprehend it may	–16	–30%	–
Douai T 1710	Marlborough	I hope	–22	–42%	–
Douai T 1710	Cadogan	in all probability will have	–14	–26%	–
Douai T 1710	Marlborough	we believe (engrs excluded)	–6	–11%	–
Douai T 1710	courrier	it is believed	–16	–30%	–
Douai T 1710	*Des Rocques*	*one can estimate till at least*	*–6*	*–10%*	–
Béthune T 1710	Vegelin	I hope	–9	–64%	–
Béthune T 1710	Marlborough	we hope	–9	–24%	–
Aire T 1710	Albemarle	I expect	21	36%	+
Aire T 1710	Cronstrom	one expects	–20	–34%	–
Aire T 1710	Cardonnel	most people	2	3%	+
Douai T 1712	Bernières	expect	–7	–28%	–

Table E.1 (*cont.*)

Siege	Author	Wording of Estimate	Error	Error as % of length	+/–
Douai T 1712	Broglie	I am persuaded	–19	–76%	–
Douai T 1712	Broglie	expect	–8	–32%	–
Douai T 1712	Destouches	don't think as long as	4	16%	+
Douai T 1712	Broglie	expect	–4	–16%	–
Douai T 1712	Broglie	probably less than	2	8%	+
Douai T 1712	Broglie	I am persuaded	–1	–2%	–
Douai T 1712	Alègre	those who know (not me)	4	16%	+
Douai F 1712	Contades	expect	–1	–8%	–
Douai F 1712	Broglie	expect	2	12%	+
Le Quesnoy T 1712	Alègre	expect	7	44%	+
Le Quesnoy T 1712	Alègre	the most common opinion	1	6%	+
Le Quesnoy T 1712	Alègre	I'll be happy if by	6	38%	+
Le Quesnoy T 1712	*Alègre*	*Valory promises by*	*4*	*25%*	+
Le Quesnoy T 1712	Alègre	I still say by	6	38%	+
Le Quesnoy T 1712	Contades	at latest	–1	–6%	–
Bouchain T 1712	Alègre	most common expect	6	60%	+
Bouchain T 1712	Alègre	hope by	2	20%	+

Table Sources

Venlo 30 August 1702: Royal Commission on Historical Manuscripts, *Report on the Manuscripts of the Earl of Mar and Kellie*, 225, Harry Erskine to the Earl of Mar, Breda.
Venlo 14 September 1702: Snyder (ed.), *Marlborough-Godolphin Correspondence*, 1:111 #97 Marlborough to Godolphin 3 September 1702 O.S.
Venlo 18 September 1702: Veenendaal, Jr. (ed.), *Briefwisseling Heinsius*, 1:429 Wassenaar-Obdam to Heinsius, Venlo.
Venlo 21 September 1702: Snyder (ed.), *Marlborough-Godolphin Correspondence*, 1:112 #99 Marlborough to Godolphin, Zutendaal 10 September 1702 O.S.
Liège 5 October 1702 O.S.: Snyder (ed.), *Marlborough-Godolphin Correspondence*, vol. 1:121 #106 Marlborough to Godolphin, Liège.
Limburg ? 1703: *The Daily Courant*, 20 September 1703 O.S.
Limburg 16 September 1703: Snyder (ed.), *Marlborough-Godolphin Correspondence*, 1:241 #238 Marlborough to Godolphin 5 September 1703 O.S.
Limburg 20 September 1703: Snyder (ed.), *Marlborough-Godolphin Correspondence*, 1:242 #240 Marlborough to Sarah Churchill, Verviers 9 September 1703 O.S.
Limburg 21 September 1703: Murray (ed.), *Letters and Dispatches*, 1:184 Marlborough to Ouwerkerk, Verviers.
Limburg 24 September 1703: Van 't Hoff (ed.), *Marlborough-Heinsius Correspondence*, p. 92 Marlborough to Heinsius, Verviers.
Limburg ? 1703: *Daily Courant*, 20 September 1703 O.S.
Huy 16 August 1703: Snyder (ed.), *Marlborough-Godolphin Correspondence*, 1:231 #225 Marlborough to Sarah Churchill 5 August 1703 O.S.
Huy 20 August 1703: Royal Commission on Historical Manuscripts, *The Manuscripts of the Earl Cowper*, 3:25 Hans Hamilton to Coke, Huy 9 August 1703 O.S.
Huy 23 August 1703: Veenendaal, Jr. (ed.), *Briefwisseling Heinsius*, 2:416 #1066 Jacob Hop to Heinsius.
Huy 23 August 1703: Royal Commission on Historical Manuscripts, *The Manuscripts of the Duke of Beaufort, K.G., The Earl of Donoughmore, and Others*, 96, T. Welby to the Earl of Coventry, Huy 12 August 1703 O.S.

Huy 30 May 1705 (French siege): AG A[1] 1835, #193 Valory to Chamillart, Huy.
Huy 4 June 1705: AG A[1] 1835, #222 Villeroi, Vignamont.
Huy 5 June 1705: AG A[1] 835, #232 Valory to Chamillart, Huy.
Oostende 1 July 1706: Snyder (ed.), *Marlborough-Godolphin Correspondence*, 2:594 #606 Marlborough to Godolphin, Rosselare.
Menin 2 August 1706: Marlborough complained of the delays: "I should be happy if we might be sure of being masters of it by the end of this month." Snyder (ed.), *Marlborough-Godolphin Correspondence*, 2:631 #638 Marlborough to Godolphin, Helchin.
Menin 5 August 1706: Snyder (ed.), *Marlborough-Godolphin Correspondence*, 2:633 Marlborough to Godolphin, Helchin.
Menin 12 August 1706: Snyder (ed.), *Marlborough-Godolphin Correspondence*, 2:641 Marlborough to Sarah Churchill, Helchin.
Menin 16 August 1706: Snyder (ed.), *Marlborough-Godolphin Correspondence*, 2:644 #651 Marlborough to Godolphin, Helchin.
Menin 19 August 1706: Veenendaal, Jr. (ed.), *Briefwisseling Heinsius*, vol. 5, p. 477 #918 Geldermalsen to Heinsius, Menin.
Dendermonde 26 August 1706: Snyder (ed.), *Marlborough-Godolphin Correspondence*, 2:652 #660 Marlborough to Godolphin, Helchin.
Dendermonde 27 August 1706: Murray (ed.), *Letters and Dispatches*, 3:106 Marlborough to Hop, Helchin.
Dendermonde 4 September 1706: Veenendaal, Jr. (ed.), *Briefwisseling Heinsius*, vol. 5, p. 511 Ouwerkerk to Heinsius, Helchin.
Ath 18 September 1706: Snyder (ed.), *Marlborough-Godolphin Correspondence*, 2:672 # 679 Marlborough to Godolphin, Grandmetz.
Ath 20 September 1706: Snyder (ed.), *Marlborough-Godolphin Correspondence*, 2:673 #681 Marlborough to Godolphin, Grandmetz.
Ath 22 September 1706: Murray (ed.), *Letters and Dispatches*, 3:143 Marlborough to Prince Salm.
Ath 30 September 1706: Snyder (ed.), *Marlborough-Godolphin Correspondence*, 2:690 #697 Marlborough to Sarah Churchill, Grandmetz.
Ath 2 October 1706: BL Add MSS 61174, f. 92b Geldermalsen to Marlborough, Ath.
Lille town ? 1708: Sautai, *Le siège de la ville et de la citadelle de Lille en 1708*, pp. 91–92.
Lille town 15 October 1708: Royal Commission on Historical Manuscripts, *Report on the Manuscripts of the Earl of Mar and Kellie*, p. 465 Stair to Mar, Rosselare.
Lille citadel 4 December 1708: BL Add MSS 61312, f. 196 Des Roques to Marlborough, Lille.
Tournai town 3 July 1709: Veenendaal, Jr. (ed.), *Briefwisseling Heinsius*, vol. 9, p. 10 Hop to Heinsius, Lille.
Tournai town 18 July 1709: Murray (ed.), *Letters and Dispatches*, 4:549 Marlborough to Boyle, Tournai.
Tournai town 21 July 1709: Veenendaal, Jr. (ed.), *Briefwisseling Heinsius*, vol. 9, p. 79 Vegelin van Claerbergen to Heinsius, Tournai.
Tournai town 25 July 1709: Veenendaal, Jr. (ed.), *Briefwisseling Heinsius*, vol. 9, p. 94 Goslinga to Heinsius, Tournai.
Tournai citadel 1 August 1709: Royal Commission on Historical Manuscripts, *Report on the Manuscripts of Mrs. Frankland-Russell-Astley of Chequers Court, Bucks*, (London, 1900), p. 198 Revett to Pollhill, Tournai.
Tournai citadel 12 August 1709: Van 't Hoff (ed.), *Marlborough-Heinsius Correspondence*, p. 454 Marlborough to Heinsius, Orchies.
Tournai citadel 16 August 1709: Royal Commission on Historical Manuscripts, *The Manuscripts of His Grace the Duke of Portland*, (London, 1891), vol. 4, p. 497 Cranstoun to Cunningham, Tournai 5 August 1709 O.S.
Tournai citadel 24 August 1709: Veenendaal, Jr. (ed.), *Briefwisseling Heinsius*, vol. 9, p. 194 Albemarle to Heinsius, Tournai.

Tournai citadel 30 August 1709: AG A^1 2154, #274 conseiller de Parlement de Tournay.
Tournai citadel 31 August 1709: PRO SP 77/58, f. 196.
Douai 25 April 1710: Royal Commission on Historical Manuscripts, *The Manuscripts of the Marquess Townshend*, 64, dated 25 April 1710.
Douai 1 May 1710: AG A^1 2225, #133 Bernières to Voysin.
Douai 8 May 1710: PRO SP 77/59, f. 115–116b Cadogan to Sunderland, Douai.
Douai 12 May 1710: BL Add MSS 61401, f. 25 Cardonnel to Walpole.
Douai 16 May 1710: AG A^1 2220, #123 Sailly.
Douai 1710: *Postman*, From the confederate camp near Doway 20 May 1710 O.S.
Douai 22 May 1710: Murray (ed.), *Letters and Dispatches*, vol. 5, p. 30 Marlborough to Boyle, Douai.
Douai 22 May 1710: Snyder (ed.), *Marlborough-Godolphin Correspondence*, vol. 3, p. 1490 Marlborough to Godolphin, Douai.
Douai 24 May 1710: Veenendaal, Jr. (ed.), *Briefwisseling Heinsius*, vol. 10, p. 378 Hooft to Heinsius, Lallaing.
Douai 27 May 1710: Snyder (ed.), *Marlborough-Godolphin Correspondence*, vol. 3, p. 1496 Godolphin to Marlborough, 16 May 1710 O.S.
Douai 2 June 1710: Murray (ed.), *Letters and Dispatches*, vol. 5, p. 39 Marlborough to Boyle, Douai.
Douai 2 June 1710: PRO SP 77/59, ff. 138–138b Cadogan to Sunderland, Douai.
Douai 5 June 1710: Snyder (ed.), *Marlborough-Godolphin Correspondence*, vol. 3, p. 1506 Marlborough to Godolphin, Douai.
Douai June 1710: *London Gazette*, p. 2, Hague 25 May 1710 O.S.
Douai 12 June 1710: Veenendaal, Jr. (ed.), *Briefwisseling Heinsius*, vol. 10, p. 430 des Rocques to Heinsius, Douai.
Béthune 7 August 1710: Veenendaal, Jr. (ed.), *Briefwisseling Heinsius*, vol. 11, p. 29 Vegelin van Claerbergen to Heinsius, Bevri.
Béthune 7 August 1710: Van 't Hoff (ed.), *Marlborough-Heinsius Correspondence*, 512 #882 Marlborough to Heinsius, Villers Brulin.
Aire 31 August 1710: Veenendaal, Jr. (ed.), *Briefwisseling Heinsius*, vol. 11, p. 107 Albemarle to Heinsius, Tournai.
Aire 29 September 1710: Veenendaal, Jr. (ed.), *Briefwisseling Heinsius*, vol. 11, p. 197 Cronström to Heinsius, Saint-Venant.
Aire 7 November 1710: BL Add mss 61401, f. 124 Cardonnel to Lumley, Aire.
Bouchain 7 September 1711: Van 't Hoff (ed.), *Marlborough-Heinsius Correspondence*, 563 #985 Marlborough to Heinsius, Bouchain.
Douai town 16 August 1712: AG A^1 2382, #7 Bernières.
Douai town 17 August 1712: AG A^1 2382, #22 Broglie to Voysin.
Douai town 18 August 1712: AG A^1 2382, #30 Broglie to Voysin, Auby.
Douai town 19 August 1712: AG A^1 2382, #37 Broglie.
Douai town 20 August 1712: AG A^1 2382, #54 Valory to Voysin.
Douai town 24 August 1712: AG A^1 2382, #132 Broglie.
Douai town 2 September 1712: AG A^4 Carton 8 2, 2 Alègre to Maine.
Douai town 4 September 1712: AG A^4 Carton 8 2, Alègre to Maine.
Douai fort Scarpe 22 August 1712: AG A^1 2382, #97 Contades.
Douai fort Scarpe 24 August 1712: AG A^1 2382, #132 Broglie.
Le Quesnoy 12 September 1712: AG A^4 Carton 8 2, Alègre to Maine.
Le Quesnoy 18 September 1712: AG A^4 Carton 8 2, 2 Alègre to Maine.
Le Quesnoy 18 September 1712: AG A^4 Carton 8 2, 2 Alègre to Maine.
Le Quesnoy 23 September 1712: AG A^4 Carton 8 2, 2 Alègre to Maine.
Le Quesnoy 23 September 1712: AG A^4 Carton 8 2, 2 Alègre to Maine.
Le Quesnoy 1 October 1712: AG A^1 2385, #2 Contades.
Bouchain 6 October 1712: AG A^4 Carton 8 2, 2 Alègre to Maine.
Bouchain 17 October 1712: AG A^1 2386, #31 Alègre.

APPENDIX F

SIEGE ARTILLERY

I have only included in this table sieges for which I have data on at least one of the two side's guns-26 of the sieges in the theater, including a representative sample of both short and long sieges. Generally speaking, collecting data on siege trains is extremely difficult for several reasons. Not only are sources often lacking, but those accounts that do mention numbers of cannon are usually limited to when specific batteries opened fire (whether these guns came from old, abandoned batteries or added their firepower to the older batteries remains unspecified). They rarely give the calibers of the guns, and they rarely report the several different measures of besieging artillery: the number of cannon in a single battery (and the number of batteries firing at one time), the number of cannon opening fire on a particular day, and the total number of guns in the siege train (including how many of these were used in the siege vs. in the lines of circumvallation/contravallation). The data in the table is the number of guns opening fire, unless otherwise noted in the notes. As the source notes also detail, different accounts (even from the same side) can vary wildly in the number of guns reported, and additional artillery reinforcements might be sent over the course of a long siege. Therefore, this data can do no more than give a general sense of the magnitude of train sizes.

The number of garrison pieces does not necessarily indicate whether these were all used in the siege—a lack of carriages, ammunition, battery locations, etc. could quickly put a majority of a garrison's guns out of service.

Key

c/m/h/hm: cannons/mortars/howitzers/hand-mortars.
c/m/p: cannons/mortars/*pierriers*.

Table F.1 Siege Artillery in Selected Flanders Sieges

Siege	Total Besieger (c/m/h/hm)	Total Garrison (c/m/p)
Ath 1697	42/22	32
Kaisersweert 1702	80/59/6/70	30
Venlo town 1702	70/40/108	47/10
Stevensweert 1702		21
Roermond 1702	54/44	30/2
Liège citadel 1702	72/48/0/200	42/5
Huy 1703	70/46	
Limburg 1703	49	
Huy 1705 French	32/16	
Huy 1705 Allies	74/34	
Zoutleeuw 1705	16	
Oostende 1706	37/18/20	
Menin 1706	60/44	65/6
Dendermonde 1706	32/15/10/80	22
Ath 1706	76/16	
Lille town 1708	120/40/40/?	
Tournai town 1709	130/40	
Mons 1709	105/29	
Douai 1710	288/80	
Béthune 1710	43/26/?/170	27/11
Saint-Venant 1710		7
Aire 1710	73	32
Bouchain 1711	38/8	
Le Quesnoy 1712	30/24	46/4/3
Douai town 1712	50/30/0/6	
Le Quesnoy 1712	72/30/12p	102/47
Bouchain 1712	40	23/2/2

Sources and Notes

Kaisersweert 1702: Wijn, *Het Staatsche Leger*, 1:52 (besieger). Wijn, *Het Staatsche Leger*, 1:50 (garrison).

Venlo town 1702: According to a French journal of the siege printed in *Le Mercure Galant*, octobre 1702, Relation de ce qui s'est passé à la deffense de Venloo, 327–328. For the garrison, the *Europische Mercurius* claims an enormous 160 cannon and 30 mortars, but Geldermalsen's account indicates only 47 pieces and 10 mortars, a more likely figure given his status as field deputy in charge of supplying the artillery. Wijn, *Het Staatsche Leger*, 1:151.

Stevensweert 1702: Wijn, *Het Staatsche Leger*, 1:171. The besieging commander found in the town between twenty and thirty guns. Veenendaal, Jr. (ed.), *Briefwisseling Heinsius*, 1:462 #949, Wassenaar-Obdam to Heinsius, Roermond 5 October.

Roermond 1702: When the besiegers first opened fire, in Wijn, *Het Staatsche Leger*,

1:172. For the garrison, see Murray (ed.), *Letters and Dispatches*, 1:45, bulletin from Zutendaal 9 October.

Liège citadel 1702: Murray (ed.), *Letters and Dispatches*, 1:47 Marlborough, Liege 19 October. The Allies initially opened fire with between 40 and 50 cannon and 28 mortars. Wijn, *Het Staatsche Leger*, 1:185–186. An English army bulletin says they initially opened fire with 50 battering pieces in addition to smaller 12-pdrs and 8–pdrs, as well as 48 mortars and 200 handmortars. Murray (ed.), *Letters and Dispatches*, 1:47, Liège 19 October. For the garrison, Wijn, *Het Staatsche Leger*, 1:187.

Huy 1703: Indicative of the imprecision of these numbers, Rorive's exhaustive study of the sieges of Huy indicate anywhere from 45 to 70 cannon and 30 to 46 mortars. Rorive, *La guerre de siège*, 217 and 254. According to the *Relation de ce qui s'est passé à l'attaque de la ville et château de Huy* written by the commander, 38 cannon and 40 mortars opened fire against Ft. St. Joseph; Huy's château was fired on by 46 (later 56) cannon. Raoul de Linière, "M. de Millon et Marlborough aux Sièges de Liege et de Huy, 1702–3," *Revue historique et archéologique du Maine*, 54 (1903), 255–256. Vault and Pelet mention 70 cannon and 46 mortars.

Limburg 1703: Snyder (ed.), *Marlborough-Godolphin Correspondence*, 1:241 #238, Marlborough to Godolphin, 5 September.

Huy 1705 French: Rorive, *La guerre de siège*, 217–218 and 254. Valory mentions 22 cannon and 10 mortars planning to open fire. AG A[1] 1835, #193, Valory to Chamillart, Huy 30 May. The siege commander claims 24 24-pdrs and 6 mortars were to be used against the château. AG A[1] 1835, #235 Villeroi to Chamillart, Vignamont 6 June.

Huy 1705 Allies: Rorive, *La guerre de siège*, 217–218 and 254.

Zoutleeuw 1705: ARA collectie Wassenaer 5, Mémoire touchant le siège de Leewe, 27 August.

Oostende 1706: This is the artillery available when the siege first began. BL Add MSS 61179, ff. 151–151b, Ouwerkerk to Marlborough, Oostende 22 June. See also A.J. Veenendaal Jr., "De Mémoires van Sicco van Goslinga," 22. To the Allied siege train we must include the naval fleet that helped bombard the town (Wijn, *Het Staatsche Leger*, 2:86 note 1). Before the town was invested Goslinga's memoirs claimed the garrison had 80 cannon. When the besieger's guns opened fire, he only mentions two enemy pieces responding. A.J. Veenendaal Jr., "De Mémoires van Sicco van Goslinga," 22 and 23. Wijn, *Het Staatsche Leger*, 2:79.

Menin 1706: Wijn, *Het Staatsche Leger*, 2:95 gives a total of more than 102 pieces. Van Nimwegen, *De subsistentie*, 186 says 80 to 100 24-pdrs and 20–30 howitzers. On 9 August, sixty cannon and forty-four mortars opened fire according to BL Add MSS 4742, f. 68b, Journal of the siege of Menin, Helchin 9 August. 40 cannon and 40 mortars according to Vault and Pelet (eds.), *Mémoires militaires*, 6:542 Journal du siége de Menin 9 August. According to AG A[1] 1988, #57 Relation ou journal du siège de Menin 1706, 4 August, 70 24-pdrs and smaller pieces opened fire on 9 August. For the garrison, see *Daily Courant*, 26 August 1706 O.S. Helchin 31 August. The *Antwerp Courant* claims 118 cannon and 59 mortars.

Dendermonde 1706: BL Add MSS 61335, f. 104, Mémoire de ce qui sera nécessaire pour le siège de Termonde. Marlborough indicated that they were to open fire with 36 cannon and 15 mortars—the *Europische Mercurius*, September 1706 p. 224 says 36 breaching guns and twenty smaller calibers. Murray (ed.), *Letters and Dispatches*, 3:115 Marlborough to Hedges, Dendermonde, 3 September. For the garrison's guns, see Wijn, *Het Staatsche Leger*, 2:47.

Ath 1706: Veenendaal, Sr. (ed.), *Het Dagboek van Gisbert Cuper*, 256 #11, Cuper to van den Bergh, Velaines 25 September. On the discrepancies among sources, see Wijn, *Het Staatsche Leger*, 2:124–125.

Lille town 1708: Wijn, *Het Staatsche Leger*, 4:342, firing from the first parallel around 7 September.

Tournai town 1709: PRO SP 77/58, f. 127b, John Laws' newsletter from Brussels 4 July. AG A[1] 2159, #413, Nebermont, Tournai 30 July reports the town being battered by 109 cannon and 40 mortars on 30 July.

Mons 1709: Wijn, *Het Staatsche Leger*, 2:550, 552–553.

Douai 1710: The total number of guns. Allard, "Relation du siege de Douay en 1710," *Le Spectateur militaire*, 16 1834, 354. Wijn, *Het Staatsche Leger*, 2:605 gives opening fire with 70 heavy cannon and 80 hand-mortars.

Béthune 1710: Wijn, *Het Staatsche Leger*, 2:605. ARA CvdH 20 mentions 170 hand-mortars. The 1 July (O.S.) *Postman* reports the arrival of an 80-piece convoy on 6 July. From the camp near Doway 7 July. For the defenders, see Wijn, *Het Staatsche Leger*, 2:651.

Saint-Venant 1710: The garrison's guns. Wijn, *Het Staatsche Leger*, 2:668.

Aire 1710: Opening fire 18 September. Wijn, *Het Staatsche Leger*, 2:673. See Wijn, *Het Staatsche Leger*, 2:687 for the defender's pieces.

Bouchain 1711: Firing 30 August. Wijn, *Het Staatsche Leger*, 3:60.

Le Quesnoy 1712: Firing 26 June. Wijn, *Het Staatsche Leger*, 3:160–161. Millner gives 100 cannon, 40 mortars and howitzers, with a garrison artillery of 40 guns and 8 mortars. For the garrison, see Wijn, *Het Staatsche Leger*, 3:165.

Douai town 1712: The total. AG A[1] 2382, #33, Destouches 18 August. Initially the French opened fire with only 26 cannon and 20 mortars due to a shortage of horse teams. AG A[1] 2382, #18, Bernières to Voysin 17 August.

Le Quesnoy 1712: Vault and Pelet (eds.), *Mémoires militaires*, 11:110–116. A French account lists 56 cannon, 30 mortars and 12 *pierriers* opening fire. AG A[1] 2384, #78, Vallière 22 September and #94, Hermand to Voysin, Quesnoy 24 September, f. 2. AG Article 15 Section 2 §1 Douai folder, #19 Siège du Quesnoy en 1712, f. 17 says they opened fire with 50 cannon and 18 mortars. For the garrison, see Vault and Pelet (eds.), *Mémoires militaires*, 11:527 Etat de l'artillerie et des munitions trouvées dans le Quesnoy 4 October. The marquis de Quincy gives 50 heavy cannon and 20 mortars. Marquis de Quincy, *Histoire militaire du règne de Louis le Grand*, 3:184–197.

Bouchain 1712: For both the besieger's and garrison's pieces, consult AG Article 15 Section 2 §1 Douai folder, #19 Siége de Bouchain, 20.

BIBLIOGRAPHY

Archives

Service historique de l'armée de terre, Archives de Guerre (AG), Vincennes
Series A[1]
Series A[4] carton 8
Letters addressed to the Duc du Maine
1 K 299 Papiers Montmorency-Tingry
1M (was MR) 95 Relations
1M (was MR) 126–1 Mémoires historiques
1M (was MR) 126–3 Mémoires historiques
Article 8 Section 1 Places françaises
Article 15 Section 2 Histoire militaire, campagne et sièges: Sièges des places françaises
British Library (BL), London
Additional manuscripts (Add MSS), Blenheim Papers
Public Record Office (PRO), Kew
State Papers (SP)
Algemeen Rijksarchief (ARA)—now Nationaal Archief, The Hague
Familie-Archief Surendonck
Familie-Archief Wassenaar
Staten-Generaal (SG)
Raad van State (RvS)
Ryksargyf Friesland (RA), Leeuwarden
Familie-Archief Eysinga-Vegelin van Claerbergen (EVC)

Published Works

A.P. [A. Preux?], "Mémoires inédits de Ch. Caudron sur le siège de Douai en 1710." *Souvenirs de la Flandre wallonne* 10 (1870), 100–119.

Allard, "Appendice à la Relation du siège de Douay." *Le Spectateur militaire* 16 (1834), 609–626.

——, "Relation du siége de Douay en 1710." *Le Spectateur militaire* 16 (1834), 353–390, 492–529.

Allard, N., "Note sur Vauban: Sur les changements apportés par Vauban dans l'art de l'attaque et de la défense des places," *Le Spectateur militaire*, 18 (1835), 437–443.

Allent, *Histoire du corps impérial du génie . . . depuis l'origine de la fortification moderne jusqu'à la fin du règne de Louis XIV*, (Paris, 1805).

Anderson, M.S., *War and Society in Europe of the Old Regime 1618–1789*, (New York, 1988).

Anquetil, "Suite des mémoires rédigée par Anquetil," in A. Petitot and Monmerqué, (eds.), *Collection des Mémoires pour servir à l'historie de France*, vol. 69 (Paris, 1828).

Arnold, Thomas, "Fortifications and the Military Revolution: The Gonzaga Experience, 1530–1630," in C. Rogers (ed.), *The Military Revolution Debate: Readings on the Military Transformation of Early Modern Europe*, (Boulder, CO, 1995), 201–226.

——, *The Renaissance at War*, (London, 2001).

Atkinson, C.T., "Marlborough's Sieges," *Journal of the Society for Army Historical Research*, 13 (1935), 195–205.

Atkinson, Christopher, "Gleanings from the Cathcart MSS," *The Journal of the Society for Army Historical Research*, 29 (1951), 20–25, 64–68, 97–103.
——, *Marlborough and the Rise of the British Army*, (New York, 1921).
Augoyat, Antoine-Marie, "Siège du château d'Alicante et surprise de Tortose pendant la guerre de la Succession d'Espagne," *Le Spectateur militaire*, 43(254) 1847, 129–145.
——, *Aperçu historique sur les fortifications, les ingénieurs et sur le corps du Génie en France*, (Paris, 1860) vol. 1.
Bachrach, Bernard, "Medieval Siege Warfare: A Reconnaissance," *The Journal of Military History*, 58(1) 1994, 119–133.
Baudrillart, Alfred and Lecestre, Léon (eds.), *Lettres du duc de Bourgogne au roi d'Espagne Philippe V*, 3 vols. (Paris, 1912).
Baurin, Georges, "Les sièges de Namur de 1692 et 1695," in F. and Ph. Jacquet-Ladrier (eds.), *Assiégeants et assiégés au coeur de l'Europe, Namur 1688–1698*, (Brussels, 1991), 74–106.
Bély, Lucien, *Espions et ambassadeurs au temps de Louis XIV*, (Paris, 1990).
Bertin, Pierre, "Guerre de Trente Ans: Le siège de Dôle en 1636," in *Revue historique de l'armée*, (1970), 7–33.
Black, Jeremy, *European Warfare, 1660–1815*, (New Haven, 1994).
Blackley, W. (ed.), *The diplomatic correspondance of the Right Hon. Richard Hill . . . envoy extraordinary from the court of St. James to the duke of Savoy in the reign of Queen Anne: from July 1703 to May 1706*, (London, 1845).
Blanchard, Anne, *Les ingénieurs du "Roy" de Louis XIV à Louis XVI: Etude du corps des fortifications*. (Montpellier, 1979).
——, *Vauban*, (Paris, 1996).
Bland, Humphrey, *A Treatise of Military Discipline; in which is laid down and explained the duty, of the officer and soldier, thro the several branches of the service. By Humphrey Bland, Esq; Lieutenant-Colonel of His Majesty's Own Regiment of Horse*, (London, 1727).
Blay, Michel, "Le développement de la balistique et la pratique du jet des bombes en France à la mort de Colbert," in L. Godard de Donville (ed.), *De la mort de Colbert à la révocation de l'édit de Nantes, un monde nouveau? XIV*[e] *colloque au Centre méridional de rencontre sur le XVII*[e] *siècle*, (Marseilles, 1984), 33–50.
Blomfield, Reginald, *Vauban*, (New York, 1971 reprint of 1938 edition).
Bluche, François, "Laparra de Fieux" in F. Bluche (ed.), *Dictionnaire du Grand Siècle*, (Paris, 1990), 828–829.
Bodart, G., *Militär-historisches kriegs-lexicon (1618–1905)*, (Vienna, 1907–1908).
Bois, Jean-Pierre, "Armes, tactiques et batailles d'Azincourt à Fontenoy" in *Armes et Alliances en Europe*, (Nantes, 1992), 27–65.
——, *Maurice de Saxe*, (Paris, 1992).
Bonnefoy, François, "Louvois et la politique d'armement des troupes" *Histoire, Economie et Société* 15(1) (1996), 95–103.
Bonney, Richard, "The Eighteenth Century II: The Struggle for Great Power Status and the End of the Old Fiscal Regime," in R. Bonney (ed.), *The rise of the fiscal state in Europe, c. 1200–1815*, (Oxford, 1995), 315–392.
——, *Jean-Roland Malet: premier historien des finances de la monarchie française*. (Paris, 1993).
Bradbury, Jim, *The Medieval Siege*, (Woodridge, 1992).
Bragard, Philippe, "Un exemple de la transformation de la poliorcétique au XVe siècle: le premier siege au canon à Namur (1488)," in *Revue belge d'histoire militaire*, (1994), 117–152.
Brewer, John, *The Sinews of Power: War, Money and the English State, 1688–1783*, (New York, 1989).
Brice, Martin, *Forts and Fortresses: From the Hill Forts of Prehistory to Modern Times*, (London, 1990/1999).

Buisseret, David, "Monarchs, Ministers and Maps in France before the Accession of Louis XIV," in D. Buisseret (ed.), *Monarchs, Ministers and Maps: The Emergence of Cartography as a Tool of Governance in Early Modern Europe* (Chicago, 1992), 99–123.
Burke, James, "Siege Warfare in Seventeenth Century Ireland," in Pádraig Lenihan (ed.), *Conquest and Resistance: War in seventeenth-century Ireland*, (Leiden, 2001), 257–291.
——, "The New Model Army and the problems of siege warfare, 1648–51," *Irish Historical Studies* 27, no. 105 (May 1990), 7–29.
Burton, Ivor, *The Captain-General: The Career of John Churchill, Duke of Marlborough, from 1702 to 1711*, (London, 1968).
Campbell, John, *The military history of the late Prince Eugene of Savoy, And of the Late John, Duke of Marlborough, Including a particular Description of the several Battles, Sieges, etc. in which either or both of those generals commanded. Collected from the best authors in all languages.* 2 vols. (London, 1737).
Cardini, Franco, *La culture de la guerre X^e^–XVIIIe siècles*, French translation of 1982 Italian original (Paris, 1992).
Carle, Lucia (ed.), *Situazioni d'Assedio. Cities under Siege*, (Florence, 2002).
Carlton, Charles, *Going to the Wars: The Experience of the British Civil Wars, 1638–1651*, (New York, 1992).
Carpi-Mailly, Olivia, "Amiens au XVIe siècle: Le destin d'une ville frontière," in P. Nivet (ed.), *Picardie, terre de frontière: actes du colloque, Amiens, 26 avril 1997*, (Amiens, 1998), 53–72.
Chandler, David, *The Campaigns of Napoleon: The Mind and Methods of History's Greatest Soldier*, (New York, 1966).
——, (ed.), *Robert Parker and Comte de Mérode-Westerloo: The Marlborough Wars.* (Hamden, CT, 1968 reprinted 1998).
——, *Marlborough as Military Commander*, (New York, 2000 reprint of 1973 original).
——, *The Art of Warfare in the Age of Marlborough*, (London, 1994 reprint of 1976 original).
—— (ed.), *A Journal of Marlborough's campaigns during the War of the Spanish Succession 1704–1711.* (London, 1984).
—— (ed.), *The Military Maxims of Napoleon*, (New York, 1988).
Childs, John, "The Military Revolution I: The Transition to Modern Warfare," in Charles Townshend (ed.), *The Oxford Illustrated History of Modern War* (Oxford, 1997), 19–34.
——, *The Nine Years' War and the British Army, 1688–1697: The Operations in the Low Countries* (Manchester, 1991).
Churchill, Winston, *Marlborough: His Life and Times*, (New York, 1947).
Claessens, P.-E., "A propos d'ingénieurs et d'établissements d'instruction dans les Pays-Bas espagnols au XVIIe siècle," *Intermédiaire des Généalogistes*, 88 (1960), 186–191.
Cocula, Anne-Marie (ed.), *Mémoires de Monsieur de La Colonie: maréchal de camp des armées de l'Electeur de Bavière*, (Paris, 1992).
Cra'aster. H.H.E., "Letters of the first Lord Orkney during Marlborough's campaigns," *English Historical Review*, (1904), 307–321.
Crichton, Andrew (ed.), *The Life and Diary of Lieut.-Col. J. Blackadder, of the Cameronian Regiment, and Deputy Governor of Stirling Castle, who served with distinguished honour in the wars under King William and the Duke of Marlborough, and afterwards in the Rebellion of 1715 in Scotland.* (Edinburgh, 1824).
Dalton, Charles, *English Army Lists and Commission Registers 1661–1714*, 6 vols. (London, 1898).
De Fer, Nicolas, *Les forces de l'Europe ou description des principales villes avec leurs fortifications*, (Paris, 1705).
De Jongste, Jan A.F. (ed.), *Anthonie Heinsius and the Dutch Republic 1688–1720. Politics, Finance and War*, (The Hague, 2002).

De May [Du Mée], Sieur, *An exact journal of the siege of Lisle, from the opening of the trenches, to the entire reduction of the said place under the Command of his Highness Prince Eugene of Savoy; with a list of the kill'd, wounded and prisoners and Private men on both sides: Nicely calculated from the best Accounts: Together with the Numbers of the Besiegers, and the Besieged, at the Opening of the Trenches; with other Remarakable Passages relating to the said siege by the Sieur d'Mey*, (1708).

De Ville, Anthoine, *Les fortifications du chevalier Antoine de Ville*, (Lyon, 1628).

——, *De la charge des Gouverneurs des Places*, (Paris, 1640).

Decker, Michel, "Louvois, l'artillerie et les sièges," *Histoire, économie et société* 15(1) 1996, 75–94.

Deidier, abbé, *Le parfait ingénieur francois, ou la fortification offensive et défensive, contenant la construction, l'attaque et la défense des places régulières et irrégulières, selon les méthodes de Monsieur de Vauban, et des plus habiles auteurs de l'Europe, qui ont écrit sur cette science*, (Paris, 1757).

DeJean, Joan, *Literary Fortifications: Rousseau, LaClos, Sade*, (Princeton, 1984).

Delorme, "Le journal d'un capitaine de mineurs (1711)," *Carnet de la Sabretache*, 4–6 (1896–1898), 368–382, 422–431, 233–249.

DeVries, Kelly, "The Impact of Gunpowder Weaponry on Siege Warfare in the Hundred Years War," in I. Corfis and M. Wolfe (eds.), *The Medieval City under Siege*, (Woodbridge, 1995), 227–244.

——, *Joan of Arc: A Military Leader*, (Gloucestershire, 1999).

Dickinson, H.T., "The Earl of Peterborough's Campaign in Valencia 1706," *The Journal of the Society for Army Historical Research*, 45(181) 1967, 35–52.

——, "The Richards Brothers: Exponents of the Military Arts of Vauban," *The Journal of the Society for Army Historical Research*, 46(186) (Summer 1986), 76–86.

Dickson, P.G.M. and Sperling, John, "War Finance, 1689–1714," in J.S. Bromley (ed.), *The New Cambridge Modern History, Vol. 6: The Rise of Great Britain and Russia, 1688–1715/25* (Cambridge, 1971), 284–315.

Duffy, Christopher, *Fire and Stone: The Science of Fortress Warfare, 1660–1860* (London, 1975).

——, *Siege Warfare: The Fortress in the Early Modern World 1494–1660* (London, 1979).

——, *The Fortress in the Age of Vauban and Frederick the Great 1660–1789*, (London, 1985).

——, *The Military Experience in the Age of Reason*, (New York, 1988).

DuMoulin, *Campagne de Monsieur le maréchal de Villars en Allemagne l'an MDCCCIII. Contenant les lettres de ce maréchal & celles de plusieurs autres officiers-généraux au roi & à Mr. de Chamillart, ministre au Département de la guerre, avec les réponses du roi, & de ce ministre. Recueil . . . formé sur les originaux qui se trouvent en dépôt au Bureau de la guerre de la Cour de France*, 2 vols. (Amsterdam, 1762).

——, *Campagne de Monsieur le maréchal de Tallard en Allemagne l'an MDCCIV contenant les lettres de ce maréchal & celles de plusieurs autres officiers-généraux au roi & à Mr. de Chamillart . . . avec les réponses . . . Recueil formé sur les originaux qui se trouvent en dépôt au Bureau de la guerre de la Cour de France*, 2 vols. (Amsterdam, 1763).

Duvivier, J. and J.W. Wijn, "Documents hollandais rélatifs à la guerre de succession d'Espagne dans le Nord de la France." *Revue du Nord* 39(156) October-December 1957, 243–253. [duplicates of letters in *Briefwisseling Heinsius*]

Eltis, David, *The Military Revolution in Sixteenth-Century Europe*, (London, 1995).

Errard Bar-le-Duc, Jean, *La fortification demonstrée et réduicte en art par feu I. Errard de Bar le Duc Ingénieur du Treschrestienne Roy de France et de Navarre*, (Paris, 1622).1

Esnault, G., *Correspondance et papiers inedits recueillis et publies par l'abbe G. Esnault*, (Geneva, 1970). [Chamillart papers.]

Europische Mercurius, behelzende de voornaamste zaken van Staat en Oorlog, voorgevallen in alle de Koningryken en Heerschappyen van Europa; benevens eenige meldenswaardige tydingen uit verscheide andere deelen der waerled. Historischer wyze, onverward en klaar beschreven, en met

de noodige authentyke stukken voorzien. (Leyden and Amsterdam, 1710). [contemporary semi-official Dutch newspaper]
Evans, John (ed.), *The Works of Sir Roger Williams*, (Oxford, 1972).
Faucherre, Nicholas and Prost, Philippe, *Le triomphe de la methode: le traité de l'attaque des places*, (Paris, 1992).
Francis, David, *The First Peninsular War, 1701–1713*, (London, 1975).
Frey, Linda and Frey, Marsha (eds.), *The Treaties of the War of the Spanish Succession: An historical and critical dictionary*, (Westport, CT, 1995).
Galland-Seguela, Martine, "Introduction à l'étude du corps des ingénieurs militaires espagnols au XVIIe siècle," *Histoire, économie et société*, 8 (1989), 551–562.
Garland, John, "Irish Officers in the Bavarian Service During the War of the Spanish Succession," *The Irish Sword*, 14(56) 1981, 240–255.
Gat, Azar, *The Origins of Military Thought from the Enlightenment to Clausewitz*, (Oxford, 1991).
Glover, Michael, *Wellington as Military Commander*, (London, 1968).
Goënaga, Jean-Marie, "Sièges" in F. Bluche (ed.), *Dictionnaire du Grand Siècle*, (Paris, 1990), 1448–1449.
Gould, Stephen Jay, *Full House: The Spread of Excellence from Plato to Darwin*, (New York, 1996).
Goulon, Charles, *Memoirs of Monsieur Goulon, being a Treatise on the Attack and Defence of a Place. To which is added, a journal of the siege of Ath, in the year 1697, under the conduct of Monsieur de Vauban. For Use of the Young Gentlemen of the Royal Academy at Woolwich*, (London, 1745).
Guerlac, Henry, "Vauban: The Impact of Science on War," in Peter Paret (ed.), *Makers of Modern Strategy from Machiavelli to the Nuclear Age* (Princeton, 1986), 64–90.
Guilmartin Jr., John, *Gunpowder and galleys: changing technology and Mediterranean warfare at sea in the sixteenth century*, (New York, 1974).
Hahn, Roger and Taton, René (eds.), *Écoles techniques et militaires au XVIIIe siècle*, (Paris, 1986).
Hale, John R., "A humanistic visual aid. The military diagram in the Renaissance," *Renaissance Studies*, 2(2) 1988, 280–298.
——, "The Early Development of the Bastion: An Italian Chronology c. 1450–1534" in J.R. Hale (ed.), *Europe in the Late Middle Ages*, (Evanston, IL, 1965), 466–494.
——, "Tudor Fortifications: The Defence of the Realm, 1485–1558," reprinted in Hale (ed.), *Renaissance War Studies*, (London, 1983), 61–97.
Hall, A.R., "Gunnery, Science, and the Royal Society," in J. Burke (ed.), *The Uses of Science in the Age of Newton* (Berkeley, 1983), 111–141.
Hall, Bert, "The Changing Face of Siege Warfare: Technology and Tactics in Transition," in I. Corfis and M. Wolfe (eds.), *The Medieval City under Siege*, (Woodbridge, 1995), 257–275.
Hall, Bert S., *Weapons and Warfare in Renaissance Europe*, (Baltimore, 1997).
Hanlon, Gregory, *The Twilight of a Military Tradition: Italian Aristocrats and European Conflicts, 1560–1800*, (New York, 1998).
Harrington, Peter, "English Civil War Fortifications," *Fort: The International Journal of Fortification and Military Architecture* 15 (1987), 39–60.
Harsin, Paul, "Vauban à Liège en 1702," *Bulletin de la Société royale du vieux Liège* 104–105 (1954), 307–313.
Hasselt, W.J.C. van, "Brieven geschreven in het jaar 1710 uit het Nederlandsche leger in Frankrijk." *De Militaire Spectator* (1860), 502–510.
Hebbert, F.J. and Rothrock, George, *Soldier of France: Sebastien Le Prestre de Vauban, 1633–1707*, (New York, 1989).
——, "The Memoirs of Monsieur Goulon," *The Journal of the Society for Army Historical Research*, 69(279) 1991, 161–165.
Hipssich, C., *Spanischer Successions-krieg. Feldzüge 1710.* (Vienna, 1887). Volume 12 of Kriegsarchiv, *Feldzüge des Prinz Eugen von Savoyen.* 20 vols. (Vienna, 1876–91).

Hugill, J.A.C., *No Peace without Spain*, (Oxford, 1991).
Hutton, Ronald and Reeves, Wylie, "Sieges and Fortifications," in J. Kenyon and J. Ohlmeyer, (eds.), *The Civil Wars: A Military History of England, Scotland and Ireland, 1638–1660*, (Oxford, 1998), 195–233.
Jadin, Louis (ed.), *Correspondance du Baron Karg de Bebenbourg, chancelier du Prince-Eveque de Liège Joseph-Clément de Bavière, Archeveque Électeur de Cologne, avec le Cardinal Paolucci, secrétaire d'état (1700–1719)*. 2 vols. (Brussels, 1968), vol. 1.
Johnston, S.F.H. (ed.), "Letters of Samuel Noyes, Chaplain of the Royal Scots, 1703–4" *The Journal of the Society for Army Historical Research*, 37 (March-December 1959), 33–40, 67–71, 128–135, 145–152.
Jones, Archer, *The Art of War in the Western World* (New York, 1987).
Jones, Dwyryd W., *War and Economy in the Age of William III and Marlborough*, (Oxford, 1988).
Jones, J.R., *Marlborough*, (Cambridge, 1993).
Jones, Richard, "Fortifications and Sieges in Western Europe c. 800–1450" in M. Keen (ed.), *Medieval Warfare: A History*, (Oxford, 1999), 163–185.
Keen, Maurice H., *The Laws of War in the late Middle Ages* (London, 1968),
Kennett, Lee, "Tactics and Culture: The Eighteenth-Century Experience," *International Colloquy on Military History, No. 5* (1981), 152–159.
Kern, Paul Bentley, *Ancient Siege Warfare*, (Bloomington, IN, 1999).
Kuypers, F.H.W., *Geschiedenis der Nederlandsche artillerie van af de vroegste tijden tot op heden*, (Nijmegen, 1869).
La Vergne, Jacob de, *De l'utilité d'avoir un bon ingénieur, ou directeur général des fortifications dans un État: Joint à un bon corps d'ingénieurs, sur lesquels il aye le commandement absolu, et la manière de le composer, sans augmenter les fonds, qui sont déjà fait; Avec un petit Traitté demonstratif de leurs devoirs, chacun en particulier, par lesquels l'on pourroit éviter les grosses confusions et fautes qui sont infallibles dans un corps qui n'a point de commandant tant aux approches qu'aux fortifications ordinaire. Par Jacob de la Vergne Ingenieur en Chef, et Capitaine au Service de Sa Majesté Imperiale*, (Vienne, 1698).
——, *Nouveau exercice du gabion, et de la fascine*, (Vienne, 1698).
Landosle, Hyrvoix de, *Vauban: Lettres intimes (inédites) addressées au marquis de Puyzieulx (1699–1705)*, (Paris, 1924).
Landsberg, J.-H.-D., *Nouvelle manière de fortifier les places, qui demontre en même tems les défauts qui se rencontrent dans la construction des ouvrages faits par les plus fameux ingénieurs modernes, et les moyens faciles & assurez pour les corriger dans toutes les parties des fortifications, tant des villes, que des citadelles, par une nouvelle méthode qui n'a point été mise au jour.* (The Hague, 1712).
Langins, Janis, *Conserving the Enlightenment: French Military Engineering from Vauban to the Revolution*. (Cambridge, MA, 2003).
Lazard, Paul, *Vauban, 1633–1707*. (Paris, 1934).
Le cabinet du Roi: recueil d'estampes de differents auteurs concernant les bâtimens, les tapisseries, tableaux, conquêtes et autres sujets qu'on trouve dans les maisons royales, (Paris, 1679–1743).
Le Mercure Galant. (Paris, 1710). [semi-official contemporary newspaper]
Le Michaud d'Arçon, Jean-Claude-Eléonore, *Considerations sur l'influence du génie de Vauban dans la balance des forces de l'État*, (Strasbourg, 1786).
Lecestre, Léon (ed.), *Mémoires du chevalier de Quincy*. 3 vols. (Paris, 1898–1901). vol. 3
—— (ed.), *Mémoires de Saint-Hilaire*, (Paris, 1903).
Lecomte, Charles Albert Samuel, "Du service des ingénieurs militaires en France pendant le règne de Louis XIV," *Revue du génie militaire*, 25–26 (1877), vol. 25: 105–124, 193–222, 289–322, 473–500; vol. 26: 1–36, 159–173.
——, *Les ingénieurs militaires en France pendant la règne de Louis XIV*, (Paris, 1904).
Lenihan, Pádraig, "Conclusion: Ireland's Military Revolution(s)," in P. Lenihan (ed.), *Conquest and Resistance: War in seventeenth-century Ireland*, (Leiden, 2001), 345–369.
—— (ed.), *Conquest and Resistance: War in seventeenth-century Ireland*, (Leiden, 2001).

Linière, Raoul de, "M. de Millon et Marlborough aux Sièges de Liege et de Huy, 1702–3," *Revue historique et archéologique du Maine*, 54 (1903), 201–273.

Loeber, Rolf and Parker, Geoffrey, "The military revolution in seventeenth century Ireland," in J. Ohlmeyer (ed.), *Ireland from Independence to Occupation 1641–1660*, (Cambridge, 1994), 66–88.

Luh, Jürgen, *Ancien Regime Warfare and the Military Revolution*, (Groningen, 2000).

Lund, Erik, *War for the Every Day: Generals, Knowledge and Warfare in Early Modern Europe, 1680–1740*, (Westport, CT, 1999).

Lynn, John, "Recalculating French Army Growth During the Grand Siècle, 1610–1715," *French Historical Studies* 18(4) Autumn 1994, reprinted in C. Rogers (ed.), *The Military Revolution Debate: Readings on the Military Transformation of Early Modern Europe*, (Boulder, CO, 1995), 117–148.

——, "Vauban," *MHQ: The Quarterly Journal of Military History*, 1(2) 1989, 51–61.

—— (ed.), *Feeding Mars: Logistics in Western Warfare from the Middle Ages to the Present*, (Boulder, CO, 1993)

——, *Bayonets of the Republic: Motivation and Tactics in the Army of Revolutionary France, 1791–94*, (Boulder, CO, 1996).

——, *Giant of the Grand Siècle: The French Army, 1610–1715*, (Cambridge, 1996).

Mallet, Alain Manesson, *Les travaux de Mars, ou La fortification nouvelle tant reguliere, qu'irreguliere. Divisée en trois parties. La première contient la construction des places regulières, citadelles & dehors . . . La seconde partie donne & examine les diverses constructiones ou methodes de fortifier toutes sortes de villes . . . La dernière donne la connoissance des materiaux qui servent à l'éleevation des ramparts, parapets, & chemises des villes . . .*, 3 vols. (Amsterdam, 1672).

Malo, Henri (ed.), "Journal du siège de Lille rendu aux Alliés le 23 octobre 1708," *Revue du Nord*, 8 (1922)., 279–290.

Menning, Bruce, "Train Hard, Fight Easy: The Legacy of A.V. Suvorov and his 'Art of Victory'," *Air University Review*, 38 (1) 1986, 79–88.

Michel, Georges, *Histoire de Vauban*, (Paris, 1879).

Millner, J.A., *A Compendious Journal of all the Marches, Famous Battles, Sieges And other most noteworthy, heroical and ever memorable Actions of the Triumphant Armies, Of the ever glorious Confederate High Allies, In their late and victorious War Against the Powerful Armies of proud and lofty France, In and on the Confines of Holland, Germany, & Flanders, So far as our successful British Troops extended in Conjunction therein. Digested into Twelve Campaigns, begun AD 1701, and ended 1712. All but the first and last, the Grand Confederate Armies were under the conduct and command of our Honourable & much Honor-worthy, ever-renown'd, graceful, & excellent war-like Hero John, duke of Marlborough, Prince of the Holy Empire, etc. Truly & punctually collected, form'd, compos'd & written in the Time of the said War, By John Millner, Serjeant in the Honourable Royal Regiment of Foot of Ireland. Having been therewith during the War an Eye-witness of the most of the following Marches & Actions of the said War.* (London, 1733).

Milot, Jean, "Un problème opérationnel du XVIIe siècle illustré par un cas régional," *Revue du Nord* 53, no. 209 (1971), 269–285

Monck, George, duke of Albemarle, *Observations upon Military and Political Considerations*, (London, 1671).

Muller, Josy, "Les ingénieurs militaires dans les Pays-Bas espagnols (1500–1715), " *Revue internationale d'histoire militaire*, 20 (1959), 467–478.

Mulloy, Sheila, "French engineers with the Jacobite Army in Ireland, 1688–1691," *The Irish Sword*, 15(16) 1983, 222–232.

Murray, George (ed.), *The Letters and Dispatches of John Churchill, First Duke of Marlborough from 1702 to 1712.* 5 vols. (London, 1845), vols. 4–5.

Naudin, M., *L'ingénieur françois: contenant la geometrie pratique sur le papier & sur le terrain, avec le toisé des travaux & des bois, la fortification reguliere & irreguliere, sa construction effective, l'attaque & la défense des places: avec la methode de Monsieur de Vauban, & l'explication de son nouveau systeme*, (Paris, 1695).

Ostwald, Jamel, "The 'Decisive' Battle of Ramillies, 1706: Prerequisites for Decisiveness in Early Modern Warfare," *Journal of Military History* 64 (July 2000), 649–677.
Ouverleaux, Emile, *Mesgrigny, ingénieur militaire, lieutenant-général des armées du Roi, 1630–1720*, (Brussels, 1928).
Pagan, *Les Fortifications de Monsieur le Comte de Pagan avec ses theorems sur la fortification*, (Brussels, 1668).
Papillon, *A Practicall Abstract of the Arts of Fortification and Assailing . . .*, (London, 1645).
Parent, Michel, and Verroust, Jacques, *Vauban*, (Paris, 1971).
Parker, Geoffrey, "The 'Military Revolution,' 1560–1660—a Myth?" *Journal of Modern History* 48 (June 1976), 195–214.
——, *The Army of Flanders and the Spanish Road, 1567–1659*, (New York, 1972).
——, *The Grand Strategy of Philip II*, (New Haven, 1998).
——, *The Military Revolution: Military Innovation and the Rise of the West, 1500–1800*, (New York, 1988, second edition 1996).
Parrott, David, "Strategy and Tactics in the Thirty Years War: The 'Military Revolution'," *Militargeschichtliche Mitteilungen*, 18 (2) (1983), 7–25.
——, *Richelieu's Army: War, Government and Society in France, 1624–1642*, (Cambridge, 2001).
Pelet (eds.), *Atlas des Mémoires militaires relatifs à la succession d'Espagne sous Louis XIV . . .*, (Paris, 1862).
Pepper, Simon and Nicholas Adams, *Firearms and Fortifications: Military Architecture and Siege Warfare in Sixteenth-Century Siena*. (Chicago, 1986).
Pepper, Simon, "Castles and cannon in the Naples campaign of 1494–95," in D. Abulafia (ed.), *The French Descent into Renaissance Italy, 1494–95. Antecedents and Effects*, (Aldershot, 1995), 263–293.
——, "Military Architecture in Baroque Europe," in H. Millon (ed.), *The Triumph of the Baroque: Architecture in Europe, 1600–1750* (New York, 1999), 333–347.
——, "Artisans, Architects and Aristocrats: Professionalism and Renaissance Military Engineering," in D. Trim (ed.), *The chivalric ethos and the development of military professionalism*. (Leiden, 2003), 118–147.
Perjés, Géza, "Army Provisioning, Logistics and Strategy in the Second Half of the 17th Century," *Acta Historica Academiae Scientiarum Hungaricae* 16 (1970), 7–51
Pernot, Jean-François, "Vauban et la guerre des mines" *Histoire et Défense. Sapes et contre-sapes: la guerre des mines à l'époque moderne et contemporaire* 32 (1995), 73–104.
——, "Vauban, le siège devenu réglé ou l'économie des vies militaires," in A. Corvisier et al, *Les malheurs de la guerre I: De la guerre à l'ancienne à la guerre réglée*, (Paris, 1996), 247–261.
Petrie, Charles, *The Marshal Duke of Berwick: The Picture of an Age*, (London, 1953).
Picard, Ernest, *L'Artillerie française au XVIII*[e] siècle (Paris, 1906).
Pollak, Martha, "Military Architecture and Cartography in the Design of the Early Modern City," in D. Buisseret (ed.), *Envisioning the City*, (Chicago, 1998), 109–124.
——, *Military Architecture, Cartography and the Representation of the Early Modern European City: A Checklist of Treatises on Fortification in the Newberry Library*, (Chicago, 1991).
Pontal, Edouard (ed.), *Mémoires du marquis de Sourches sur le règne de Louis XIV*, (Paris, 1882).
Porter, Whitworth, *History of the Corps of Royal Engineers*, (London, 1889).
Puységur, *Les mémoires de messire Jacques de Chastenet, chevalier, seigneur de Puységur, colonel du regiment de Piedmont, et lieutenant général des armées du roy. Sous les règnes de Louis XIII et de Louis XIV*, 2 vols. (Amsterdam, 1690).
Quincy, Charles Sevin marquis de, *Histoire militaire du règne de Louis le Grand, Roy de France, où on trouve un détail de toutes les Batailles, Sièges, Combats particuliers, et généralement de toutes les actions de Guerre qui se sont passées pendant le cours de son Règne, tant sur Terre que sur Mer, Enrichie des plans nécessaires. On y a joint un traité particulier de Pratiques et de Maximes de l'Art Militaire. Par M. le marquis de Quincy, Brigadier des*

Armées du Roy, Lieutenant Général de l'Artillerie, Lieutenant pour le Roy au Gouvernment d'Auvergne, Chevalier de l'Ordre Militaire de S. Louis, (Paris, 1726).
Relation de la campagne de l'année 1710, contenant un journal de tout ce qui s'est passé aux sièges de Douay, Béthune, St-Venant et d'Aire . . . (La Haye, 1711).
Relation de la campagne de Tannières, contenant un journal exact & fidel de ce qui s'est passé au siège de la ville & citadelle de Tournai, à la battaile de Blaugis, ou Malplacquet, & au siège de Mons, avec quelques autres particularités, & les plans. En l'an 1709, (La Haye, 1709).
Ringoir, *Afstammingen en voortzettingen der génie en trein*, ('s-Gravenhage, 1980).
Rochas d'Aiglun, Albert de, "La campagne de la Succession d'Espagne dans les Alpes (1707–13)," *Bulletin de la Société d'études des Hautes-Alpes*, 7 (1888), 101–113.
Rochas d'Aiglun (ed.), Albert de, *Vauban, sa famille et ses écrits, ses oisivetés et sa correspondance*, 2 vols. (Paris, 1910).
Rogers, Clifford, "The Military Revolution of the Hundred Years War," in C. Rogers (ed.), *The Military Revolution Debate: Readings on the Military Transformation of Early Modern Europe*, (Boulder, CO, 1995), 55–94.
Rorive, Jean-Pierre, *La guerre de siège sous Louis XIV en Europe et à Huy*, (Brussels, 1997).
Rothrock, George (ed. and trans.), *A Manual of Siegecraft and Fortification*, (Ann Arbor, 1968).
Rousset, *Histoire de Louvois et de son administration politique et militaire*, (Paris, 1864–1864).
Royal Commission on Historical Manuscripts, *Report on the Manuscripts of Mrs. Frankland-Russell-Astley of Chequers Court, Bucks*, (London, 1900).
——, *Report on the Manuscripts of the Earl of Mar and Kellie*, (London, 1904).
——, *The Manuscripts of the Earl Cowper*, 3 vols. (London, 1888).
——, *The Manuscripts of the Marquess Townshend*, (London, 1887).
——, *The Manuscripts of the Duke of Beaufort, K.G., The Earl of Donoughmore, and Others*, (London, 1891).
——, *The Manuscripts of the Duke of Athole, K.T., and of the Earl of Home*, (London, 1891).
——, *The Manuscripts of His Grace the Duke of Portland*, vol. 8 (London, 1891).
Russell Weigley, *The Age of Battles: The Quest for Decisive Victory from Breitenfeld to Waterloo*, (Bloomington, 1991).
Saint-Martin, Y.J., "Le Pseudo-Siège de Toulon en 1707," *Provence historique* 176 (1994), 199–209.
Saint-Rémy, Surirey de, *Mémoires d'artillerie où il est traité des mortiers, pétards, arquebusies à croc, mousquets, fusils, et de ce qui peut servir à l'execution et au service de toutes ces Armes; des bombes, carcasses, grenades, etc . . .*, 2 vols. (The Hague, 1741).
Salamagne, Alain, "L'attaque des places-fortes au XVe siècle à travers l'exemple des guerres anglo- et franco-bourguignonnes," *Revue historique*, 585 (1993), 65–113.
——, "Vauban et les fortifications du Quesnoy," *Revue historique des armées* 1986, 45–51.
Satterfield, George, *Princes, Posts and Partisans: The Army of Louis XIV and Partisan Warfare in the Netherlands (1673–1678)*. (Leiden, 2003).
Sautai, Maurice, "Une lettre inédite du Gouverneur de Béthune du Puy-Vauban," *Bulletin de la commission historique du département du Nord* 28 (1911), 518–519.
——, *Le siège de la ville et de la citadelle de Lille en 1708*, (Lille, 1899).
Scholten, F.W.J., *Militaire topografische kaarten en stadsplattegronden van Nederlanden, 1579–1795*, (Alphen aan den Rijn, 1989).
Schulenburg, Johann Mathias reichsgrafen von der, *Leben und Denkwürdigkeiten Johann Mathias Reichsgrafen von der Schulenburg, Erbherrn auf Emden und Delifs, Feldmarschalls in Diensten der Republik Venedig, Zus Original-Quellen bearbeitet.* 2 vols. (Leipzig, 1834), vol. 1.
Seymour, William, *Great Sieges of History*, (London, 1991).

Showalter, Dennis, *The Wars of Frederick the Great*, (London, 1996).
Sicard, "Récherches historiques sur les guerres, sièges et traités de paix depuis l'origine du nom français," *Le Spectateur militaire* 7 (1829), 619–649.
Snyder, Henry (ed.), *The Marlborough-Godolphin Correspondence*. 3 vols. (Oxford, 1975).
Soulié (ed.), *Journal du marquis de Dangeau publié en entier pour la première fois . . . avec les additions inédits du Duc de Saint-Simon*, 19 vols. (Paris, 1857).
Steele, Brett, "Muskets and Pendulums: Benjamin Robin, Leonhard Euler and the Ballistic Revolution," *Technology and Culture* 35 (1994), 348–382.
Sturgill, Claude, *Marshal Villars and the War of the Spanish Succession*, (Lexington, KY, 1965).
Susanne, Louis, *Histoire de l'artillerie française*, (Paris, 1874).
Tallett, Frank, *War and Society in Early Modern Europe, 1495–1715*, (London, 1992).
The Daily Courant. (London). [contemporary newspaper]
The Post Boy. (London). [contemporary newspaper]
The Postman. (London). [contemporary newspaper]
Tracy, James (ed.), *City Walls—The Urban Enceinte in Global Perspective*, (Cambridge, 2000).
Trotter, Ben Scott, *Marshal Vauban and the Administration of Fortifications under Louis XIV (to 1691)*, Ohio State University Ph.D. Dissertation, Columbus, OH (1993).
Van 't Hoff, B. (ed.), *The Correspondence 1701–1711 of John Churchill, First Duke of Marlborough, and Anthonie Heinsius, Grand Pensionary of Holland*. (The Hague, 1951).
Van Creveld, Martin, *Supplying War: Logistics from Wallenstein to Patton*, (New York, 1977)
Van Nimwegen, Olaf, "Kannonen en houwitsers. De Staatsh veldartillerie in de eerste helft van de achttiende eeuw," *Armamentaria* 32 (1997), 50–66.
——, "Maurits van Nassau and siege warfare (1590–1597)," in M. van der Hoeven (ed.), *Exercise of Arms: Warfare in the Netherlands, 1568–1649*, (Leiden, 1997), 113–132.
Van Nimwegen, Olaf, *De subsistentie van het leger: Logistiek en strategie van het Geallieerde en met name het Staatse leger tijdens de Spaanse Successieoorlog in de Nederlanden en het Heilige Roomse Rijk (1701–1712)*, (Amsterdam, 1995).
Van Sypesteyn, J.W., *Het leven van Menno Baron van Coehoorn*, (Leeuwarden, 1860).
Vauban, Sébastien le Prestre de, *De l'attaque et de la défense des places* (ed.) Antoine-Marie Augoyat, 2 vols. (Paris, 1828).
Vault and J. Pelet (eds.), *Mémoires militaires relatifs à la succession d'Espagne sous Louis XIV, extraits de la correspondance de la cour et des généraux par le lieutenant général de Vault, directeur du Dépot de la Guerre, mort en 1790* . . . 11 vols. (Paris, 1836–42).
Vauvilliers, L.-H.-C., *Recherches historiques sur le rôle et l'influence de la Fortification* (Paris, 1845).
Veenendaal Jr., A.J. (ed.), *De Briefwisseling van Anthonie Heinsius, 1702–1720*. 19 vols. (The Hague, 1976–2001).
Veenendaal Sr., A.J. (ed.), *Het Dagboek van Gisbert Cuper, gedeputeerde te velde, gehouden in de Zuidlijke Nederlanden in 1706*, ('s-Gravenhage, 1950).
Vérin, Hélène, *La gloire des ingénieurs: L'intelligence technique du XVIe au XVIIIe siècles*, (Paris, 1993).
Virgin, Jean-Bernard, *La défense des places, mise en équilibre avec les attaques savantes et furieuses d'aujourd'hui*, (Stockholm, 1781).
Virol, Michèle, *Les Oisivetés de Monsieur de Vauban* (Paris, 2000).
Vogel, H., "Arms production and exports in the Dutch Republic, 1600–1650," in M. van der Hoeven (ed.), *Exercise of arms: warfare in the Netherlands, 1568–1648* (Leiden, 1997), 197–210.
Vogüé, Marquis de (ed.), *Le duc de Bourgogne et le duc de Beauvillier: Lettres inédites, 1700–1708*, (Paris, 1900).
—— (ed.), *Mémoires du Maréchal de Villars publié d'après le manuscrit original*, 6 vols. (Paris, 1889).
Watson, Bruce Allen, *Sieges: A Comparative Study*, (Westport, CT, 1993).

Wauwermans, H., "Le marquis de Verboom ingénieur militaire flamand au service d'Espagne au XVIIe siècle," *Annales de l'Académie d'archéologie de Belgique*, 4th series, 10 (1891), 276–317.
Wheeler, James Scott, "Logistics of the Cromwellian Conquest of Scotland, 1650–1651," *War and Society*, 10(1) (1992), 1–18.
Wijn, J.W., *Het krijgswesen in den tijd van Prins Mauritz*, (Utrecht, 1934).
——, *Het Staatsche Leger: Het Tijdperk van de Spaanse Successieoorlog, 1702–1715*, vol. 8, 3 parts (The Hague, 1959–1964).
——, "Une bataille classique du XVIII[e] siecle: Almansa1707 25 avril 1707," *Revue internationale d'histoire militaire*, 22 (1960), 92–104.
Wolf, John, "Louis XIV, Soldier-King," in J. Rule (ed.), *Louis XIV and the Craft of Kingship* (Columbus, OH, 1969), 196–223.
——, *Louis XIV*, (New York, 1968).
Wolfe, Michael, "Writing the City under Attack: Siege Warfare as Vicarious Experience during the French Wars of Religion," in L. Carle (ed.), *Situazioni d'Assedio. Cities under Siege*, (Florence, 2002), 197–203.
——, "Walled towns during the French wars of religion (1560–1630)," in J. Tracy (ed.), *City Walls: The Urban Enceinte in Global Perspective*, (Cambridge, 2000), 317–348.
Wood, James, *The King's Army: Warfare, Soldiers and Society during the Wars of Religion in France, 1562–1576*, (Cambridge, 1996).
Ziegler, François, *Villars: Le centurion de Louis XIV*, (Paris, 1996).

INDEX

Italics—alternate spelling

Aalst (town, *Alost*), 100
Act of Settlement (1701), 94
Adams, Nicholas, 6
Aire, 1710 siege of, 103, 122, 170, 206, 301–302, 308, 343, 361, 363, 368
Albemarle, Arnold Joost van Keppel Earl van, 104, 115, 178, 179, 196, 200, 233–234, 262
Albergotti, François-Zénobie-Philippe comte de, 229, 257, 299
Alègre, Yves marquis de Tourzel et d', 175, 249, 266
Amiens, 1597 siege of, 59
Anhalt-Zerbst, Anton Günther Prince of, 23
Antwerp (town), 98, 100
Ardennes Mountains, 98, 261
Armstrong, John, 133
Arnold, Thomas, 4
Arras (town), 222, 234
Arras, 1654 siege of, 60
artillery
  duties of, 159
  independence from engineers, 158–166
  failed 1755 union with engineers, 160
  siege park sizes, 266–270, 367–370
  munitions, 295–296
  brute force siege tactics, 257–297
  indirect fire, 31–32, 33–34, 65, 88, 265, 284, 287–290
  ricochet fire, 31–33, 37, 64–67, 282–284
  see also: Bélidor, Board of Ordnance, Coehoorn, Goor, Gribeauval, IJssel, La Frézelière, Maine, Saint-Hilaire, Saint-Rémy, Vallière, Vigny
Asfeld, Claude-François chevalier d', 176
Ath (town), 21
Ath, 1697 siege of
  town, 21
  fortifications, 25–28, 40–41
  expectations for, 202
  investment, 22–25
  civilian labor in, 24, 36
  opening the trenches, 28–29, 193
    trenchwork, 29–35, 42–45
  capturing the covered way, 32–33, 242
  capturing the Brabançon ravelin, 33–34
  breaching the bastions, 34–36, 274–275
  capitulation, 36
  artillery at, 30–35, 43–45, 66, 91, 161–162, 268, 280, 282, 288, 290–291, 368
  as model siege, 37–38
  engineers in, 137, 139
Ath, 1706 siege of, 100, 121, 142, 176, 193, 246, 263–264, 277, 342, 363, 368
Athlone, Godaert van Reede-Ginkel graaf van, 97
Augoyat, Antoine Marie, 38
Austria, 93, 99, 103, 104, 133–135, 148, 157

Barbesieux, Louis-François-Marie Le Tellier marquis de, 125n6, 137, 212, 228
Barcelona, 1697 siege of, 36, 140, 175, 188, 286
Barcelona, 1706 siege of, 100, 342
Barcelona, 1714 siege of, 104
battle,
  decisive, 304–305, 333
  see also: Denain, Ekeren, Malplaquet, Oudenaarde, Ramillies, Stekene, Wijnendael
Bavaria (Electorate of), 99
Bavaria, Maximilien II Emmanuel Wittelsbach, Elector of, 93, 99, 129
Beauvillier, Paul duc de, 280
Bedmar, Isidoro Juan Hose Domingo de la Cueva y Benavides, Marques de, 113
Belgrade, 1717 siege of, 135

Bélidor, Bernard Forest de, 313
Bergeyck, Jan van Brouchoven graaf van, 101, 109
Berwick, James Fitz-James Stewart Duke of, 163, 174
Béthune, 1710 siege of, 102–103, 122, 145, 169–170, 179–180, 196, 200, 234–235, 239, 257, 265, 273, 275, 283, 289, 301, 343, 363, 368
bicoque, *see*: fortifications
Blackadder, John, 168, 223–224, 300–301
Blainville, Jules-Armand Colbert, marquis de, 243
Bland, Humphrey, 138
Blaregnies (town), 102
Blenheim, 1704 battle of, 99
blockade, 16, 225–226, 351–352, 354–359
Blomfield, Reginald, 9–10
Blood, Holcroft, 133, 236, 294
Board of Ordnance (England), 132–133
Bois, Jean-Pierre, 314
bombardment, 16, 224–225, 290–293, 349, 351–352, 354–359
Bonaparte, Napoleon, 13, 317, 319
Bonn (town), 106
Bonn, 1703 siege of, 98, 156, 237–238, 261, 282, 291, 341
Bosch, Zas van den, 146
Bouchain, 1711 siege of, 103, 122, 145, 235, 265, 269, 275, 303, 343, 368
Bouchain, 1712 siege of, 104, 190, 249, 256, 266, 343, 364, 368
Boufflers, Louis François duc de, 23, 28, 97, 107, 243
Bourgogne, *see*: Burgundy
Bousmard, Henri Jean Baptiste, 38
Brabant, Lines of, 98–99, 107, 120
Brandenburg, Frederick Hohenzollern Elector of, 225
Breisach (Old, *Alt*), 1703 siege of, 46, 82, 113, 162, 167, 228, 239–240, 280–281, 285, 292, 341
Brest (town), 9
Broglie, François-Marie comte de, 187, 190, 206–207
Bruges (town, *Brugge*), 104, 224
Bruges, 1708 surprise of, 101, 110, 223
Bruijn, 146
Brussels (town), 23, 36, 108, 110, 224–225
Brussels, 1695 bombardment of, 290
Brussels, 1708 surprise/siege of, 101, 122, 343
Burgundy (*Bourgogne*), Louis Bourbon duke of, 50, 82, 101, 113, 151, 167, 228, 280, 292

Cadogan, William, 196, 226, 234
Cambon, François Du Puy heer van, 130
Cambrai, 1677 siege of, 268
Cambrin, Lines of, 102, 120
Candia (Crete), 1648–1669 siege of, 58
Caraman, Pierre-Paul Riquet comte de, 255
Cardonnel, Adam, 178–179, 199, 237
Carlos II (Habsburg), King of Spain, 92–93
Carlton, Charles, 7
Carnot, Lazare, 38
Catinat, Nicolas, 22, 28
Chamillart, Michel, 107, 125n6, 151, 307
Chamlay, Jules-Louis Bolé, marquis de, 264
Chandler, David, 7, 10, 15, 328–337
Charlemont-Givet (town), 92
Charleroi (town), 99
Charleroi, 1693 siege of, 39, 147, 161, 167, 192, 193, 212, 228, 247, 288
Charles (Habsburg), Archduke (Charles III of Spain), 93, 100
Charles XII (Vasa), King of Sweden, 311
Cherbourg (town), 9
Childs, John, 7, 11, 221, 329
Chivasso, 1705 siege of, 188
Churchill, Charles (brother of John), 226
Clausewitz, Carl von, 182, 297, 317–319
Clerville, Louis-Nicolas chevalier de, 8, 149, 150
Coehoorn, Menno van, 97, 114–116, 156, 179, 187, 189, 209, 235–238, 261, 279, 282, 285–287, 311
Colbert, Jean-Baptiste, 18, 125
Cologne (*Köln*), Joseph Clemens, Bishop of Liège and Elector of, 93, 97–98
Condé (town), 194
Condé, Louis II de Bourbon prince de, 60, 86, 136
Coni, 1691 siege of, 175
Contades, Erasme, 152–153

Cormontaigne, Louis de, 313
Cronström, Isaac, 361
Croxton, Derek, 6
Cuper, Gisbert, 277

Danton, Georges, 317
Danube River, 164
De Ville, Anthoine, 52, 61, 80, 267
Defoe, Daniel, 303
Denain, 1712 battle of, 104, 189–190, 265, 299, 303
Dender River, 24, 165
Dendermonde, 1706 bombardment/blockade of, 165, 225–226
Dendermonde, 1706 siege of, 100, 121, 226, 342, 363, 368
Des Forges, 265, 283
Des Rocques, Guillaume le Vasseur, 139n48, 144–146, 148, 157–158, 176, 180, 183, 184, 189, 194, 196, 204, 206, 208–209, 232–234, 239, 244, 246–247, 294, 300
Desmaretz, Nicolas, 307
Diest, 1705 siege of, 120, 342
Dinant (town), 92
Dixmuide, 1647 siege of, 60
Dôle, 1636 siege of, 60
Donauwörth (town), 164
Dopff, Daniël Wolf, baron von, 302
Douai (town), 198–199
Douai, 1667 siege of, 139
Douai, 1710 siege of, 102, 122, 144–145, 168–169, 178–179, 180, 183, 189, 194–196, 200, 202, 220, 222, 233–234, 239, 247, 283–284, 293–294, 301–302, 343, 363, 368
Douai, 1712 siege of, 104, 122, 140, 154, 163, 175, 187, 190, 197–198, 210, 228–229, 248–249, 255, 257, 266, 281, 283–284, 293, 298–299, 343, 363–364, 368
Du Mée, Lucas, 106, 139n48, 143, 144, 158, 184, 189, 212–213, 232–234, 246, 257, 275, 294, 300
Duffy, Christopher, 5, 12, 55, 59–60, 181, 250, 274, 336, 348
DuPuyVauban, Antoine le Prestre, 139, 189, 275
Dutch Brabant, 225
Dutch Revolt (Eighty Years War), 5, 54, 59, 105, 326–327, 335, 348
Dutch War (1672–1679), 13, 49, 171, 189, 221, 267

Ekeren, 1703 battle of, 98, 120
Elixhem (town), 99
engineers
  in 16th century, 55–56, 267
  in 18th century, 312–313
  French, 125, 128–129
  Spanish, 128–129
  Bavarian, 129
  Dutch, 129–131, 147
  German, 130, 135–136
  British, 131–133, 147
  Austrian, 133–135, 148
  limited knowledge about, 126–127
  professionalism of, 111–113, 123–126
  education of, 47, 123–126
  efficiency and, 47 *passim*
  and siege journals, 183–184
  in sieges
  outside of chain of command, 148
  numbers and casualties, 36, 136–146
  disputes between, 156–158
  threats of violence towards, 149, 153
  criticized for conduct of, 173–180, 185–186, 232–235
  see also: Asfeld, Bélidor, Blood, Bosch, Bousmard, Bruijn, Cambon, Carnot, Clerville, Coehoorn, Cormontaigne, De Ville, Des Forges, Des Rocques, Du Mée, DuPuyVauban, Errard Bar-le-Duc, Filley, Flamand, Fourcroy de Ramecourt, Goulon, Hertel, L'Espinasse, La Colonie, La Vergne, Landsberg, Lapara des Fieux, Le Michaud d'Arçon, Lloyd, Marsigli, Mesgrigny, Naudin, Navarro, Pagan, Papillon, Richards, Richerand, Trognée, Valory, Vauban, Verboom, Villars-Lugein
English Channel, 21
English Civil Wars (1640–1651), 7–8, 131
Errard Bar-le-Duc, Jean, 61
Escaut River, *see*: Scheldt River
Eugene, François-Eugène Prince of Savoy, 95, 101, 153, 179, 246, 265, 287

Fagel, Francis Nicolaas, 110, 234
Feuquières, Antoine de Pas marquis de, 174–175, 192, 241, 259, 273, 286

field deputies, Dutch
on campaign, 98, 114–115, 156, 231
see also: Cuper, Geldermalsen, Goslinga, Vegelin van Claerbergen
Filley, Louis, 187
Flamand, Claude, 72
Folard, Jean Charles chevalier de, 314–315, 318
Fontenelle, Bernard le Bovier de, 37, 78, 298
fortifications
bicoque defined, 105
status of, 105–111
pré carré, 9, 39, 21, 92, 95, 100, 101, 102, 104, 110, 114, 142, 262–263, 276
fortified lines, 120–122, *see also*: Brabant, Cambrin, Ne Plus Ultra
see also: individual cities
Fourcroy de Ramecourt, Charles-René, 47, 313
Frederick II (Hohenzollern), King of Prussia, 312, 319
Freiburg (town), 151, 240
Freiburg, 1713 siege of, 153, 188, 207, 250, 343
French Wars of Religion, 5, 59

Galway, Henri de Massue marquis de Ruvigny, Earl of, 133
Gat, Azar, 12
Geertruidenberg, negotiations at, 102, 103
Geldermalsen, Adriaen van Borssele, heer van, 115, 141, 164
Gelders (town, *Guelder*), 106, 107
Gelders, 1702 bombardment/blockade of, 225–226, 256, 358
generals
in sieges, 149–156 *passim*
criticizing engineers, 173–180, 185–186, 232–238
and cult of vigor, 215 *passim*
acceptance of casualties, 297–305
Genoa, 1684 bombardment of, 224
Germany
German theater, 99, 104, 174, 266, 331
see also: Blenheim, Breisach, Donauwörth, Landau, Metz, Moselle, Rhine
Ghent (town), 100, 104
Ghent, 1708 siege of, 101, 122, 143, 177–178, 256, 292–293, 343
Ghent, 1708 surprise of, 101, 110, 223, 358
Godolphin, Sidney, 103, 231
Goor, Johan Wijnand van, 164
Goslinga, Sicco van, 109, 142–144, 158, 164–165, 210, 231, 233, 244, 246, 275
Goulon, Charles, 260
Grave, 1602 siege of, 59
grenadiers, 33, 250
Gribeauval, Jean-Baptiste de, 313
Groningen, 1594 siege of, 59
Grovestins, Frederik Sirtema van, 276
Guerlac, Henry, 11, 13
Guiche, Antione de Grammont duc de, 207, 210
Guignard, 163, 223
Gustavus Adolphus (Vasa), King of Sweden, 55

Harley, Robert, 103, 237, 239
Heinsius, Anthonie, 96, 141–144, 157, 177, 183, 184, 209, 230, 232n55
Hertel, Christiaan Frederik, 146, 158, 196, 208–209, 234
Hompesch, Reynout Vincent, baron van, 310
Huguenot engineers
in sieges, 130–132, 181
see also: Cambron, Des Rocques, Du Mée, L'Espinasse, Nantes
Hulst, 1702 siege of, 46, 97, 113, 120, 341
Huy (town), 106
Huy, 1703 siege of, 98, 106, 120, 261, 284, 288, 341, 362, 368
Huy, 1705 Allied siege of, 99, 120, 262, 284, 288, 342, 362, 368
Huy, 1705 French siege of, 99, 120, 163, 261–262, 284, 288, 342, 362, 368

IJssel, Willem, 115, 164
Isabella, 1704 siege of Fort, 98, 120, 341
Israel, Jonathan, 326–327, 348
Italy
Italian theater, 2, 93, 100, 151, 331
see also: Coni, Ivrea, Nice, Savoy, Toulon, Turin
Ivrea (*Ivrée*), 1704 siege of, 188

James II (Stuart), King of England, 93, 148–150, 220
James III the 'Old Pretender' (Stuart), chevalier de Saint-George, 93

Joseph I (Habsburg), Holy Roman Emperor, 103
Jülich (*Juliers*), 1610 siege of, 59

Kaisersweert, 1702 siege of, 97, 120, 141–142, 164, 188, 243, 341, 368
Kara Mustafa, 59
Kehl, 1703 siege of, 174, 193, 341

L'Espinasse, Charles Du Puy de, 130
La Colonie, Jean de, 129
La Feuillade, Louis d'Aubusson, duc de, 151, 175, 228, 286–287, 298
La Frézelière, Jean-Angélique Frézeau, 162–163
La Rochelle, 1573 siege of, 59
La Vergne, Jacob de, 51, 134, 139, 148, 157, 168, 241–242
Landau, 1702 siege of, 115, 187–188, 341
Landau, 1703 siege of, 113, 174, 186, 188, 341
Landau, 1713 siege of, 188, 230, 250, 343
Landrecies, 1712 siege of, 104, 122, 265, 303, 343
Landsberg, Johan, 169, 177, 180, 209, 262
Langins, Janis, 12, 83, 313
Lapara des Fieux, Louis, 175, 188
Lawe River, 234
Lazard, Paul, 12
Le Michaud d'Arçon, Jean-Claude-Eléonore, 48
Le Peletier de Souzy, Michel, 77, 137, 175
Le Quesnoy, 1712 Allied siege of, 104, 122, 146, 265, 343, 368
Le Quesnoy, 1712 French siege of, 104, 122, 141, 152, 187, 210, 229–230, 249, 269–270, 281, 284, 343, 364, 368
Lefebvre d'Orval, Jean-Robert, 175
Leffinghe, 1708 siege of, 122, 343
Leuven (town, *Louvain*), 108
Leuze River, 24
Liège (bishopric), 97–98, 106
Liège (town, *Luik, Lüttich*), 106
Liège, 1702 siege of, 97, 115, 120, 261, 282, 289, 306, 337, 341, 368
Liège, 1705 siege of, 120, 342
Lier (town, *Lierre*), 108
Lille (town, *Rijssel*), 194
Lille, 1667 siege of, 72, 78–79, 338
Lille, 1708 siege of, 101, 122, 143, 158, 176–177, 184, 189, 199, 204, 210, 212, 232, 246–248, 257, 264, 273, 275, 283, 292, 294, 300, 310, 342, 363, 368
Limburg (town), 107
Limburg, 1703 siege of, 98, 107, 120, 261, 341, 362, 368
Lloyd, Henry, 314
logistics
  administrators, 69, 156
  impact on campaigns, 1–4, 101, 216–217, 218n8, 296
  impact on sieges, 237–238
Lorraine, 99
Louis XIII (Bourbon), King of France, 6
Louis XIV (Bourbon), King of France, *appears throughout this study; omitted here*
Louvois, François-Michel Le Tellier marquis de, 49, 68, 82, 125, 159–160, 167, 203
Luxembourg, 1684 siege of, 65, 203
Luxembourg, François Henri de Montmorency-Bouteville duc de Piney de, 192
Lynn, John, 7, 304, 306, 329–332, 335–338

Maas River, *see*: Meuse River
Maastricht (town), 97, 106, 225
Maastricht, 1673 siege of, 57, 58, 61, 111, 236
Madrid (town), 100
Maine, Louis-Auguste de Bourbon duc de, 161, 175
Maintenon, Françoise d'Aubigné marquise de, 112
Mallet, Alain Manesson, 203
Malplaquet, 1709 battle of, 102, 122, 304–305
Mannheim, 1688 siege of, 139
Marchiennes, 1712 siege of, 104, 122, 137–138, 190, 206–207, 250, 265, 343
Marchiennes, 1712 storm of, 359
Marlborough, John Churchill Earl then Duke of,
  campaigns, 97–103, 325
  stripped of command and exiled, 103
  appears throughout this study
Marsigli, Luigi Ferdinando, 87, 135n37
Marsin, Ferdinand comte de, 28

Maurits of Nassau-Orange, 55
Mechelen (town, *Malines*), 108
Menin, 1706 siege of, 100, 109, 121, 142, 152, 157–158, 176, 189, 193, 231–232, 238–239, 245–246, 255–257, 263, 273, 277, 286, 292, 294, 342, 362, 368
Meredith, Thomas, 165
Mesgrigny, Jean de, 22, 28
Metz (town), 135
Meuse River, 21, 97–99, 106, 236
Millner, John, 144, 145, 271
miners
  duties of, 33, 159, 166–168, 212, 272–273
  see also: Goulon, Mesgrigny
Mirandola, 1705 siege of, 188, 342
Monck, George, 2
Mons (town, *Bergen*), 285
Mons, 1691 siege of, 288
Mons, 1709 siege of, 102, 122, 165–166, 168, 293, 300–301, 343, 368
Montauban, 1621 siege of, 60
Montesquiou, Pierre d'Artaignan comte de, 163–164
Morvan (Burgundy), 8, 86, 87
Moselle River, 99
Mouzon (Mousson), 1650 siege of, 148–150

Namur (town, *Namen*), 99, 109–110
Namur, 1692 siege of, 39, 139, 288
Namur, 1695 siege of, 39, 80, 114, 225, 242, 268, 285, 287
Nantes, Edict of, revoked (1685), 130
Napoleon, *see*: Bonaparte, Napoleon
Nassau-Saarbrücken, Walrad vorst von, 261
Naudin, M., 223
Naulet, Frédéric, 268, 285
Navarro, Pedro, 267
Ne Plus Ultra, Lines of, 103, 122
Newton, Isaac, 11
Nice, 1705 siege of, 188, 342
Nijmegen, 1702 pursuit to, 97, 130
Nijmegen, Treaty of (1678), 21, 68
Nine Years' War (1688–1697)
  campaigns of, 7, 13, 21, 39, 128, 135, 189, 224, 268, 307
  see also: Ath, Charleroi, Coni, Mannheim, Mons, Namur, Philippsbourg, Rijswijk
Obdam, Jacob van Wassenaer, heer van, 106, 164, 237
Oostende (town, *Ostend*), 101
Oostende, 1706 siege of, 100, 121, 164, 231, 244–245, 263, 292, 342, 368
Ormonde, James Butler Duke of, 103
Orrery, Roger Boyle, 1st Earl of, 7
Oudenaarde (town, *Audenarde*), 23, 36
Oudenaarde, 1677 siege of, 10
Oudenaarde, 1708 battle of, 101, 122
Ouwerkerk, Hendrik van Nassau, heer van, 164–165, 244, 246, 292

Pagan, Blaise-François comte de, 56
Papillon, David, 60, 61, 63–64
Parker, Geoffrey, 5, 7, 326, 348–349
Parliament (England), 93, 97
Parrott, David, 6, 258
Pepper, Simon, 6
Perjés, Géza, 15
Philippe V (Bourbon), duc d'Anjou, 93, 102, 104
Philippsbourg, 1688 siege of, 64, 139
Pollak, Martha, 11
Portland, Hans Willem Bentinck, 1st Earl of, 36
pré carré, see: fortifications
Puységur, Jacques de Chastenet marquis de, 58
Puységur, Jacques François Chastenet marquis de, 105, 107

Quincy, Charles Sevin marquis de, 38
Quincy, Joseph Sevin chevalier de, 37, 92, 137, 155–156, 176, 207, 265, 299

Raad van State (Dutch Council of State), 98, 69, 183
Ramillies, 1706 battle of, 99, 108–109, 121, 304
Rheinberg, 1702–3 siege/blockade of, 97, 225, 341, 358
Rhine River, 6, 97, 135
Richards, Jacob, 133
Richards, John, 133
Richards, Michael, 133
Richerand, Guy Creuzet de, 188
Rijswijk, Treaty of (1697), 21, 37, 92, 94
Roberts, Michael, 5
Roermond (town), 106
Roermond, 1702 siege of, 97, 120, 237, 261, 282, 341, 368
Roeux, Ferdinand-Gaston-Lamorald de Croy, comte de, 23, 24

Rogers, Clifford, 326
Rothrock, George, 6–7
Ryswick, Treaty of, *see*: Rijswijk, Treaty of

Saint-Donas, 1702 siege of Fort, 97, 120, 341
Saint-Ghislain, 1708 siege of, 122, 343
Saint-Hilaire, Armand de Mormes sieur de, 195, 288
Saint-Malo (town), 224
Saint-Michel, 1702 storm of Fort, 115, 244
Saint-Omer (town), 303
Saint-Rémy, Pierre Surirey de, 159–160, 268, 270, 279–280, 288
Saint-Venant, 1710 siege of, 122, 256, 301, 343, 368
Salisch, Ernst Willem van, 176, 231
Santa Cruz de Marcenado, Alvar de Navia Osoris de, 260
Satterfield, George, 227
Savoy, Vittorio II Amadeo, Duke of, 21
Saxe, Maurice de, 311, 315
Scarpe River, 194
Scheldt River (*Escaut*), 22, 102, 165, 177
Schulenburg, Johann Matthias count von der, 177, 179, 212, 233, 275, 283, 311
Seignelay, Jean-Baptiste Colbert marquis de, 125n6
siege tactics
  artillery and, 200–206, *see also*: artillery
  as scientific, 72–89
  mapping of, 71–72
  planning of, 22–25, 67–78, 155
  choice of approaches, 152–153, 190–199
  trenchworks, described, 28–35
  trench parallels, 28–30, 32, 42–45, 48, 57–62, 166, 245, 254–256, 258
  bombardment as, *see*: bombardment
  sorties, 24, 57, 315–316
  trench cavaliers (*cavaliers de tranchée*), 32, 48, 62–64, 65, 256–257
  storming the covered way, 62, 115, 241–251, *see also*: storm
  countermines, 178, 207, 232–233, 247, 249
  breaching walls, 34, 271–273
  retrenchments behind a breach, 35–36, 273–276
  cavalry and, 24, 193, 197n66, 217n6
Sluis, 1604 siege of, 274
soldiers
  as workmen in sieges, 166–170, 212–213
  demoralized by sieges, 299–303
Spain
  campaigns in, 2, 100, 133, 331
  see also: Barcelona, Madrid, Philippe V
Spanish Brabant, 20, 98–100, 101, 107–110
Spanish Flanders, 20, 101
Spanish Guelders, 106, 225
Stair, John Dalrymple 2nd Earl of, 199
States-General (Netherlands), 93, 97, 141–142, 157, 183, 207, 231
Stekene, 1703 battle of, 98, 120
Stevensweert (town), 106, 237
Stevensweert, 1702 siege of, 97, 120, 341, 368
storm
  tactic of, 16, 56, 223, 350–352, 356–359
  see also: Marchiennes, Saint-Michel
surprise tactic, 56, 221–223, 351–352
Suvorov, Aleksandr Vasil'evich, 317

Tallard, Camille d'Hostun, 174
Tavannes, Gaspard de Saulx, seigneur de, 55
Tessé, René de Froullay comte de, 28, 36
Thionville (town), 135
Thirty Years' War, 6, 54
Toulon, 1707 siege of, 100, 342
Tournai (town, *Doornik*), 24
Tournai, 1709 siege of, 102, 122, 143–144, 153, 158, 165, 178, 184, 199–200, 212, 232–233, 247–248, 256, 264–265, 276, 300–301, 343, 363, 368
Trognée, Jean Gérard Baron de, 130
Turenne, Henri de La Tour d'Auvergne vicomte de, 149, 153, 310
Turin, 1706 siege of, 100, 151, 193, 228, 285–287, 298, 342

Utrecht, Peace of (1713), 103–104, 135

Valenciennes (town), 24, 102
Vallière, Jean-Florent, 160, 210, 240, 281, 298, 313

Valory, Charles-Guy, 153, 183, 188–189, 197, 210, 229–230, 248–249
Vauban, Sébastien le Prestre de,
  early life, 8, 86–87
  injured, 33, 139
  semi-retirement in 1703, 113–114
  criticized, 173 *passim*, 311
  criticisms of artillery, 160–163
  and humanitarianism, 50–52, 227, 290–291
  and Science, 83–89
  Mémoire pour servir d'instruction dans la conduite des siéges et dans la défense des places, 49, 56–57, 82–83, 170–171
  Traité de l'attaque des places, 49–50, 56–57, 82–83, 85, 113–114
  Traité de la défense des places, 114
  see also: sieges of Ath, Breisach, Charleroi, Hulst, Luxembourg, Mons, Namur, Oudenaarde
  also appears throughout this study
Vegelin van Claerbergen, Philip Frederick, 144, 158, 165, 300
Vendôme, Louis-Joseph duc de, 100, 101, 315
Venlo (town), 106, 107
Venlo, 1702 siege of, 46, 97, 115–116, 120, 236–237, 244, 256, 260–261, 276, 281–282, 289, 341, 362, 368
Verboom, Georges-Prospero, 129
Vercelli, 1704 siege of, 188
Verrua, 1704–1705 siege of, 188, 286
Vienna, 1683 siege of, 59
Vigny, Jean-Baptiste de, 30
Villars, Louis-Hector duc de, *appears throughout this study, omitted here*
Villars-Lugein, Louis-Joseph de Plaibault de, 138, 188
Ville, Anthoine de, *see*: De Ville
Villeroi, François de Neufville duc de, 23, 99, 100, 107, 108
Voysin, Daniel-François, 198, 228

War of Devolution (1667–1668), 13, 21, 111
War of the Reunions (1683–1684), 13
Webb, John, 101
Wijnendael, 1708 battle of, 101, 120
William III (Orange), *Stadhouder* of Netherlands, King of England, 35, 94, 96, 98, 114, 131, 209, 225, 268
Williams, Sir Roger, 274
Wood, James, 5

Ypres (town), 222, 303
Ypres, 1647 siege of, 60
Yvoy heer van Binderen, Frederick Thomas d'Hangest-Genlis, 196

Zandvliet, 1703 siege of, 99, 120, 342
Zoutleeuw (*Léau*), 1705 siege of, 99, 120, 342, 368

HISTORY OF WARFARE

History of Warfare *presents the latest research on all aspects of military history. Publications in the series will examine technology, strategy, logistics, and economic and social developments related to warfare in Europe, Asia, and the Middle East from ancient times until the early nineteenth century. The series will accept monographs, collections of essays, conference proceedings, and translation of military texts.*

1. HOEVEN, M. VAN DER (ed.). *Exercise of Arms.* Warfare in the Netherlands, 1568-1648. 1997. ISBN 90 04 10727 4
2. RAUDZENS, G. (ed.). *Technology, Disease and Colonial Conquests, Sixteenth to Eighteenth Centuries.* Essays Reappraising the Guns and Germs Theories. 2001. ISBN 90 04 11745 8
3. LENIHAN P. (ed.). *Conquest and Resistance.* War in Seventeenth-Century Ireland. 2001. ISBN 90 04 11743 1
4. NICHOLSON, H. *Love, War and the Grail.* 2001. ISBN 90 04 12014 9
5. BIRKENMEIER, J.W. *The Development of the Komnenian Army: 1081-1180.* 2002. ISBN 90 04 11710 5
6. MURDOCH, S. (ed.). *Scotland and the Thirty Years' War, 1618-1648.* 2001. ISBN 90 04 12086 6
7. TUYLL VAN SEROOSKERKEN, H.P. VAN. *The Netherlands and World War I.* Espionage, Diplomacy and Survival. 2001. ISBN 90 04 12243 5
8. DEVRIES, K. *A Cumulative Bibliography of Medieval Military History and Technology.* 2002. ISBN 90 04 12227 3
9. CUNEO, P. (ed.). *Artful Armies, Beautiful Battles.* Art and Warfare in Early Modern Europe. 2002. ISBN 90 04 11588 9
10. KUNZLE, D. *From Criminal to Courtier.* The Soldier in Netherlandish Art 1550-1672. 2002. ISBN 90 04 12369 5
11. TRIM, D.J.B. (ed.). *The Chivalric Ethos and the Development of Military Professionalism.* 2003. ISBN 90 04 12095 5
12. WILLIAMS, A. *The Knight and the Blast Furnace.* A History of the Metallurgy of Armour in the Middle Ages & the Early Modern Period. 2003. ISBN 90 04 12498 5
13. KAGAY, D.J. & L.J.A. VILLALON (eds.). *Crusaders, Condottieri, and Cannon.* Medieval Warfare in Societies Around the Mediterranean. 2002. ISBN 90 04 12553 1
14. LOHR, E. & M. POE (eds.). *The Military and Society in Russia: 1450-1917.* 2002. ISBN 90 04 12273 7
15. MURDOCH, S. & A. MACKILLOP (eds.). *Fighting for Identity.* Scottish Military Experience c. 1550-1900. 2002. ISBN 90 04 12823 9
16. HACKER, B.C. *World Military History Bibliography.* Premodern and Nonwestern Military Institutions and Warfare. 2003. ISBN 90 04 12997 9
17. MACKILLOP, A. & S. MURDOCH (eds.). *Military Governors and Imperial Frontiers c. 1600-1800.* A Study of Scotland and Empires. 2003. ISBN 90 04 12970 7

18. SATTERFIELD, G. *Princes, Posts and Partisans.* The Army of Louis XVI and Partisan Warfare in the Netherlands (1673-1678). 2003. ISBN 90 04 13176 0
20. MACLEOD, J. & P. PURSEIGLE (eds.). *Uncovered Fields.* Perspectives in First World War Studies. 2004. ISBN 90 04 13264 3
21. WORTHINGTON, D. *Scots in the Habsburg Service, 1618-1648.* 2004. ISBN 90 04 13575 8
22. GRIFFIN, M. *Regulating Religion and Morality in the King's Armies, 1639-1646.* 2004. ISBN 90 04 13170 1
23. SICKING, L. *Neptune and the Netherlands.* State, Economy, and War at Sea in the Renaissance. 2004. ISBN 90 04 13850 1
24. GLOZIER, M. *Scottish Soldiers in France in the Reign of the Sun King.* Nursery for Men of Honour. 2004. ISBN 90 04 13865 X
25. VILLALON, L.J.A. & D.J. KAGAY (eds.). *The Hundred Years War.* A Wider Focus. 2005. ISBN 90 04 13969 9
26. DEVRIES, K. *A Cumulative Bibliography of Medieval Military History and Technology, Update 2004.* 2005. ISBN 90 04 14040 9
27. HACKER, B.C. *World Military History Annotated Bibliography.* Premodern and Nonwestern Military Institutions (Works Published before 1967). 2005. ISBN 90 04 14071 9
28. WALTON, S.A. (ed.). *Instrumental in War.* Science, Research, and Instruments. Between Knowledge and the World. 2005. ISBN 90 04 14281 9
29. STEINBERG, J.W., B.W. MENNING, D. SCHIMMELPENNINCK VAN DER OYE, D. WOLFF & S. YOKOTE (eds.). *The Russo-Japanese War in Global Perspective.* World War Zero, Volume I. 2005. ISBN 90 04 14284 3
30. PURSEIGLE, P. (ed.). *Warfare and Belligerence.* Perspectives in First World War Studies. 2005. ISBN 90 04 14352 1
31. WALDMAN, J. *Hafted Weapons in Medieval and Renaissance Europe.* The Evolution of European Staff Weapons between 1200 and 1650. 2005. ISBN 90 04 14409 9
32. SPEELMAN, P.J. (ed.). *War, Society and Enlightenment.* The Works of General Lloyd. 2005. ISBN 90 04 14410 2
33. WRIGHT, D.C. *From War to Diplomatic Parity in Eleventh-Century China.* Sung's Foreign Relations with Kitan Liao. 2005. ISBN 90 04 14456 0
34. TRIM, D.J.B. & M.C. FISSEL (eds.). *Amphibious Warfare 1000-1700.* Commerce, State Formation and European Expansion. 2006. ISBN 90 04 13244 9
35. KENNEDY, H. (ed.). *Muslim Military Architecture in Greater Syria.* From the Coming of Islam to the Ottoman Period. 2006. ISBN 90 04 14713 6
36. HALDON, J.F. (ed.). *General Issues in the Study of Medieval Logistics.* Sources, Problems and Methodologies. 2006. ISBN 90 04 14769 1
37. CHRISTIE, N. & M. YAZIGI (eds.). *Noble Ideals and Bloody Realities.* Warfare in the Middle Ages. 2006. ISBN 90 04 15024 2
38. SHAW, C. (ed.). *Italy and the European Powers.* The Impact of War, 1500–1530. 2006. ISBN-13: 978 90 04 15163 5, ISBN-10: 90 04 15163 X
39. BIGGS, D. *Three Armies in Britain.* The Irish Campaign of Richard II and the Usurpation of Henry IV, 1397-99. 2006. ISBN-13: 978 90 04 15215 1, ISBN-10: 90 04 15215 6
40. WOLFF, D., MARKS, S.G., MENNING, B.W., SCHIMMELPENNINCK VAN DER OYE, D., STEINBERG, J.W. & S. YOKOTE (eds.). *The Russo-Japanese War in Global Perspective.* World War Zero, Volume II. 2007. ISBN-13: 978 90 04 15416 2, ISBN-10: 90 04 15416 7

41. OSTWALD, J. *Vauban under Siege.* Engineering Efficiency and Martial Vigor in the War of the Spanish Succession. 2007. ISBN-13: 978 90 04 15489 6, ISBN-10: 90 04 15489 2

ISSN 1385–7827